ADAPTIVE ROBUST CONTROL SYSTEMS

Edited by **Le Anh Tuan**

Adaptive Robust Control Systems
http://dx.doi.org/10.5772/intechopen.68813
Edited by Le Anh Tuan

Contributors
Krishna Bhavithavya Kidambi, William MacKunis, Mahmut Reyhanoglu, Jaime Rubio Hervas, Andrea L'Afflitto, Keyvan Mohammadi, Silvio Simani, Paolo Castaldi, Nikhil Angad Bakshi, Marian Gaiceanu, Seung-Bok Choi, Pham Tam Thanh, Dao Phuong Nam, Cong Nho Luong, Konrad Gromaszek, Andrzej Kotyra, Antonio Rosales, Iván Villanueva, Pedro Ponce, Arturo Molina, Steven Su, Yi Zhang, Branko Celler, Hung Nguyen, Winnie Pang, Qin Yang, Andrey V. Savkin, Kairui Guo, Jian Chen, Yuan Tian, Zhang Ren, Chen Bai, Peng Li, Gangbing Song, Hidetoshi Oya, Shunya Nagai, Yiming Wu, Tong Yang, He Chen, Ramiro Ibarra, Gerardo Romero, Aldo Jonathan Muñoz, David Lara, Silvia Florida, Tung Lam Nguyen, Minh Duc Duong, Nga Thi - Thuy Vu, Thanh Pham, Phuoc Nguyen, Pham Xuan Duong

Published by InTech
Janeza Trdine 9, 51000 Rijeka, Croatia

Publishing Process Manager Slobodan Momcilovic
Technical Editor SPi Global
Cover InTech Design team

First published March, 2018
Printed in Croatia
Legal deposit, Croatia: National and University Library in Zagreb

Additional hard copies can be obtained from orders@intechopen.com

Adaptive Robust Control Systems, Edited by Le Anh Tuan
p. cm.
Print ISBN 978-953-51-3796-2
Online ISBN 978-953-51-3797-9

Meet the editor

Le Anh Tuan received his BE and ME degrees in Mechanical Engineering and Marine Machinery from the Vietnam Maritime University in 2003 and 2007, respectively. He received his PhD degree in Mechanical Engineering from the Kyung Hee University, South Korea, in 2012. He was a research fellow at the Nanyang Technological University, Singapore; a visiting scholar at the University of Birmingham, UK; and an endeavor research fellow at the University of Technology Sydney, Australia. He is a contract research professor at the Center of Wind Energy System, Kunsan National University, Korea, since 2016. He is currently an associate professor at the Department of Automotive Engineering of the Vietnam Maritime University. His research interests are nonlinear robust adaptive control, dynamics, vibration, and control of industrial machines.

Contents

Preface

We would like to introduce to the readers a monograph related to *robust adaptive control and its application*. The book presents the recent studies on applications of robust and adaptive control approaches to practical systems. Being a generalized concept, robust adaptive control indicates the control techniques that make the systems stable, work well, and robust despite the actions of uncertainties and disturbances. In addition, the controlled systems tend to the adaptive features, in which the systems show the ability in self-changing and self-adjusting the control structures to adapt to the influences of uncertainties. Robust control requires the information of all plant parameters, and these parameters must lie in the deterministic boundaries. Meanwhile, adaptive control does not need to know many system parameters even the system structure because the adaption mechanisms are integrated for dealing with parametric uncertainties, parameter estimation, and structural approximation. Robust control simply keeps the consistence of system responses, while adaptive control makes the systems to self-learn or self-train to achieve the control objectives. In this regard, adaptive control is more superior to robust control. Otherwise, robust control holds the strong points that adaptive control does not have. Robust control can treat well the problems of quick varying parameters, while adaptive control is only effective in case of slow parametric uncertainties. In addition, robust control is better for control problems with disturbances and modeling imprecision. The combination of robust control methods and adaptive control techniques leads to the *robust adaptive control systems* having two key advantages:

(I) *Robustness*: The control system is robust with the large variation in plant parameters and disturbances.

(II) *Adaptation*: The control system does not need the knowledge of plant parameters even the plant model. The adaptation behavior will automatically find the proper estimated parameters even synthetizing the approximated model to support the controller working well.

Tending to robust adaptive control systems, the studies in this book consisting of 17 chapters are classified into four groups as follows:

(1) *Advanced sliding mode control* (SMC): As a key approach of robust controls, SMC has been largely applied in designing the control systems with robust features. SMC does not require much accuracy of plant model. It treats well the systems having widely varying parameters and disturbances. However, chattering phenomenon of system responses is a challenge that needs to be solved completely in conventional SMC. Chapter 1 integrates the fuzzy approach into two robust control methods composed of SMC and H-infinity to construct two adaptive robust controllers for reducing the vibration of vehicle seat suspension systems under the excitation of several road profiles. Chapter 2 designs the guidance system for aircrafts based

on SMC and simulates the two complex operation cases. Chapter 3 proposes an SMC controller for multimotor drive systems, integrates the high-gain observer, and considers the influence of nonlinearities such as backlash, friction, and elasticity. Chapter 4 utilizes the stator flux–oriented SMC to regulate torque and reactive power of doubly fed induction generator.

(2) *Model reference adaptive control* (MRAC): Dissimilar to SMC whereas the control structure is fixed, MRAC systems tend to adaptation behavior in terms of parametric uncertainties by self-varying the control structures. In fact, the control structures are parameterized with respect to varying plant parameters, and the adaptation mechanisms are constituted to approximate these uncertainties. Chapter 5 develops a MRAC-based robust tracking system and an autopilot for quadrotors considering the parametric uncertainties composed of translational mass and inertial mass with the presence of wind disturbances. Chapter 6 applies MRAC to construct an attitude control system for unmanned quadcopters when faced with unknown plant parameters, the action of disturbances, and the influence of nonlinearities in actuators. Chapter 7 designs the control algorithms for DC motors on the basis of compound MRAC, whereas the system robustness is investigated by using simulation. Chapter 8 enhances MRAC for quadrotor unmanned aerial vehicles with the foundation of neural networks and machine learning. On the basis of improved MRAC, Chapter 9 analyzes and designs an adaptive controller including SMC observers and parametric estimators for thermo-acoustic oscillations of Rijke-type systems with the presence of dynamic model uncertainty and unknown disturbances.

(3) *The other robust adaptive control approaches*: In this part, the other adaptive techniques such as gain-scheduling and fuzzy logic together with robust controls such as H-infinity are untitled for analyzing and designing the control systems. Chapter 10 synthesizes the robust adaptive controllers for a class of mechanical linear systems with adjustable time-varying parameters taking uncertainties and perturbation into account. Chapter 11 provides two kinds of adaptive robust control strategies for wind turbines using fuzzy logic, data-driven, and model-based approaches fully considering the stochastic disturbances and load uncertainties. Using model predictive control together with MRAC, Chapter 12 develops the process control system for biomass fuel cocombustion. Chapter 13 improves an adaptive embedded control system for measuring yield strength of plate-formed materials, in which a cerebellar model articulation controller (CMAC) is integrated in feedforward loop, and a proportional-derivative structure is equipped on feedback loop for training CMAC. With high applicability in medical practice and sport science, Chapter 14 proposes an adaptive control system for regulating the heart rate during treadmill exercises.

(4) *Control of overhead cranes—the underactuated systems*: We usually face with underactuated systems in control engineering practice. Overhead crane is an underactuated system, whereas the number of actuators is lesser than that of outputs. For 2D motion, only two actuators composed of trolley-moving motors and cargo-hosting motors are utilized for controlling three outputs. For 3D motion, three motors are applied for driving five outputs composed of trolley motion, bridge motion, cargo hoisting displacement, and two cargo swing angles. Control of such a system is harder than full-actuator systems and meets many challenges. This part introduces the robust and adaptive control techniques applied for overhead cranes and the symbolic underactuated mechanical systems. Chapter 15 proposes a robust nonlinear controller integrating state observer by using Lyapunov-based design, in which controllability and observability are also investigated. Chapter 16 constitutes a distributed mass model of overhead crane with flexible handling cable and proposes two nonlinear control-

lers on the basis of Lyapunov candidate and its barrier version. Finally, Chapter 17 uses the input-to-state stable method together with Lyapunov stability for constructing the robust control algorithms. The robustness of crane system when facing with parametric uncertainties and disturbances is also analyzed and investigated.

This book is formed by the recent studies of many authors around the world. As an editor, I would like to thank all the authors for their excellent contributions to the book. I am also sincerely grateful to Mr. Slobodan at IntechOpen who helped me to manage the editorial process positively and effectively. Hopefully, the readers will find many useful information and professional knowledge in this book.

Le Anh Tuan
Associate Professor at Automotive Engineering Department
Vietnam Maritime University, Haiphong, Vietnam
Research Professor at Center of Wind Energy System
Kunsan National University, Gunsan, South Korea

Section 1

Sliding Mode Based Controls

Chapter 1

Robust Adaptive Controls of a Vehicle Seat Suspension System

Do Xuan Phu, Ta Duc Huy and Seung Bok Choi

Additional information is available at the end of the chapter

http://dx.doi.org/10.5772/intechopen.71422

Abstract

This work proposes two novel adaptive fuzzy controllers and applies them to vibration control of a vehicle seat suspension system subjected to severe road profiles. The first adaptive controller is designed by considering prescribed performance of the sliding surface and combined with adaptation laws so that robust stability is guaranteed in the presence of external disturbances. As for the second adaptive controller, both the H-infinity controller and sliding mode controller are combined using inversely fuzzified values of the fuzzy model. In order to evaluate control performances of the proposed two adaptive controllers, a semi-active vehicle suspension system installed with a magneto-rheological (MR) damper is adopted. After determining control gains, two controllers are applied to the system and vibration control performances such as displacement at the driver's position are evaluated and presented in time domain. In this work, to demonstrate the control robustness two severe road profiles of regular bump and random step wave are imposed as external disturbances. It is shown that both adaptive controllers can enhance ride comfort of the driver by reducing the displacement and acceleration at the seat position. This excellent performance is achieved from each benefit of each adaptive controller; accurate tracking performance of the first controller and fast convergence time of the second controller.

Keywords: adaptive fuzzy control, sliding mode control, H-infinity control, prescribed performance of the sliding surface, vibration control, seat suspension system

1. Introduction

Nowadays, modern control-based technical devices such as robotics, assistive machines and home appliances are popularly used to improve the level of human being's life. In these devices, control algorithm is one of the most important components which brings comfortable requirements to the consumer. The development of control algorithms in recent years is abundantly being undertaken from the aspect of classical control to salient characteristics of intelligent control. The classical control methods are frequently combined with modern control technique to resolve parameter uncertainties and disturbances those are existed in most of control devices. A controller which is formulated using more than two different control schemes is called "a hybrid controller" or "composite controller" [1, 2]. Among many candidates of the hybrid controller, the type of hybrid adaptive controller is the most popular since its structure is relatively simple and its control performance is very robust against the uncertainties or/and external disturbances. A hybrid adaptive control with fuzzy model and wavelet neural networks was presented in [1, 3] in which the sliding mode control was used to connect the parameters of the fuzzy model and the neural networks. This method is the typical model to develop the adaptive control in the last few years. Besides of uncertain nonlinear system, the problem of unknown input nonlinearity such as dead-zone or backlash-like hysteresis was also studied through the hybrid adaptive control [4]. It has been also shown that the neural works can be designed for a good performance of the hybrid adaptive control to deal with the uncertain system [5]. A hybrid adaptive controller possessing the robustness against input and parameter uncertainties was studied using the sliding mode controller associated with the fuzzy model [6, 7]. When a hybrid adaptive controller is formulated, in general the adaptation laws are simultaneously calculated. Furthermore, the back-stepping method was integrated with the fuzzy mode to achieve high performance of the hybrid adaptive controller [8].

As mentioned earlier, both the fuzzy model and the neural networks model are frequently used for the formulation of high performance of a hybrid adaptive controller [9]. Recently, a modified type of the fuzzy model called interval type 2 was combined with the back-stepping method to design of a hybrid adaptive control [10, 11]. It is remarked that the fixed fuzzy model always provides a safe choice in design of a hybrid adaptive control. However, this choice may cause a large error in finding the final values. To resolve this problem, an adaptive interval type 2 fuzzy neural network was developed on the basis of the online technique which can strengthen the flexibility of design parameters against the uncertainties [12]. Besides the above, there are many approaches to formulate new hybrid adaptive controllers such as output feedback control approach to take account for unknown hysteresis [13]. From the aspect of experimental implementation of hybrid adaptive controllers, several dynamic systems featuring magneto-rheological (MR) mount and MR damper are adopted for vibration control [2, 14–18]. Most of hybrid adaptive controllers used in these experimental realizations have been formulated by combining the models of interval type 2 fuzzy and interval type 2 fuzzy neural networks, and the control techniques of H-infinity control and sliding mode control. The advantage of using the interval type 2 fuzzy model is its flexibility in which optimized fuzzy values can be achieved unlike the classical fuzzy rule with the fixed value [19]. In order to improve the fuzzy model, clustering method [20] and data-driven for fuzzy rules [21] were also introduced.

As a subsequent work to develop a new hybrid adaptive controller, in this work two different new hybrid adaptive controllers are developed and their control performances are evaluated by investigation on vibration control of a semi-active seat suspension system installed with MR damper. The first hybrid adaptive controller is designed by combing online interval type 2 fuzzy neural networks model and prescribed performance of the sliding surface associated with adaptation laws to guarantee robust stability (HAC-PP in short). The second hybrid adaptive controller is formulated by combining inversely fuzzified value with H-infinity control to minimize computational cost algorithm (HAC-IFV in short).The stability of both adaptive controllers are rigorously proved based on the Lyapunov stability and appropriate control gains are determined to evaluate vibration control performance. It is shown that both proposed adaptive controllers are very effective and robust for controlling unwanted vibrations or excitations from the road profiles. These are validated by presenting control results showing significant reduction of both the displacement and acceleration at the seat position subjected to external excitations.

2. Formulation of HAC-PP

As mentioned in Introduction, the online interval type 2 fuzzy neural networks (OIT2FNN in short) model is used to formulate two adaptive controllers. The rule base of OIT2FNN can be expressed as follows [22].

$$R_f^j : \text{If}\, h_1\, \text{is}\, \mathbf{H}_{f1}^{\mathbf{j}}\, \text{and}\ldots\text{and}\, h_n\, \text{is}\, \mathbf{H}_{f\mathbf{n}}^{\mathbf{j}}\, \text{Then}\, g\, \text{is}\, a_0^j + \sum_{i=1}^{n} a_i^j h_i \tag{1}$$

where, $\mathbf{H}_{f\mathbf{i}}^{\mathbf{j}}(i = 1, \ldots, n; j = 1, \ldots, m)$ are fuzzy sets, m is the number of rules, and a_i^j are interval sets. The calculation process of OIT2FNN is clearly explained in [22]. The defuzzified output is then determined by

$$g_f = \frac{g_l + g_r}{2} = \frac{\boldsymbol{\theta}_l^T \xi_l^f + \boldsymbol{\theta}_r^T \xi_r^f}{2} \tag{2}$$

In the above, $\boldsymbol{\theta}_l^T = \left[w_1^l\ w_2^l\ w_3^l \ldots w_n^l\right]$ and $\boldsymbol{\theta}_r^T = \left[w_1^r\ w_2^r\ w_3^r \ldots w_n^r\right]$ are the weighting vectors, which symbolize the relation of the rule layer and type-reduction, and the weighted firing strength vectors given by

$$\xi_l^f = \left[\frac{\underline{f}_1}{\sum_{i=1}^{n} \underline{f}_i}\ \frac{\underline{f}_2}{\sum_{i=1}^{n} \underline{f}_i}\ \frac{\underline{f}_3}{\sum_{i=1}^{n} \underline{f}_i} \cdots \frac{\underline{f}_n}{\sum_{i=1}^{n} \underline{f}_i}\right]^T, \xi_r^f = \left[\frac{\bar{f}_1}{\sum_{i=1}^{n} \bar{f}_i}\ \frac{\bar{f}_2}{\sum_{i=1}^{n} \bar{f}_i}\ \frac{\bar{f}_3}{\sum_{i=1}^{n} \bar{f}_i} \cdots \frac{\bar{f}_n}{\sum_{i=1}^{n} \bar{f}_i}\right]^T$$

As a problem formulation, consider a single-input and single-output (SISO) nonlinear system governed by the following equation:

$$\dot{\mathbf{x}} = \mathbf{f}(\mathbf{x}) + \mathbf{g}(\mathbf{x})u(t) + \mathbf{d}(\mathbf{t}) \tag{3}$$

where $\mathbf{f}(\mathbf{x}) \in R^n$ and $\mathbf{g}(\mathbf{x}) \in R^n$ are two unknown non-linear function vectors, $u(t) \in R^1$ is control function, $\mathbf{d}(\mathbf{t}) \in R^n$ is an external disturbance vector, $|\mathbf{d}(\mathbf{t})| \leq \boldsymbol{\delta}\mathbf{d}$ where $\boldsymbol{\delta}\mathbf{d} \in R^n$ is upper bound of $\mathbf{d(t)}$, $\mathbf{x} = [x_1, x_2, \ldots, x_n] = \left[x_1, \dot{x}_1, \ldots, x_1^{(n-1)}\right]^T \in R^n$ is the state vector of the system. The first sliding surface s_s is defined as follows:

$$s_s = k_1x_1 + k_2x_2 + k_3x_3 + \ldots + k_nx_n = \sum_{i=1}^{n} k_ix_i \tag{4}$$

where, $\mathbf{K} = [k_n, k_{n-1}, k_{n-2}, \ldots, k_1]$ is defined as the coefficients such that all of the roots of the polynomial $\sigma^n + k_{n-1}\sigma^{n-1} + k_{n-2}\sigma^{n-2} + \ldots + k_1$ are in the open left-half complex plane. The sliding surface (4) is rewritten using the state variables as follows:

$$x_n = -k_1x_1 - k_2x_2 - k_3x_3 - \ldots - k_{n-1}x_{n-1} + s_s \tag{5}$$

A new vector $\tilde{\mathbf{x}}$ is defined by $\tilde{\mathbf{x}} = [x_1\ x_2\ x_3 \ldots x_{n-1}]^T$, and thus the system (3) is rewritten as follows:

$$\dot{\tilde{\mathbf{x}}} = \mathbf{S}_1\tilde{\mathbf{x}} + \mathbf{S}_2^T s_s \tag{6}$$

where,

$$\mathbf{S}_1 = \begin{bmatrix} 0 & 1 & 0 & \ldots & 0 \\ 0 & 0 & 1 & \ldots & 0 \\ \cdot & \cdot & \cdot & \ldots & \cdot \\ -k_1 & -k_2 & -k_3 & \ldots & -k_{n-1} \end{bmatrix}, \mathbf{S}_2 = \begin{bmatrix} 0 \\ 0 \\ \cdot \\ 1 \end{bmatrix}$$

The tracking error is defined as $e = x_1 - x_d$ with the desired states of x_d. Then the error performance function is defined as follows [23]:

$$\lambda(t) = (\lambda(0) - \lambda_\infty)e^{-lt} + \lambda_\infty \tag{7}$$

where, $l > 0, 0 < |e(0)| < \lambda(0), \lambda_\infty > 0, \lambda_\infty < \lambda(0)$ then $\lambda_t > 0$ and $\lambda(t)$ tend to λ_∞ exponentially. In order to guarantee fast convergence of tracking error, and obtain a certain convergence accuracy, the tracking error is set as follows:

$$e(t) = \lambda(t)S(\phi) \tag{8}$$

In the above, the prescribed error performance function $S(\phi)$ found as follows:

$$S(\phi) = \frac{e(t)}{\lambda(t)} \tag{9}$$

The function $S(\phi)$ must satisfy the following conditions.

(i) $S(\phi)$ is smooth continuous and monotone increasing function

$$\text{(ii)} -1 < S(\phi) < 1 \tag{10}$$

(iii) $\lim_{\phi \to +\infty} S(\phi) = 1$ and $\lim_{\phi \to -\infty} S(\phi) = -1$

From the above conditions (10), the function $S(\phi)$ can be determined as follows:

$$S(\phi) = \frac{e^{\phi} - e^{-\phi}}{e^{\phi} + e^{-\phi}} \tag{11}$$

Then using Eq. (8), the tracking error is obtained by

$$-\lambda(t) < \lambda(t)S(\phi) < \lambda(t) \Leftrightarrow -\lambda(t) < e(t) < \lambda(t) \tag{12}$$

Hence, the tracking error can be summarized as $\Xi = \{e \in R : |e(t)| < \lambda \ \ \forall t \geq 0$ and $e(t) < \lambda_\infty$ for $t \to \infty\}$. On the other hand, the inverse function of (11) is expressed as:

$$\phi = \frac{1}{2}\ln\frac{1+S}{1-S} = \frac{1}{2}\ln\frac{1+(e/\lambda)}{1-(e/\lambda)} = \frac{1}{2}\ln\frac{\lambda+e}{\lambda-e} = \frac{1}{2}[\ln(\lambda+e) - \ln(\lambda-e)] \tag{13}$$

Hence, the derivatives of Eq. (13) are obtained as:

$$\dot{\phi} = \frac{1}{2}\left[\frac{\dot{\lambda}+\dot{e}}{\lambda+e} - \frac{\dot{\lambda}-\dot{e}}{\lambda-e}\right] \tag{14}$$

$$\ddot{\phi} = M_1 + M_2 + M_3\ddot{e} \tag{15}$$

where,

$$M_1 = \frac{\ddot{\lambda}(\lambda+e) - (\dot{\lambda}+\dot{e})^2}{2(\lambda+e)^2}, M_2 = -\frac{\ddot{\lambda}(\lambda-e) - (\dot{\lambda}-\dot{e})^2}{2(\lambda-e)^2}, M_3 = \left(\frac{\lambda+e}{2(\lambda+e)^2} + \frac{\lambda-e}{2(\lambda-e)^2}\right).$$

In order to realize $\phi \to 0$, the second sliding surface is defined as follows:

$$\sigma_s = \dot{\phi} + c_s\phi \tag{16}$$

where $c_s > 0$. The derivative of Eq. (16) is obtained as:

$$\dot{\sigma}_s = \ddot{\phi} + c_s\dot{\phi} = M_1 + M_2 + M_3(f(x) + g(x)u(t) + d(t) - \ddot{x}_d) + c_s\dot{\phi} \tag{17}$$

The lumped uncertainty of system is defined as:

$$w = M_3\tilde{\gamma}_f\xi_f + M_3\tilde{\gamma}_g\xi_g u + M_3 d(t) \tag{18}$$

where $\gamma_f = f(x) - f^*(x)$, $\gamma_g = g(x) - g^*(x)$. Using Eqs. (17) and (18), the derivative Eq. (17) is rewritten as:

$$\dot{\sigma}_s = M_1 + M_2 + M_3 f^*(x) + M_3 g^*(x)u(t) - M_3\ddot{x}_d + c_s\dot{\phi} + w \tag{19}$$

Based on Eq. (2), the relationship between Eq. (19) and OIT2FNN is expressed as follows:

$$\dot{\sigma}_s = M_1 + M_2 + M_3\theta_f{}^*\xi_f + M_3\theta_g{}^*\xi_g u - M_3\ddot{x}_d + c_s\dot{\phi} + w \tag{20}$$

where

$\theta_f^* = \arg\min_{\theta_f\in\Delta\theta_f}\left[\sup_{x\in\Delta x}|f(x) - f^*(x)|\right]$, $\theta_g^* = \arg\min_{\theta_g\in\Delta\theta_g}\left[\sup_{x\in\Delta x}|g(x) - g^*(x)|\right]$, $\Delta\theta_f = \{\theta_f\in R^n, \|\theta_f\|\le\Theta_f\}$, $\Delta\theta_g = \{\theta_g\in R^n, \|\theta_g\|\le\Theta_g\}$, $\Delta x = \{x\in R^n, \|x\|\le\Theta_x\}$

Now, an equivalent control is determined from Eq. (20) based on the assumption $\dot{\sigma}_s\approx 0$:

$$u_1 = \frac{1}{M_3\widehat{\theta}_g\xi_g}\left(-M_1 - M_2 - M_3\widehat{\theta}_f\xi_f + M_3\ddot{x}_d - c_s\dot{\phi}\right) \tag{21}$$

The equivalent control u_1 cannot control the system because it cannot compensate the error from the fuzzy approximation. To guarantee the robustness and stability in control, a robust control part u_2 should be introduced as follows:

$$u_2 = \frac{1}{M_3\widehat{\theta}_g\xi_g}\left(-\sum_{i=1}^{n-1} P_{(n-1)i}x_i - \frac{\sigma_s}{\beta} + \frac{1}{2}M_3\Gamma\xi_z\tilde{\mathbf{x}}\mathbf{P}\mathbf{S}_2\mathbf{S}_2^T\mathbf{P}\tilde{\mathbf{x}}^T\right) \tag{22}$$

Then, the total control u of the system is determined as follows:

$$u = u_1 + u_2 \tag{23}$$

The control u_2 is the combination of two sliding surfaces s_s and σ_s. The value Γ is the adaptive parameter where its boundary is given by $\Delta\Gamma = \{\Gamma\in R, \|\Gamma\|\le\Theta_\Gamma, \sigma_s\Gamma\xi_{fz}\le\rho\}$, and Θ_Γ is constant boundary. The matrix $\mathbf{P} = \mathbf{P}^T\ge 0$ in which its result is a solution of Riccati-like equation given by

$$\mathbf{PS_1} + \mathbf{S_1^T P} + \mathbf{Q} - \sigma_s\Gamma\xi_z\mathbf{PS_2S_2^TP} + \rho\mathbf{PS_2S_2^TP} = 0 \tag{24}$$

where, $\rho\ge\sigma_s\Gamma\xi_z$, ρ is the prescribed attenuation level, $\mathbf{Q} = \mathbf{Q^T}\ge 0$, ξ_z is consequent membership value of the OIT2FNN. When the value $\rho = \sigma_s\Gamma\xi_z$, the Riccati-like equation is rewritten as:

$$\mathbf{PS_1} + \mathbf{S_1^T P} + \mathbf{Q} = 0 \tag{25}$$

Now, Eq. (20) can be analyzed as follows:

$$\dot{\sigma}_s = M_1 + M_2 + M_3\tilde{\gamma}_f\xi_f + M_3\tilde{\gamma}_g\xi_g u - M_3\ddot{x}_d + c_s\dot{\phi} + w + \left[M_3\widehat{\theta}_f\xi_f + M_3\widehat{\theta}_g\xi_g u\right] \tag{26}$$

where $\tilde{\gamma}_f = \theta_f{}^* - \widehat{\theta}_f, \tilde{\gamma}_g = \theta_g{}^* - \widehat{\theta}_g$. Using Eqs. (23) and (26), Eq. (26) is rewritten by

$$\dot{\sigma}_s = \left[-\sum_{i=1}^{n-1} P_{(n-1)i}x_i - \frac{\sigma_s}{\beta} + \frac{1}{2}M_3\Gamma\xi_z\tilde{\mathbf{x}}\mathbf{P}\mathbf{S}_2\mathbf{S}_2^T\mathbf{P}\tilde{\mathbf{x}}^T\right] + \left[M_3\tilde{\gamma}_f\xi_f + M_3\tilde{\gamma}_g\xi_g u + w\right] \tag{27}$$

Now, the stability of the proposed adaptive control system can be solidly proved with Eqs. (21)–(23) and adaptation laws as follows:

$$\dot{\tilde{\gamma}}_f = -\mu_1 M_3 \sigma_s \xi_f; \ \dot{\tilde{\gamma}}_g = -\mu_2 M_3 \sigma_s \xi_g u; \ \dot{\Gamma} = -\mu_3 M_3 \sigma_s \xi_z \tilde{\mathbf{x}} \mathbf{P} \mathbf{S}_2 \mathbf{S}_2^T \mathbf{P} \tilde{\mathbf{x}}^T \tag{28}$$

In order to make a proof, in this work the following Lyapunov function candidate is proposed.

$$L_v = \frac{1}{2}\sigma^2{}_s + \frac{1}{2}\tilde{\mathbf{x}}\mathbf{P}\tilde{\mathbf{x}}^T + \frac{1}{2\mu_1}\tilde{\gamma}^2{}_f + \frac{1}{2\mu_2}\tilde{\gamma}^2{}_g + \frac{1}{2\mu_3}\Gamma^2 \tag{29}$$

The derivative of Eq. (29) is then obtained by

$$\dot{L}_v = \sigma_s \dot{\sigma}_s + \frac{1}{2}\dot{\tilde{\mathbf{x}}}\mathbf{P}\tilde{\mathbf{x}}^T + \frac{1}{2}\tilde{\mathbf{x}}\mathbf{P}\dot{\tilde{\mathbf{x}}}^T + \frac{1}{\mu_1}\tilde{\gamma}_f\dot{\tilde{\gamma}}_f + \frac{1}{\mu_2}\tilde{\gamma}_g\dot{\tilde{\gamma}}_g + \frac{1}{\mu_3}\Gamma\dot{\Gamma} \tag{30}$$

Substituting Eq. (27) into Eq. (30), Eq. (30) is rewritten as follows:

$$\begin{aligned} \dot{L}_v = & \left[M_3\sigma_s\tilde{\gamma}_f\xi_f + \frac{1}{\mu_1}\tilde{\gamma}_f\dot{\tilde{\gamma}}\right] + \left[M_3\sigma_s\tilde{\gamma}_g\xi_g u + \frac{1}{\mu_2}\tilde{\gamma}_g\dot{\tilde{\gamma}}_g\right] \\ & + \left[M_3\Gamma\xi_z\sigma_s\tilde{\mathbf{x}}\mathbf{P}\mathbf{S}_2\mathbf{S}_2^T\mathbf{P}\tilde{\mathbf{x}}^T + \frac{1}{\mu_3}\Gamma\dot{\Gamma}\right] + \left[\sigma_s w - \frac{\sigma^2{}_s}{\beta} - \frac{1}{2}\rho\mathbf{P}\mathbf{S}_2\mathbf{S}_2^T\mathbf{P} - \frac{1}{2}\tilde{\mathbf{x}}^T\mathbf{Q}\tilde{\mathbf{x}}^T\right] \end{aligned} \tag{31}$$

It is noted that Eq. (24) is used in finding Eq. (31). Substituting Eq. (28) into Eq. (31), the following is achieved.

$$\dot{L}_v = \left[-\frac{1}{2}\tilde{\mathbf{x}}^T\mathbf{Q}\tilde{\mathbf{x}}^T - \frac{1}{2}\left(\frac{\sigma_s}{\sqrt{\beta}} - \sqrt{\beta}w\right)^2 + \beta w^2\right] - \frac{1}{2}\rho\mathbf{P}\mathbf{S}_2\mathbf{S}_2^T\mathbf{P} \le -\frac{1}{2}\tilde{\mathbf{x}}^T\mathbf{Q}\tilde{\mathbf{x}}^T + \beta w^2 \tag{32}$$

Eq. (32) cannot use for conclusion of stability. Hence, it will be integrated from $t=0$ to $t=T$, we have:

$$L_v(0) - L_v(T) + \beta\int_0^T w^2 dt \ge \frac{1}{2}\int_0^T \tilde{\mathbf{x}}\mathbf{Q}\tilde{\mathbf{x}}^T dt \tag{33}$$

where, $L_v(0) = \frac{1}{2}\sigma^2{}_s(0) + \frac{1}{2}\tilde{\mathbf{x}}(0)\mathbf{P}\tilde{\mathbf{x}}^T(0) + \frac{1}{2\mu_1}\tilde{\gamma}^2{}_f(0) + \frac{1}{2\mu_2}\tilde{\gamma}^2{}_g(0) + \frac{1}{2\mu_3}\Gamma^2(0)$. The value $L_v(T)$ is always positive, so Eq. (33) is determined as:

$$L_v(0) + \beta\int_0^T w^2 dt \ge \frac{1}{2}\int_0^T \tilde{\mathbf{x}}\mathbf{Q}\tilde{\mathbf{x}}^T dt \ge 0 \tag{34}$$

From Eqs. (32) and (34), the stability is guaranteed.

From the boundedness of the parameters $\tilde{\gamma}_f$ and $\tilde{\gamma}_g$, the closed sets are defined as $\Xi_1 = \left\{\tilde{\gamma}_f \middle| \left\|\tilde{\gamma}_f\right\| \le \aleph_f\right\}$, $\Xi_2 = \left\{\tilde{\gamma}_g \middle| \left\|\tilde{\gamma}_g\right\| \le \aleph_g\right\}$, $\Xi_{\delta 1} = \left\{\tilde{\gamma}_f \middle| \left\|\tilde{\gamma}_f\right\| \le \aleph_f + \delta_1\right\}$, $\Xi_{\delta 2} = \left\{\tilde{\gamma}_g \middle| \left\|\tilde{\gamma}_g\right\| \le \aleph_g + \delta_2\right\}$.

In here, $\aleph_f$, $\aleph_g$, δ_1, δ_2 are the choosing parameters. Hence, the adjusted adaptation laws are redefined as follows:

$$\dot{\tilde{\gamma}}_f = \begin{cases} -\mu_1 M_3\sigma_s\xi_f & \text{if } \|\tilde{\gamma}_f\| < \aleph_f \text{ or } \left(\|\tilde{\gamma}_f\| = \aleph_f \text{ and } M_3\sigma_s\xi_f\tilde{\gamma}_f \geq 0\right) \\ -\mu_1 M_3\sigma_s\xi_f + \mu_1 \dfrac{\left(\left\|\tilde{\gamma}_f\right\|^2 - \aleph_f\right) M_3\sigma_s\xi_f\tilde{\gamma}_f}{\delta_1\left\|\tilde{\gamma}_f\right\|^2} & \text{if } \|\tilde{\gamma}_f\| = \aleph_f \text{ and } M_3\sigma_s\xi_f\tilde{\gamma}_f < 0 \end{cases} \tag{35}$$

$$\dot{\tilde{\gamma}}_g = \begin{cases} -\mu_2 M_3\sigma_s\xi_g u & \text{if } \left\|\tilde{\gamma}_g\right\| < \aleph_g \text{ or } \left(\left\|\tilde{\gamma}_g\right\| = \aleph_g \text{ and } M_3\sigma_s\xi_g u\tilde{\gamma}_g \geq 0\right) \\ -\mu_2 M_3\sigma_s\xi_g u + \mu_2 \dfrac{\left(\left\|\tilde{\gamma}_g\right\|^2 - \aleph_g\right) M_3\sigma_s\xi_g u\tilde{\gamma}_g}{\delta_2\left\|\tilde{\gamma}_g\right\|^2} & \text{if } \left\|\tilde{\gamma}_g\right\| = \aleph_g \text{ and } M_3\sigma_s\xi_g u\tilde{\gamma}_g < 0 \end{cases} \tag{36}$$

$$\dot{\Gamma} = \begin{cases} -\mu_3 M_3\sigma_s\xi_z\tilde{\mathbf{x}}\mathbf{P}\mathbf{S}_2\mathbf{S}_2^T\mathbf{P}\tilde{\mathbf{x}}^T \text{if } \|\Gamma\| < \Theta_\Gamma \text{ or } \left(\|\Gamma\| = \Theta_\Gamma + \delta_3 \text{ and } M_3\sigma_s\xi_z\tilde{\mathbf{x}}\mathbf{P}\mathbf{S}_2\mathbf{S}_2^T\mathbf{P}\tilde{\mathbf{x}}^T\Gamma \geq 0\right) \\ -\mu_3 M_3\sigma_s\xi_z\tilde{\mathbf{x}}\mathbf{P}\mathbf{S}_2\mathbf{S}_2^T\mathbf{P}\tilde{\mathbf{x}}^T + \mu_3 \dfrac{\left(\|\Gamma\|^2 - \Theta_\Gamma\right) M_3\sigma_s\xi_z\tilde{\mathbf{x}}\mathbf{P}\mathbf{S}_2\mathbf{S}_2^T\mathbf{P}\tilde{\mathbf{x}}^T\Gamma}{\delta_3\|\Gamma\|^2} \\ \text{if } \|\Gamma\| = \Theta_\Gamma \text{ and } M_3\sigma_s\xi_z\tilde{\mathbf{x}}\mathbf{P}\mathbf{S}_2\mathbf{S}_2^T\mathbf{P}\tilde{\mathbf{x}}^T\Gamma < 0 \end{cases} \tag{37}$$

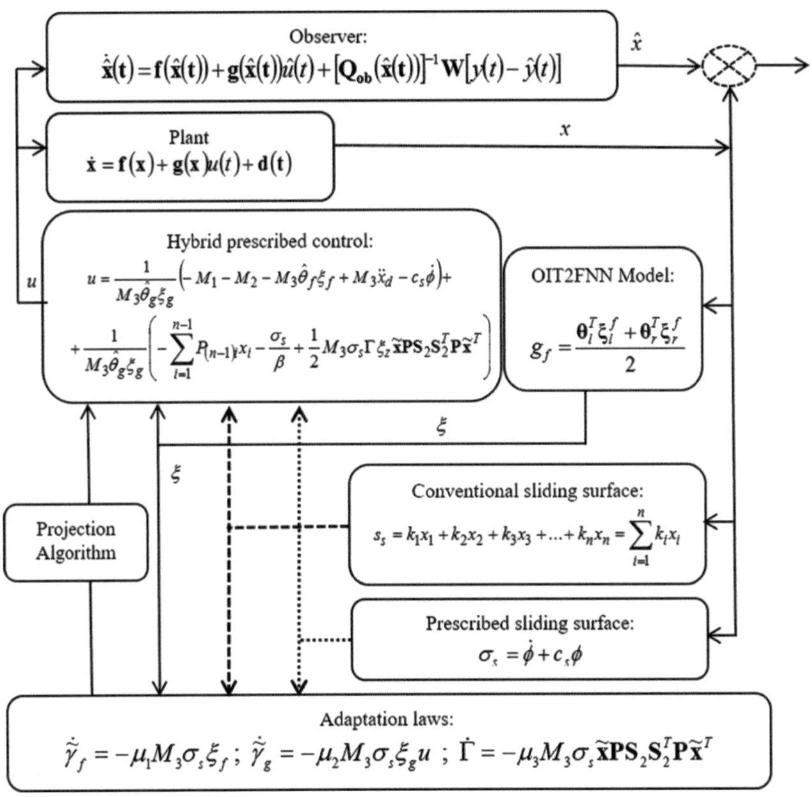

Figure 1. Control flow chart of the HAC-PP.

In the above, δ_1, δ_2 and δ_3 are choosing parameters related boundaries of $f(x)$, $g(x)$ and Γ. It is noted here that in order to utilize the states of the system, the Luenberger observer [24] has been used in this work. **Figure 1** presents a flow chart of the HAC-PP showing the combination of each controller and the prescribed performance.

3. Formulation of HAC-IFV

As a first step to design the controller, consider the system (3) rewritten by

$$\dot{\mathbf{x}} = \mathbf{f}_0(\mathbf{x}) + \mathbf{g}_0(\mathbf{x})u(t) + \mathbf{D} \tag{38}$$

where, the function $\mathbf{f}_0(\mathbf{x})$ and $\mathbf{g}_0(\mathbf{x})$ are the functions of $\mathbf{f}(\mathbf{x})$ and $\mathbf{g}(\mathbf{x})$ which are determined as:

$$\mathbf{f}(\mathbf{x}) = \mathbf{f}_0(\mathbf{x}) + \boldsymbol{\delta}\mathbf{f}(\mathbf{x}); 0 < |\boldsymbol{\delta}\mathbf{f}(\mathbf{x})| < \|\boldsymbol{\delta}\mathbf{f}\|_\infty, \mathbf{g}(\mathbf{x}) = \mathbf{g}_0(\mathbf{x}) + \boldsymbol{\delta}\mathbf{g}(\mathbf{x}); 0 < |\boldsymbol{\delta}\mathbf{g}(\mathbf{x})| < \|\boldsymbol{\delta}\mathbf{g}\|_\infty.$$

$$\mathbf{f}_0(\mathbf{x}) = \left[x_2, \dots, x_n, f_0\right]^T, \mathbf{g}_0(\mathbf{x}) = \left[0, \dots, 0, g_0\right]^T, \boldsymbol{\delta}\mathbf{f} = \left[0, 0, \dots, \delta f_0\right]^T, \boldsymbol{\delta}\mathbf{g} = \left[0, 0, \dots, \delta g_0\right]^T.$$

In the above, $\boldsymbol{\delta}\mathbf{f}$ and $\boldsymbol{\delta}\mathbf{g}$ are two positive vectors. It is noted that $\mathbf{D}=\boldsymbol{\delta}\mathbf{f}+\boldsymbol{\delta}\mathbf{g}u(t)+\mathbf{d}(\mathbf{t})$ denotes the uncertain disturbance and $\mathbf{D}=[0, 0, \dots, D_0]^T$. In order to formulate the controller, the following assumption is made: There exists a constant $g_m \in \Re^+$ to satisfy $|g(x)| > g_m$. Without loss of generality, it is assumed that the equation $g(x) > g_m$. The error between a desired output x_d and the measured output x is $e = x_d - x$. Hence, the error vector is defined by $\mathbf{E} = [e_0, e_1, e_2, \dots, e_n] = \left[e, \dot{e}, \ddot{e}, \dots, e^{(n-1)}\right]$. The sliding surface s_s can be written as $s(x,t) = \mathbf{K}^T\mathbf{E}$, and its derivative is found as $\dot{s}(x,t) = \mathbf{K}^T\dot{\mathbf{E}} = \mathbf{K}^T\dot{\mathbf{x}}_d - \mathbf{K}^T\dot{\mathbf{x}}$. Using this derivative function of the sliding surface and Eq. (38), the initial control law u is determined by:

$$u = \frac{1}{g_0(x)}\left(-f_0(x) + \dot{x}_d + \mathbf{K}^\mathbf{T}\mathbf{E} + D_0\right) \tag{39}$$

Assuming the disturbance of $\mathbf{D} \approx 0$, then Eq. (39) can be rewritten as:

$$u = \frac{1}{g_0(x)}\left(-f_0(x) + \dot{x}_d + \mathbf{K}^\mathbf{T}\mathbf{E}\right) \tag{40}$$

The relationship of Eq. (40) and OIT2FNN is expressed by

$$u = \frac{1}{g_{00}(x)}\left(-f_{00}(x) + \dot{x}_d + \mathbf{K}^\mathbf{T}\mathbf{E}\right) \tag{41}$$

where, $f_{00}(x)$ and $g_{00}(x)$ are the fuzzified functions of $f(x)$ and $g(x)$, respectively. The derivative of $\mathbf{E}$ is expressed through Eqs. (40) and (41) as follows:

$$\begin{aligned}\dot{\mathbf{E}} = \dot{\mathbf{x}}_{\mathbf{d}} - \dot{\mathbf{x}} &= (g_{00}(x) - g_0(x))u + (f_{00}(x) - f_0(x)) - \mathbf{K}^{\mathbf{T}}\mathbf{E} \\ &= \mathbf{S_1E} + \mathbf{S_2}\left[(g_{00}(x) - g(x))u + (f_{00}(x) - f(x))\right]\end{aligned} \tag{42}$$

Define the minimum approximation error due to fuzzy approximation as follows.

$$w = (f^*{}_{00}(x) - f(x)) + (g^*{}_{00}(x) - g(x))u \tag{43}$$

Substituting functions of $f_{00}(x)$, $g_{00}(x)$ and (43) into Eq. (42) yields the following equation.

$$\dot{\mathbf{E}} = \mathbf{S_1E} + \mathbf{S_2}\left[\left(\theta_f^* - \theta_f\right)\xi_f + \left(\theta_g^* - \theta_g\right)\xi_g u + w\right] \tag{44}$$

Let $\gamma_f = \left(\theta_f^* - \theta_f\right)$, $\gamma_g = \left(\theta_g^* - \theta_g\right)$. From Eq. (44), the equivalence control u_1 established without the minimum approximation error w is defined as follows:

$$u_1 = \frac{1}{\widehat{\gamma}_g \xi_g}\left(-\widehat{\gamma}_f \xi_f\right) \tag{45}$$

where, $\widehat{\gamma}_f$ and $\widehat{\gamma}_g$ are the estimates of γ_f and γ_g, respectively. The control u_1 cannot use for control the system because of the error from the fuzzy approximation. To deal with this problem, a new robust compensator based on the inversely fuzzified value is suggested as follows:

$$u_2 = -\frac{1}{\Gamma \xi_z}\mathbf{E}^{\mathbf{T}}\mathbf{P}\mathbf{S}_2 \tag{46}$$

where, Γ is a constant, and $\mathbf{P} = \mathbf{P}^{\mathbf{T}} \geq 0$ is the solution of the following Riccati-like equation.

$$\mathbf{PS_1} + \mathbf{S_1^T P} + \mathbf{Q} - \frac{1}{\Gamma \xi_z}\mathbf{PS_2S_2^TP} + \rho\mathbf{PS_2S_2^TP} = 0 \tag{47}$$

where, $\rho \geq \frac{1}{\Gamma \xi_z}$, ρ is the prescribed attenuation level, $\mathbf{Q} = \mathbf{Q}^{\mathbf{T}} \geq 0$, ξ_z is consequent membership value of the OIT2FNN. When the value $\rho = \frac{1}{\Gamma \xi_z}$, the Riccati-like equation is obtain as given in Eq. (25). It is noteworthy that Eq. (25) is objective to guarantee the stability of the system. If this condition is obtained, the fuzzy approximation error is removed, and then the control u_1 is the main controller to retain the stability of the system. From Eqs. (45) and (46), the final fuzzy control of the system is determined as follows:

$$u = u_1 + u_2 = \frac{1}{\widehat{\gamma}_g \xi_g}\left(-\widehat{\gamma}_f \xi_f\right) - \frac{1}{\Gamma \xi_z}\mathbf{E}^{\mathbf{T}}\mathbf{P}\mathbf{S}_2 \tag{48}$$

Now, substituting Eq. (48) into (44) yields he following.

$$\dot{\mathbf{E}} = \mathbf{S_1 E} + \mathbf{S_2}\left[\tilde{\gamma}_f \xi_f + \tilde{\gamma}_g \xi_g u_1 + g_o u_2 + w\right] \tag{49}$$

where, $\tilde{\gamma}_f = \gamma_f - \hat{\gamma}_f, \tilde{\gamma}_g = \gamma_g - \hat{\gamma}_g$. Consider the Lyapunov function candidate of the system as follows:

$$V = \frac{1}{2}\mathbf{E^T P E} + \frac{1}{2\alpha_1}\tilde{\gamma}_f^2 + \frac{1}{2\alpha_2}\tilde{\gamma}_g^2 \tag{50}$$

The derivative of Eq. (50), and then substituting Eq. (25) into the derivative, the result is obtained as follows:

$$\begin{aligned}\dot{V} &= -\frac{1}{2}\mathbf{E^T Q E} - \frac{g_m}{\Gamma \xi_z}\left(\mathbf{E^T P S_2}\right)^2 + \mathbf{E^T P S_2} w + \frac{1}{\alpha_1}\left(\alpha_1 \mathbf{E^T P S_2}\xi_f - \dot{\tilde{\gamma}}_f\right)\tilde{\gamma}_f \\ &+ \frac{1}{\alpha_2}\left(\alpha_2 \mathbf{E^T P S_2}\xi_g u_1 - \dot{\tilde{\gamma}}_g\right)\tilde{\gamma}_g\end{aligned} \tag{51}$$

From Eq. (51), adaptation laws are established as follows:

$$\dot{\tilde{\gamma}}_f = -\alpha_1 \mathbf{E^T P S_2}\xi_f \tag{52}$$

$$\dot{\tilde{\gamma}}_g = -\alpha_2 \mathbf{E^T P S_2}\xi_g u_1 \tag{53}$$

Applying Eqs. (52) and (53), Eq. (51) can be written as follows:

$$\begin{aligned}\dot{V} &\le -\frac{1}{2}\mathbf{E^T Q E} - \frac{g_m}{\Gamma \xi_z}\left(\mathbf{E^T P S_2}\right)^2 + \mathbf{E^T P S_2} w \\ &= -\frac{1}{2}\mathbf{E^T Q E} - \left(\sqrt{\frac{g_m}{\Gamma \xi_z}}\mathbf{E^T P S_2} - \frac{w_m}{2\rho}\right)^2 + \frac{1}{4\rho}{w_m}^2 \le -\frac{1}{2}\mathbf{E^T Q E} + \frac{1}{4\rho}{w_m}^2\end{aligned} \tag{54}$$

where, $w_m = \frac{w}{\sqrt{g_m}}$.

Now, the integration of (54) from $t = 0$ to $t = T$ yields the following equation.

$$V(0) - V(T) + \frac{1}{4\rho}\int_0^T {w_m}^2 dt \ge \frac{1}{2}\int_0^T \mathbf{E^T Q E} dt \tag{55}$$

The value of $V(T) \ge 0$, and thus Eq. (55) is rewritten as follows:

$$V(0) + \frac{1}{4\rho}\int_0^T {w_m}^2 dt \ge \frac{1}{2}\int_0^T \mathbf{E^T Q E} dt \tag{56}$$

where, $V(0) = \frac{1}{2}\mathbf{E^T}(0)\mathbf{P E}(0) + \frac{1}{2\alpha_1}\tilde{\gamma}_f^2(0) + \frac{1}{2\alpha_2}\tilde{\gamma}_g^2(0)$. Hence the H-infinity tracking performance is achieved. From the boundedness of the parameters, $\tilde{\gamma}_f$ and $\tilde{\gamma}_g$ are guaranteed by closed sets defined as $\Omega_1 = \left\{\tilde{\gamma}_f \middle| \left\|\tilde{\gamma}_f\right\| \le \Im_f\right\}$, $\Omega_2 = \left\{\tilde{\gamma}_g \middle| \left\|\tilde{\gamma}_g\right\| \le \Im_g\right\}$, $\Omega_{\delta 1} = \left\{\tilde{\gamma}_f \middle| \left\|\tilde{\gamma}_f\right\| \le \Im_f + \delta_1\right\}$, $\Omega_{\delta 2} = \left\{\tilde{\gamma}_g \middle| \left\|\tilde{\gamma}_g\right\| \le \Im_g + \delta_2\right\}$ where $\Im_f, \Im_g$, δ_1, δ_2 are the choosing parameters. Hence, the adjusted adaptation laws are redefined as follows:

$$\dot{\tilde{\gamma}}_f = \begin{cases} -\alpha_1 \mathbf{E}^{\mathbf{T}}\mathbf{P}\mathbf{S}_2\xi_f & \text{if } \|\tilde{\gamma}_f\| < \Im_f \text{ or } \left(\|\tilde{\gamma}_f\| = \Im_f \text{ and } \mathbf{E}^{\mathbf{T}}\mathbf{P}\mathbf{S}_2\xi_f\tilde{\gamma}_f \ge 0\right) \\ -\alpha_1 \mathbf{E}^{\mathbf{T}}\mathbf{P}\mathbf{S}_2\xi_f + \alpha_1 \dfrac{\left(\left\|\tilde{\gamma}_f\right\|^2 - \Im_f\right)\mathbf{E}^{\mathbf{T}}\mathbf{P}\mathbf{S}_2\xi_f\tilde{\gamma}_f}{\delta_1\left\|\tilde{\gamma}_f\right\|^2} & \text{if } \|\tilde{\gamma}_f\| = \Im_f \text{ and } \mathbf{E}^{\mathbf{T}}\mathbf{P}\mathbf{S}_2\xi_f\tilde{\gamma}_f < 0 \end{cases} \tag{57}$$

$$\dot{\tilde{\gamma}}_g = \begin{cases} -\alpha_2 \mathbf{E}^{\mathbf{T}}\mathbf{P}\mathbf{S}_2\xi_g u_1 & \text{if } \left\|\tilde{\gamma}_g\right\| < \Im_g \text{ or } \left(\left\|\tilde{\gamma}_g\right\| = \Im_g \text{ and } \mathbf{E}^{\mathbf{T}}\mathbf{P}\mathbf{S}_2\xi_f u_1\tilde{\gamma}_g \ge 0\right) \\ -\alpha_2 \mathbf{E}^{\mathbf{T}}\mathbf{P}\mathbf{S}_2\xi_g u_1 + \alpha_2 \dfrac{\left(\left\|\tilde{\gamma}_g\right\|^2 - \Im_g\right)\mathbf{E}^{\mathbf{T}}\mathbf{P}\mathbf{S}_2\xi_g u_1\tilde{\gamma}_g}{\delta_2\left\|\tilde{\gamma}_g\right\|^2} & \text{if } \left\|\tilde{\gamma}_g\right\| = \Im_g \text{ and } \mathbf{E}^{\mathbf{T}}\mathbf{P}\mathbf{S}_2\xi_g u_1\tilde{\gamma}_g < 0 \end{cases} \tag{58}$$

Figure 2 presents a flow chart of the HAC-IFV showing the combination process of each controller with the adaptation laws.

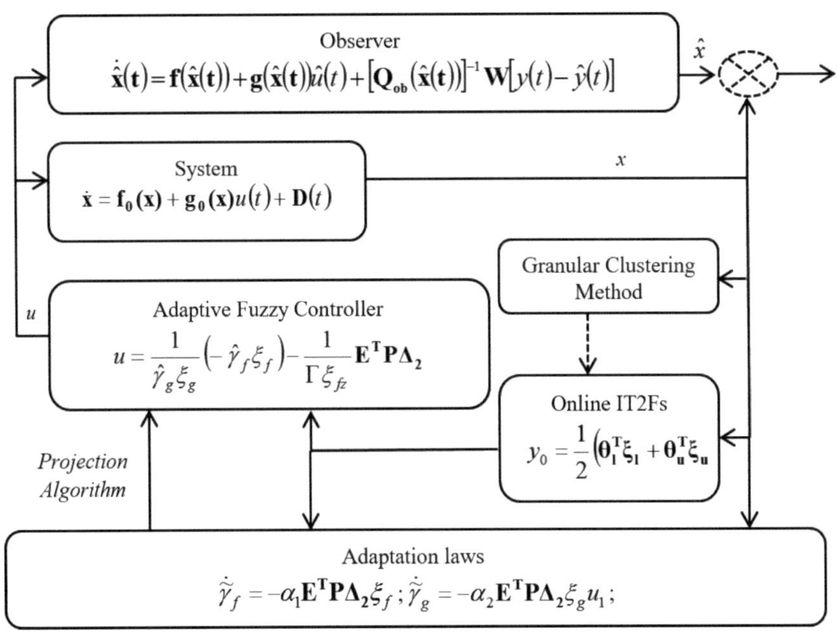

Figure 2. Control flow chart of the HAC-IFV.

4. Application to seat suspension system

4.1. Control results of the HAC-PP

In order to implement two adaptive controllers, principal parameters of the seat suspension and MR damper as shown in **Figure 3** are given in [25]. And two different road profiles of random step wave road and regular bump road are adopted to emulate severe external disturbances as shown in **Figure 4**. The first excitation is collected from the real road, and the second excitation is used same as in [25]. The process of simulation is expressed as follows: The proposed control will be simulated following an objective trajectory, which is control of [25]. Then, the outputs of the proposed control and the objective will be used for calculating error. This error will be checked by desired prescribed performance. It is remarked that the desired prescribed performance is different from the applied prescribed performance which is shown

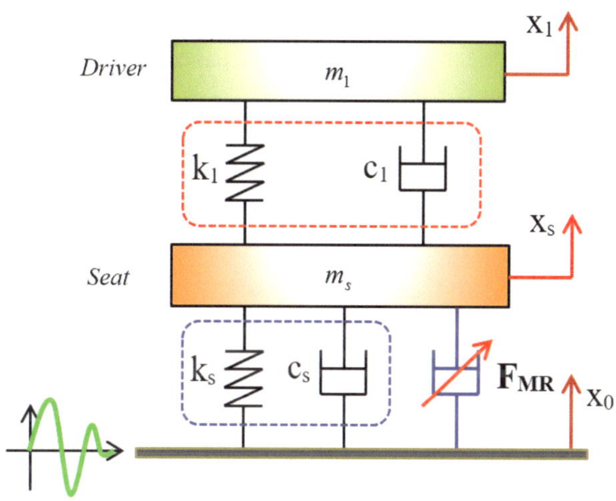

Figure 3. Mechanical model of a vehicle seat suspension system.

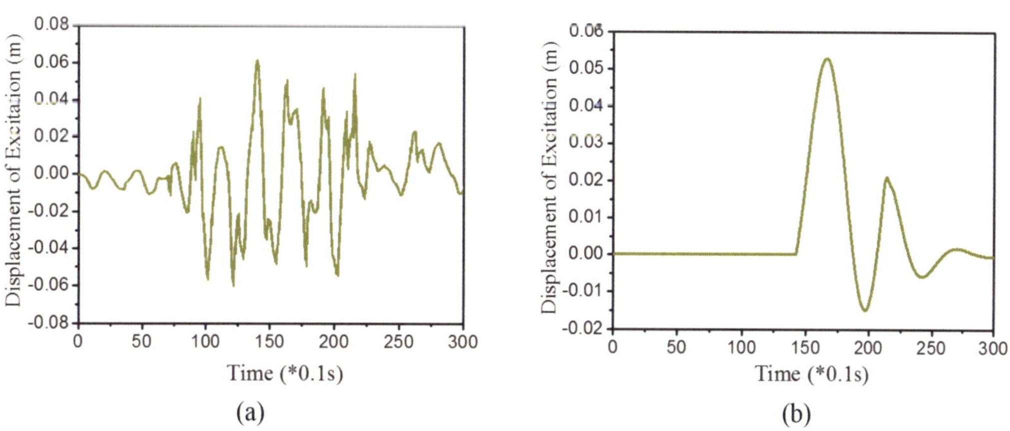

Figure 4. Road excitation signals: (a) random step wave road, (b) regular-bump road.

in Eqs. (7)–(13). The parameters of both the desired and the applied prescribed performance are listed in **Table 1**. The damping force of the MR damper is designed 1000 N (±5%) at 2 A. The fuzzy model is established based on the online model with the centroid vector as shown in [25]. It is noted that two main variables for the fuzzy models are displacement and acceleration. The fuzzy models include 6 clusters, and then the outputs of fuzzy rules become also 6. The sigma value for Gaussian function of the fuzzy model is chosen as 0.4 [22, 25], and this value is not changed through the simulation. The values of the sliding surface $[k_1, k_2]$ are chose by $[1, 20]$ for both random step wave road and regular bump road. The constant value Γ of the

Parameter	Desired prescribed performance	Applied prescribed performance
Initial value $\lambda(0)$	0.5	0.5
Infinity value λ_∞	0.001	0.001
Exponential value l	1	0.00047

Table 1. Parameters of desired prescribed performance and applied prescribed performance.

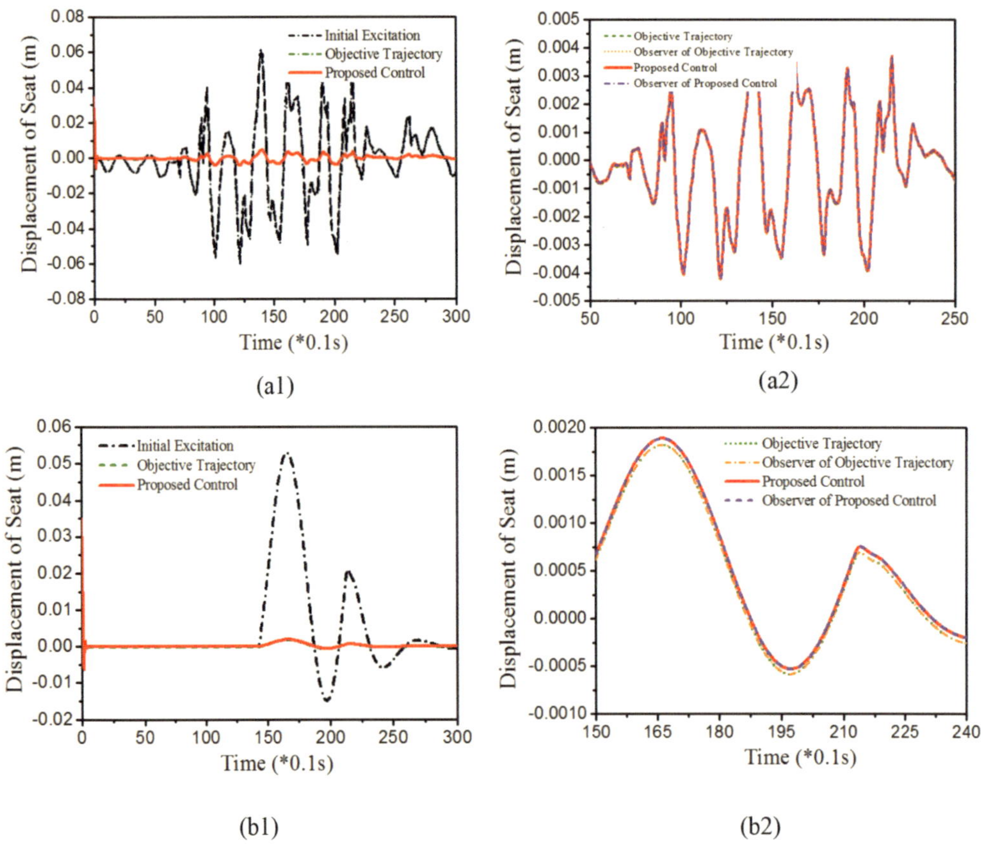

Figure 5. Control results with the HAC-PP at the seat (x_s): (a1, a2) random step wave road, (b1, b2) regular bump road.

Riccati-like equation is chosen by 10 for both roads. The constant c_s is 500 and 5000 for regular bump road and the random step wave road, respectively. In addition, the matrix **Q** of the Riccati-like equation is chosen as $\mathbf{Q}=[-2 \quad 0;0 \quad -2]$. The constants μ_1, μ_2, μ_3 of adaptation laws are chosen as 10 for two road profiles. The values of $\aleph_f$, $\aleph_g$, $\Theta\Gamma$ of the expanded adaptation laws are chosen by 0.1 and the values of $\delta_1, \delta_2, \delta_3$ are chosen by 0.1. In this simulation, the initial states for the dynamic states are used as $[0.035 \quad 2.5]$, $[0.035 \quad 2.5]$ for random regular bump, and random step wave bump, respectively. The initial states for the observer are $[0.035 \quad 0]$ for two excitations. It is noted that the observer is applied to evaluate the results of the proposed controller.

Figures 5–8 present control responses of the HAC-PP. It is clearly observed from **Figures 5** and **6** that the initial excitation has been significantly reduced by activation the proposed adaptive controller in terms of both displacement and acceleration. In addition, it is seen that the proposed control well tracks the objective trajectory which directly indicates high performance of the prescribed performance of the sliding surface. **Figure 7** presents the error of performance of the proposed adaptive controller which is always less than the boundary of the prescribed

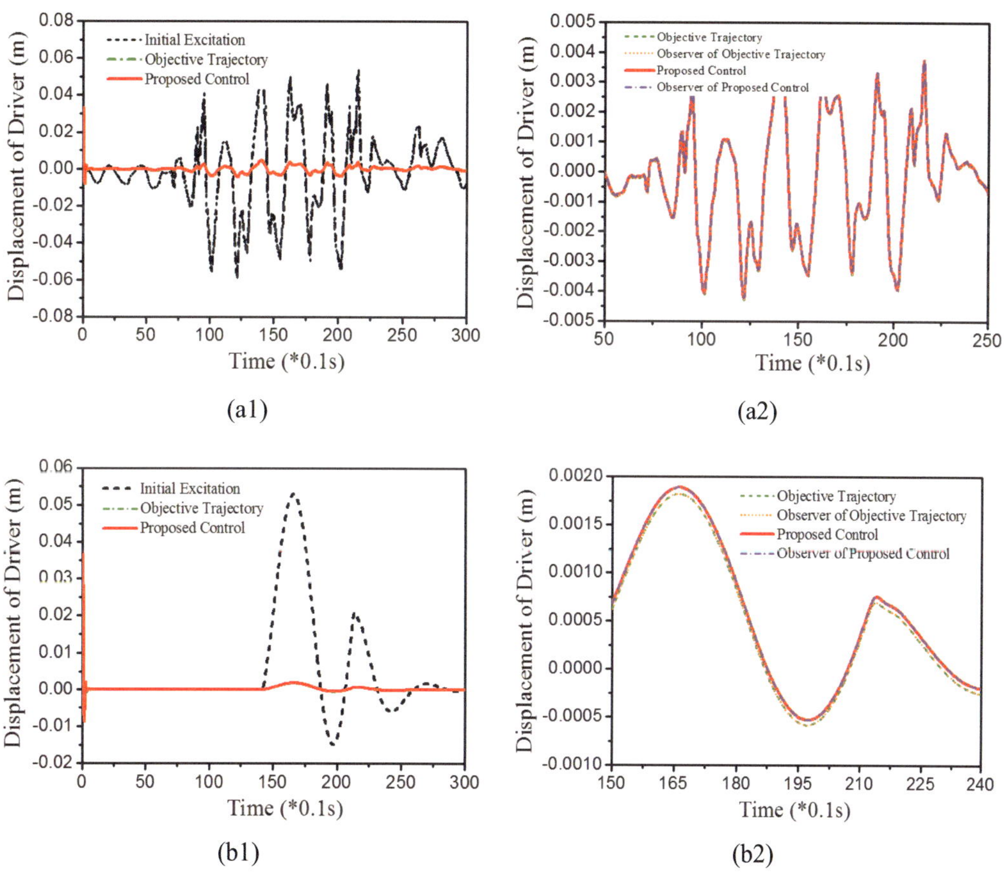

Figure 6. Control results with the HAC-PP at the driver (x_1): (a1, a2) random step wave road, (b1, b2) regular bump road.

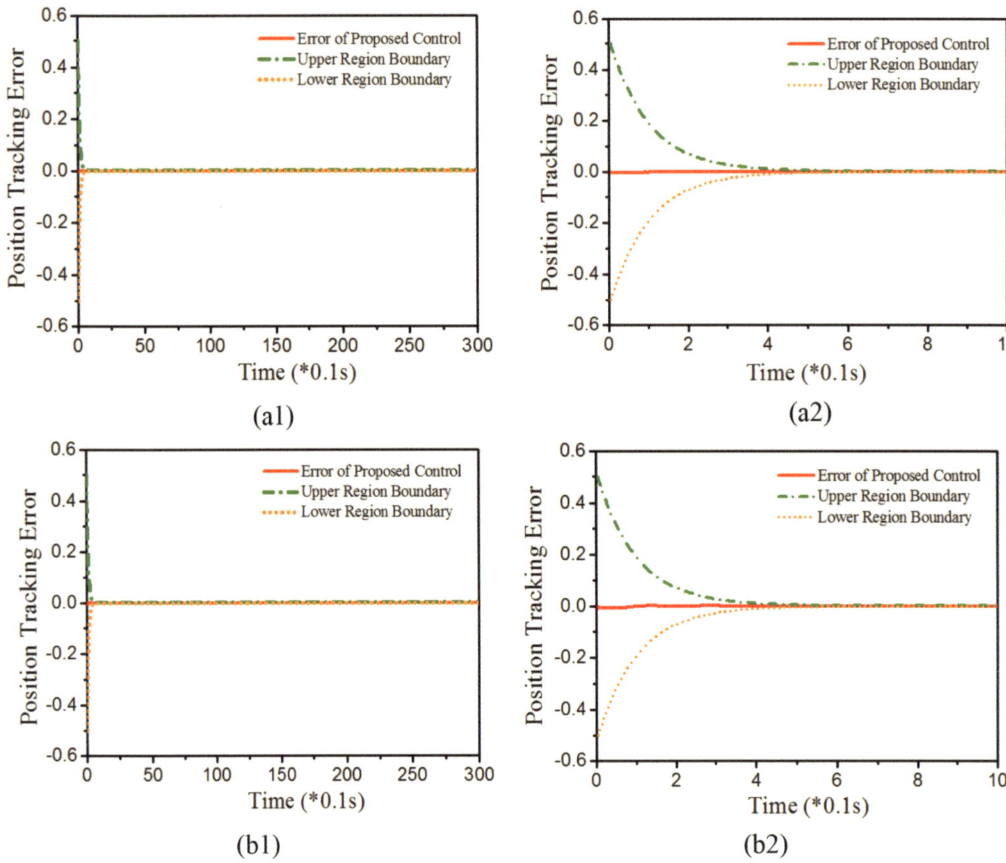

Figure 7. Tracking error with the HAC-PP: (a1, a2) random step wave road, (b1, b2) regular bump road.

performance. These results mean that the application of the prescribed performance in design of the hybrid adaptive controller can improve the quality of control with high robustness against severe excitations.

4.2. Control results of the HAC-IFV

In simulation of the HAC-IFV, the values of the sliding surface $[k_1, k_2]$ are chosen by $[1, 1.10^{-5}]$. The constant value Γ of the Riccati-like equation is chosen by 40, 10 for the regular bump road, the random step wave road, respectively. The constants α_1, α_2 of adaptation laws are chosen as 10 for all road profiles. The values of ε_f, ε_g of the expanded adaptation laws are chosen by 10 and the values of δ_1, δ_2 are chosen by 0.05. In this simulation, the initial states for the dynamic states are used as $[0.122 \quad 2.5]$, $[0.066 \quad 2.5]$, $[0.047 \quad 2.5]$ for random bump, random regular bump, and random step wave bump, respectively. The initial states for the observer are $[0.06 \quad 0]$ for two excitations. It is noted that the observer is applied to evaluate the results of the proposed controller. The parameters $[k_1, k_2]$ are chosen as $[1, 1.5]$ for random regular bump and $[1, 5]$ for random step wave bump.

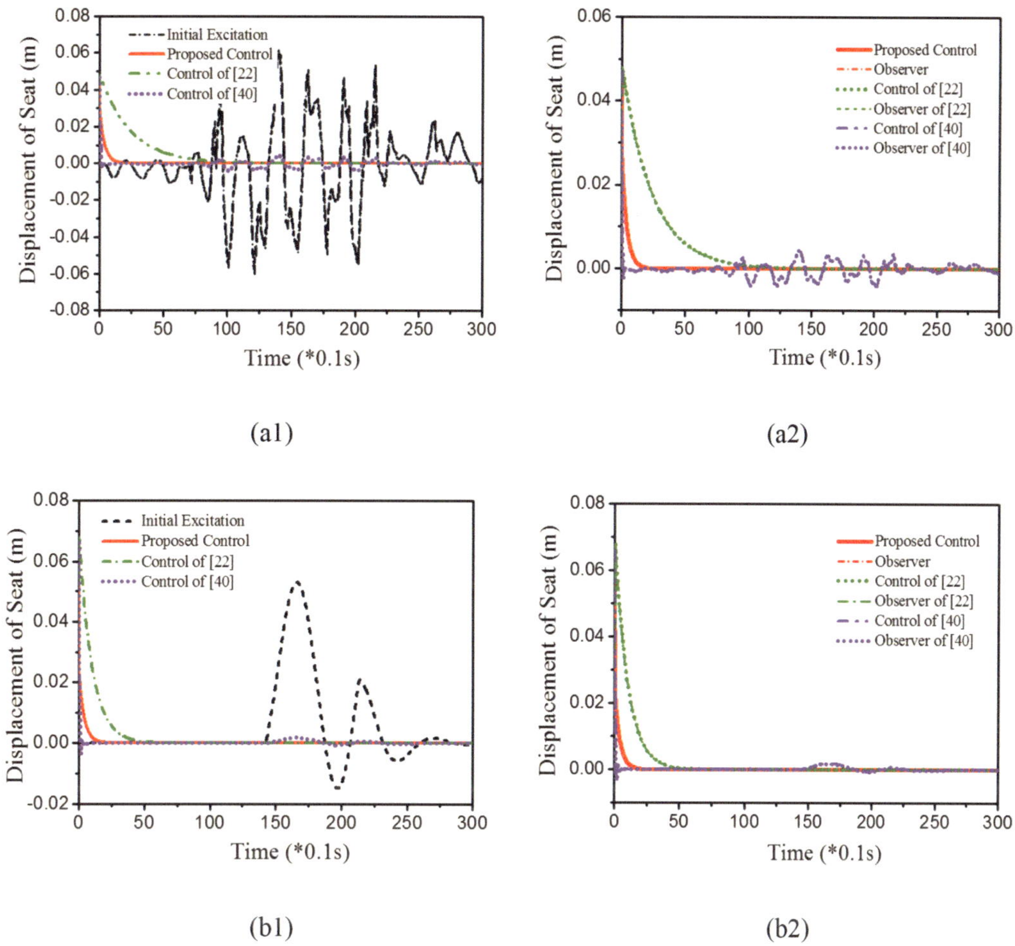

Figure 8. Control results with the HAC-IFV at the seat (x_s): (a1, a2) random step wave road, (b1, b2) regular bump road.

Figures 8–10 present control responses of the HAC-IFV. As similar to the HAC-PP, the initial excitations were remarkably reduced by applying the proposed controller. The displacements at the seat and driver positions are reduced resulting in the improvement of the ride comfort. In order to demonstrate a salient benefit of the proposed controller, its control response is compared obtained from the controller proposed in [17, 25]. It is clearly identified that the convergence time of the displacement of the proposed controller is 2 seconds for both excitations, while that is 15 seconds for the random step wave excitation, 6 seconds for regular bump excitation in [17, 25]. In **Figure 8**, the sliding surfaces of three controllers are shown. It is observed that the proposed control obtains stable motion much faster than the comparative controls at 0.1 second. It is noted here that the better control responses of the proposed controller comes from the inversely fuzzified values in given Eqs. (46)–(48). In Eq. (48), the independent of the inversely fuzzified value helps the controller to increase its robustness. This new exploration is the outstanding property of the proposed controller in the severe operation environment subjected to strong and random disturbances.

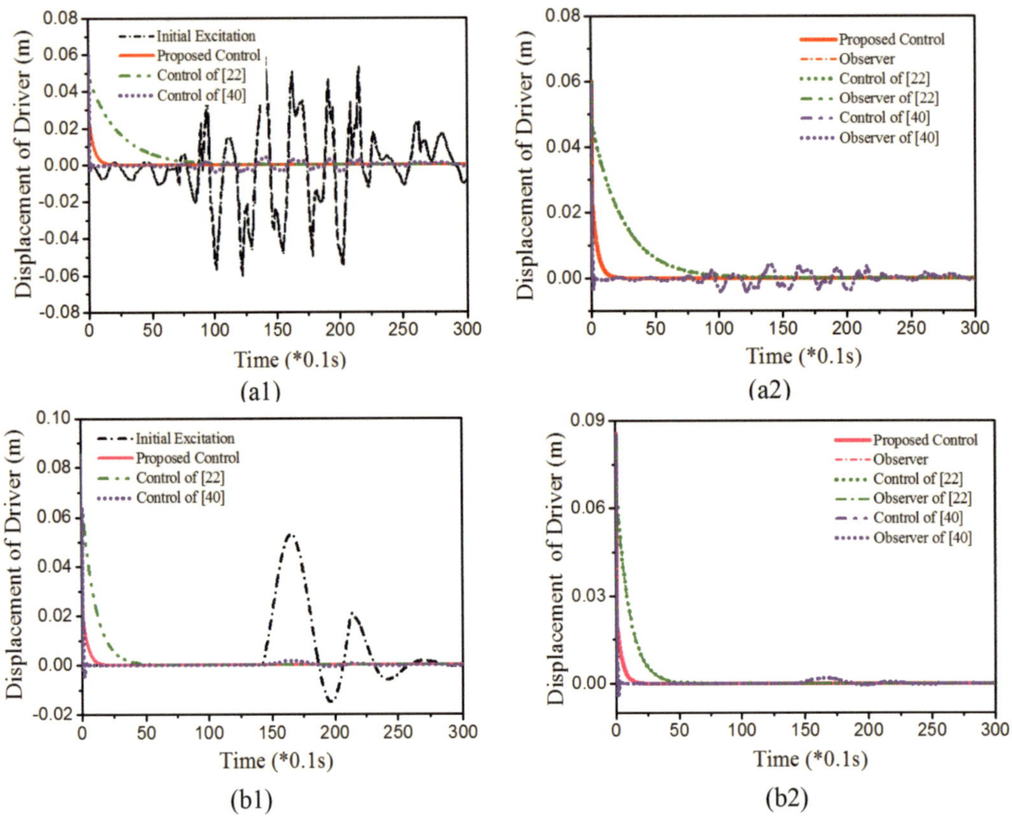

Figure 9. Control results with the HAC-IFV at the driver (x_1): (a1, a2) random step wave road, (b1, b2) regular bump road.

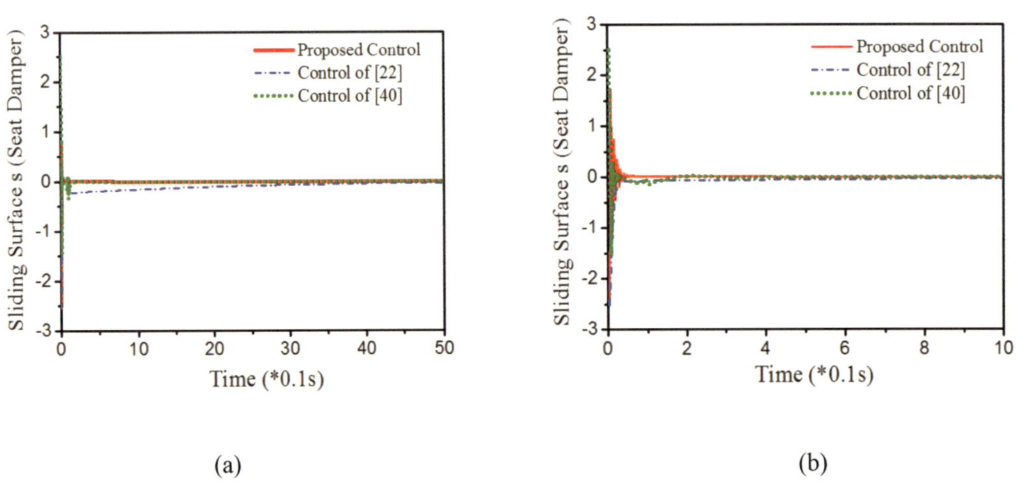

Figure 10. Sliding surface motion of the HAC-IFV (s): (a) random step wave road, (b) regular bump road.

5. Concluding remarks

In this study, two new adaptive controllers were formulated and their effectiveness was validated by applying them to vibration control of a semi-active vehicle seat suspension system featuring MR damper. The first adaptive controller includes two sliding mode controls: one for initial states of the system and the other for prescribed performance associated with the parameters of the modified Riccati-like equation. By doing this way, the tracking performance is enhanced resulting in the improved control responses. The second adaptive controller was formulated on the basis of the inversely fuzzified value with the H-infinity control to minimize computational cost algorithm. Hence, by doing this way, the convergence time can be reduced resulting in high stability of the system subjected to severe external disturbances. It has been sown that the proposed two adaptive controllers can significantly reduce the excitation from the road profiles at both the seat and driver positions. In reality, this can enhance the ride comfort of the driver. Especially, the HAC-PP provides good tracking performance with the error in range of the defined boundary and the HAC-IFV can reduce the convergence time compared with two comparative adaptive controllers. It is finally remarked that the development of a new hybrid adaptive controller needs to be connected with desired control performances to appropriately select each control scheme.

Acknowledgements

This research was funded by Vietnam National Foundation for Science and Technology Development (NAFOSTED) under grant number 107.01-2017.28. The financial support is gratefully acknowledged.

Declaration of conflicting interest

The authors declare that there is no conflict of interest.

Author details

Do Xuan Phu[1], Ta Duc Huy[1] and Seung Bok Choi[2]*

*Address all correspondence to: seungbok@inha.ac.kr

1 MediRobotics Laboratory, Department of Mechatronics and Sensor System Technology, Vietnamese-German University, Binh Duong, Vietnam

2 Smart Structures and Systems Laboratory, Department of Mechanical Engineering, Inha University, Incheon, Korea

References

[1] Shahriari Kahkeshi M, Sheikholeslam F, Zekri M. Design of adaptive fuzzy wavelet neural sliding mode controller for uncertain nonlinear systems. ISA Transactions. 2013;**52**:342-350

[2] Do XP, Shah K, Choi S-B. Damping force tracking control of MR damper system using a new direct adaptive fuzzy controller. Shock and Vibration. 2015;**2015**:947937

[3] Huang S-J, Chen H-Y. Adaptive sliding controller with self-tuning fuzzy compensation for vehicle suspension control. Mechatronics. 2006;**16**:607-622

[4] Shahnazi R. Output feedback adaptive fuzzy control of uncertain MIMO nonlinear systems with unknown input nonlinearities. ISA Transactions. 2015;**54**:39-51

[5] Cui Y, Zhang H, Wang Y, Zhang Z. Adaptive neural dynamic surface control for a class of uncertain nonlinear systems with disturbances. Neurocomputing. 2015;**165**:152-158

[6] Wu L-B, Yang G-H. Adaptive fuzzy tracking control for a class of uncertain non-affine nonlinear systems with dead-zone inputs. Fuzzy Sets and Systems. 2015;**290**:1-21

[7] Hsu C-F. Adaptive fuzzy wavelet neural controller design for chaos synchronization. Expert Systems with Applications. 2011;**38**:10475-10483

[8] Li Y, Tong S. Prescribed performance adaptive fuzzy output-feedback dynamic surface control for nonlinear large-scale systems with time delays. Information Sciences. 2015;**292**: 125-142

[9] Huang Y, Na J, Wu X, Liu X, Guo Y. Adaptive control of nonlinear uncertain active suspension systems with prescribed performance. ISA Transactions. 2015;**54**:145-155

[10] Yi-Min L, Yang Y, Li L. Adaptive backstepping fuzzy control based on type-2 fuzzy system. Journal of Applied Mathematics. 2012;**2012**:658424

[11] Gao Y, Er MJ. Online adaptive fuzzy neural identification and control of a class of MIMO nonlinear systems. IEEE Transactions on Fuzzy Systems. 2003;**11**(4):462-477

[12] Lin F-J, Chou P-H, Shieh P-H, Chen S-Y. Robust control of an LUSM-based X-Y-θ motion control stage using an adaptive interval type-2 fuzzy neural network. IEEE Transactions on Fuzzy Systems. 2009;**17**(1):24-38

[13] Wang F, Liu Z, Zhang Y, Chen CLP. Adaptive fuzzy control for a class of stochastic pure-feedback nonlinear systems with unknown hysteresis. IEEE Transactions on Fuzzy Systems. 2016;**24**(1):140-152

[14] Phu DX, Choi S-B. Vibration control of a ship engine system using high-load magnetorheological mounts associated with a new indirect fuzzy sliding mode controller. Smart Materials and Structures. 2015;**24**:025009

[15] Choi S-B, Li W, Yu M, Du H, Fu J, Do PX. State of the art of control schemes for smart systems featuring magneto-rheological materials. Smart Materials and Structures. 2016;**25**: 043001

[16] Phu DX, Shin DK, Choi S-B. Design of a new adaptive fuzzy controller and its application to vibration control of a vehicle seat installed with an MR damper. Smart Materials and Structures. 2015;**24**:085012

[17] Phu DX, Shah K, Choi S-B. Design of a new adaptive fuzzy controller and its implementation for the damping force control of a magneto-rheological damper. Smart Materials and Structures. 2014;**23**:065012

[18] Phu DX, Choi S-B, Lee Y-S, Han M-S. Vibration control of a vehicle's seat suspension featuring a magneto-rheological damper based on a new adaptive fuzzy sliding-mode controller. Proceedings of the Institution of Mechanical Engineers, Part D: Journal of Automobile Engineering. 2015;**230**(4):437-458. DOI: 10.1177/0954407015586678

[19] Fayek HM, Elamvazuthi I, Perumal N, Venkatesh B. A controller based on optimal type-2 fuzzy logic: Systematic design, optimization and real-time implementation. ISA Transactions. 2014;**53**:1583-1591

[20] Wang X, Liu X, Zhang L. A rapid fuzzy rule clustering method based on granular computing. Applied Soft Computing. 2014;**24**:534-542

[21] Mendel JM, Liu X. Simplified interval type-2 fuzzy logic systems. IEEE Transactions on Fuzzy Systems. 2013;**21**(6):1056-1069

[22] Juang C-F, Tsao Y-W. A self-evolving interval type-2 fuzzy neural network with online structure and parameter learning. IEEE Transactions on Fuzzy Systems. 2008;**16**(6):1411-1424

[23] Liu J. Sliding Mode Control Using MATLAB. Academic Press: Tsinghua University Press; 2017

[24] Ciccarella G, Dalla Mora M, Germani A. A Luenberger-like observer for nonlinear systems. International Journal of Control. 1993;**57**(3):537-556

[25] Phu DX, Choi S-M, Choi S-B. A new adaptive hybrid controller for vibration control of a vehicle seat suspension featuring MR damper. Journal of Vibration and Control. 2016: 1-22. DOI: 10.1177/1077546316629597

Adaptive Robust Guidance Scheme Based on the Sliding Mode Control in an Aircraft Pursuit-Evasion Problem

Jian Chen, Yongjun Zheng, Yuan Ren, Yuan Tian,
Chen Bai, Zhang Ren, Guangqi Wang,
Nannan Du and Yu Tan

Additional information is available at the end of the chapter

http://dx.doi.org/10.5772/intechopen.72177

Abstract

In this chapter, a robust guidance scheme utilizing a line-of-sight (LOS) observation is presented. Initial relative speed and distance, and error boundaries of them are estimated in accordance with the interceptor-target relative motion kinematics. A robust guidance scheme based on the sliding mode control (SMC) is developed, which requires the boundaries of the target maneuver, and inevitably has jitter phenomenon. For solving above-mentioned problems, an estimation to the target acceleration's boundary is developed for enhancing robustness of the guidance scheme and the Lyapunov stabilization is analyzed. The proposed robust guidance scheme's brief characteristic is to reduce the effect of relative speed and distance, to reduce the effect of target maneuverability on the guidance precision, and to strengthen the influence of line-of-sight angular velocity. The proposed scheme's performances are validated by the simulations of different target maneuvers under two worst-case conditions.

Keywords: robust guidance scheme, line-of-sight angular velocity, sliding mode control, boundary of target maneuver, Lyapunov stability

1. Introduction

The traditional Proportional Navigation Guidance (PNG) schemes including their extensions have been widely employed in interceptors because of their efficiencies and simplifications (only need line-of-sight information). PNG makes the normal load of interceptor proportional to the line-of-sight (LOS) angular velocity [1]. Nevertheless, the target can add the miss distance by acting evading maneuver because the target maneuvers are ahead of the guidance commands from PNG. For achieving desired interception performances, even for the target maneuver, it is necessary to develop advanced guidance schemes [1, 2]. These advanced guidance schemes usually require more information, like relative speed and distance, target's acceleration, or time-to-go.

For passive seekers, which are equipped with electro-optical or infrared sensors, the line-of-sight angular velocity can be observed only. If an estimation equation is assumed to estimate target kinematics, relative speed and distance and target's acceleration will be identified. The commonly used estimation model is a Kalman filter. At the same time, one must select a target's motion model, like the current statistical model, the Singer model, or the interactive multi-model scheme [3]. For polishing up the estimating performance of the target's maneuvers, observability of the interception problem with LOS angular velocity measurement is analyzed [4]. It concludes that current homing guidance schemes result in a decrease in observability of tracking the target. Because relative distance cannot be observed from the line-of-sight measurement, it is necessary to use a special type of self-motion to solve this problem. Thus, the method of introducing LOS angular oscillatory motion is presented in [4] to improve the observability. The oscillatory motion of the LOS angle improves observability, but the trajectories generated by the guidance schemes are inevitably influenced by this motion mode and affect final guidance precision. Therefore, the target maneuver estimation is constrained by a lot of practical limitations.

The sliding mode control (SMC) is robust to disturbances. Therefore, it is employed to develop adaptive guidance schemes for target's unpredictable maneuver without requirements to estimate the target's acceleration. Recently, many guidance schemes based on the SMC have been proposed, for instance, the guidance schemes based on the adaptive and optimal SMC [5–7], the high-order SMC [8, 9], SMC-based integrated guidance and control (IGC) [10, 11], and SMC with impacting angle constraint [12–14].

SMC-based optimal and adaptive guidance schemes have become a focus since the 1990s. An adaptive sliding mode guidance (ASMG) scheme is presented in [5] for target maneuvering and parameter disturbance of the guidance system. In addition, an optimal sliding mode guidance (OSMG) scheme is deduced from the ASMG, and the optimal guidance coefficients are given in [6]. In [7], the Fuzzy OSMG (FOSMG) formulated by the OSMG and PNG is stated by adjusting the weights of the OSMG and PNG using fuzzy logic. It is noted that the FOSMG owns the advantages of the PNG for nonmaneuvering targets and the OSMG for maneuvering targets. The ASMG, OSMG, and FOSMG have practical advantages of simple expressions. Nevertheless, it is essential to identify the target's normal load to adjust weights of the FOSMG.

The higher order sliding mode guidance (HOSMG) scheme is a current research highlight. While the SMC-based first-order guidance is a balance between smoothing jitter and ensuring robustness through switching frequently, the HOSMG generates control commands smoothly to systems with relative degree arbitrarily. A smooth guidance scheme based on a second-order sliding mode is developed for solving the uncertainties of the actuator and the target's maneuver [8]. In [9], a terminal guidance law with known convergent time is proposed by using the finite-time mean-square practical convergence as sliding surfaces. It validates that HOSMG is robust to stochastic noises and bounded uncertainty and does not have high-frequency jitters. The HOSMG's flaw exists in converging slowly for real time due to complex algorithms.

Integrated guidance and control (IGC) which is based on the SMC has become an unusual approach for developing a guidance system. The traditional timescale separation method splits the guidance system into an inner loop autopilot and an outer loop command system. The IGC system merges the two loops into a unique loop. Based on self-motion and relative motion states, the IGC produces commands to aero surfaces straight. Zero effort miss is used to be a sliding surface for developing the IGC [10, 11]. Due to the complex coupling between guidance and control states, this method does not spread more widely than an intuitive timescale separation method.

Based on the SMC, the guidance scheme with impacting angle constraint is designed to solve the problem of directional impacting. In practical cases, a specific impacting angle is desired to directionally hit the target or to better detect thee target. Under the impacting angle constraint, the result from ideal initial attack conditions can meet the interception requirements for nonmaneuvering and step-maneuvering targets [12–14]. However, the guidance accuracy of the target's complicated maneuver will decrease.

To sum up, although guidance schemes based on the SMC and target's maneuver estimation algorithms act excellently in simulation, the computation degrees of them have become too complex to realize. In fact, one needs a simple-expression guidance scheme based on the SMC, and if not requiring the target acceleration, it will be better. In addition, the discontinuous characteristic induced by the sliding mode part can cause jitter of guidance commands that is detrimental to the aero fins. The coefficient of sliding mode part indicates target acceleration's boundary. Actually, it is hard to get the boundary. If setting the boundary too great, the autopilot might be saturated; if setting the boundary too tiny, the sliding mode's presence cannot be guaranteed. With the target acceleration's unknown boundary, for ensuring the stability, the simplified guidance schemes based on the SMC including the FOSMG, OSMG, and ASMG have greater sliding mode part, which might cause jitter. Adaptive control offers a solution. The unknown parameters can be estimated in the online identification for the uncertain system. Nevertheless, it has no capability to suppress disturbances. Thus, the adaptive control to identify the upper boundary of the system uncertainty is merged with the SMC to suppress disturbances [15].

An interceptor-target pursuit-evasion game which only employs the line-of-sight angular velocity is under consideration in this chapter. The target maneuver is treated as a bounded perturbation. More states are demanded, including initial relative speed and distance and error boundaries of them. It is derived from recursive estimation of relative motion kinematics and obtains approximations of relative speed and distance. A simplified sliding mode guidance scheme is given, which requires target acceleration's boundary and inevitably has jitter phenomenon. For overcoming the above shortcomings, an adaptive parameter is utilized to estimate target acceleration's boundary and to adaptively adjust in terms of the line-of-sight angular velocity; moreover, the Lyapunov stabilization has been analyzed. The proposed guidance scheme's brief characteristic is to decrease the effect of relative speed and distance on the guidance precision and to strengthen the influence of line-of-sight angular velocity.

This chapter's rest part is listed hereafter. The problem statement is given in Section 2. Two robust guidance schemes based on the SMC are presented in Section 3. Section 4 carries out simulations, and conclusions are given in Section 5.

2. Pursuit-evasion game

2.1. Relative motion kinematics

The interceptor's movement consists of two orthogonal channels. The pursuit-evasion game is decomposed into two 2D channels.

Figure 1 shows the geometric diagram of the interceptor-target relative motion kinematics. A Cartesian reference system is denoted by "X-O-Y." The interceptor and target are denoted by "M" and "T". The line-of-sight angle is denoted by "q". The relative distance is denoted by "r". Flight path angles of the interceptor and target are denoted by "φ_m" and "φ_t". Velocities of the interceptor and target are denoted by "V_m" and "V_t".

Endgame relative motion kinematics are given by

$$\begin{cases} \dot{r} = V_t \cos(\varphi_t - q) - V_m \cos(\varphi_m - q), \\ r\dot{q} = V_t \sin(\varphi_t - q) - V_m \sin(\varphi_m - q). \end{cases} \tag{1}$$

Let the relative speed $v = \dot{r}$. Eq. (2) is obtained as

$$\begin{cases} \dot{v} = r\dot{q}^2 + \left[\dot{V}_t \cos(\varphi_t - q) - V_t\dot{\varphi}_t \sin(\varphi_t - q)\right] \\ \quad - \left[\dot{V}_m \cos(\varphi_m - q) - V_m\dot{\varphi}_m \sin(\varphi_m - q)\right], \\ r\ddot{q} = -2\dot{r}\dot{q} + \left[\dot{V}_t \sin(\varphi_t - q) + V_t\dot{\varphi}_t \cos(\varphi_t - q)\right] \\ \quad - \left[\dot{V}_m \sin(\varphi_m - q) + V_m\dot{\varphi}_m \cos(\varphi_m - q)\right]. \end{cases} \tag{2}$$

For simplification, we get

$$\dot{v} = r\dot{q}^2 + a_{tr} - a_{mr}, \tag{3}$$

$$\ddot{q} = -2\frac{v}{r}\dot{q} + \frac{1}{r}a_{tq} - \frac{1}{r}a_{mq}, \tag{4}$$

where acceleration components of the interceptor and target along the line-of-sight are denoted by "a_{mr}" and "a_{tr}"; acceleration components of the interceptor and target orthogonal to the line-of-sight are denoted by "a_{mq}"and "a_{tq}". The equations of them are formulated as

$$a_{tr} = \dot{V}_t \cos(\varphi_t - q) - V_t\dot{\varphi}_t \sin(\varphi_t - q), \tag{5}$$

$$a_{mr} = \dot{V}_m \cos(\varphi_m - q) - V_m\dot{\varphi}_m \sin(\varphi_m - q), \tag{6}$$

$$a_{tq} = \dot{V}_t \sin(\varphi_t - q) + V_t\dot{\varphi}_t \cos(\varphi_t - q), \tag{7}$$

$$a_{mq} = \dot{V}_m \sin(\varphi_m - q) + V_m\dot{\varphi}_m \cos(\varphi_m - q). \tag{8}$$

Assume the line-of-sight angular velocity is accurately observed at each instant. Initial relative speed and distance and error boundaries of them are obtained as

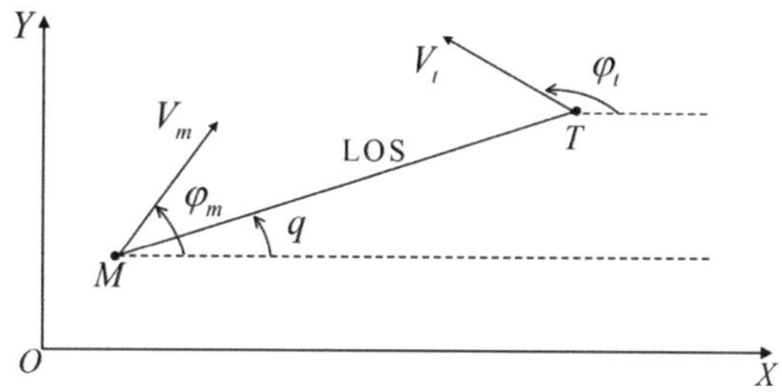

Figure 1. Relative motion kinematics.

$$\begin{cases} r_0 = \overline{r}_0 + \tilde{r}_0, & |\tilde{r}_0| \leq \delta_{r_0}, \\ v_0 = \overline{v}_0 + \tilde{v}_0, & |\tilde{v}_0| \leq \delta_{v_0}, \end{cases} \tag{9}$$

where initial relative speed and distance are denoted by "v_0" and "r_0"; observations of v_0 and r_0 are denoted by "$\overline{v}_0$" and "$\overline{r}_0$"; observation deviations of v_0 and r_0 are denoted by "$\tilde{v}_0$" and "$\tilde{r}_0$"; upper boundaries of $\tilde{v}_0$ and $\tilde{r}_0$ are denoted by "δ_{v_0}" and "δ_{r_0}".

Remark 1. $\overline{v}_0$ and $\overline{r}_0$ are detected by a radar on ground or aircraft carrier and are sent to the interceptor via the data link only once. δ_{v_0} and δ_{r_0} are treated to be maximum observation deviation of the detector.

2.2. Kinematics simplification

For successfully intercepting the target, the line-of-sight angular velocity should be constrained [5, 6]. In this chapter, seeker and autopilot loops are not considered. With this premise, relative equation of the line-of-sight angular velocity $\dot{q}$ is obtained as Eq. (4). However, $1/r$ and $2v/r$ in Eq. (4) are obtained as Eq. (3), which indicates relative speed v alters as r, v, $\dot{q}$, a_{tr}, and a_{mr} vary, and v_0 and r_0 are preset. In accordance with the characteristic of an interceptor's engine, the thrust along the line of sight almost does not change. Moreover, the target is usually escaping orthogonally to the line of sight to increase the line-of-sight angular velocity. Although acceleration component of the target along the line-of-sight is subsistent, the relative speed does not change too much with limited energy and time. Assume that acceleration components along the line of sight of the target and interceptor are zero. Simplify Eq. (3) into

$$\begin{cases} \dot{r} = v, \\ \dot{v} = r\dot{q}^2. \end{cases} \tag{10}$$

Define $z_1 = r$ and $z_2 = v$. Equation (10) becomes

$$\begin{cases} \dot{z}_1 = z_2, \\ \dot{z}_2 = z_1\dot{q}^2. \end{cases} \tag{11}$$

According to Eq. (9), initial states for Eq. (11) to get $\overline{r}_0$ and $\overline{v}_0$ are obtained as

$$z_1^{(1)}(t_0) = \bar{r}_0,\ z_2^{(1)}(t_0) = \bar{v}_0, \tag{12}$$

$$z_1^{(2)}(t_0) = \bar{r}_0 + \delta_{r0},\ z_2^{(2)}(t_0) = \bar{v}_0 + \delta_{v0}, \tag{13}$$

$$z_1^{(3)}(t_0) = \bar{r}_0 - \delta_{r0},\ z_2^{(3)}(t_0) = \bar{v}_0 - \delta_{v0}. \tag{14}$$

Equations (12)–(14) are employed to calculate Eq. (11). Boundaries of v and $r$are computed as

$$\begin{cases} |\Delta r| = \max\left\{z_1^{(2)} - z_1^{(1)}, z_1^{(1)} - z_1^{(3)}\right\}, \\ |\Delta v| = \max\left\{z_2^{(2)} - z_2^{(1)}, z_2^{(1)} - z_2^{(3)}\right\}. \end{cases} \tag{15}$$

3. Guidance scheme design

3.1. Guidance scheme based on the sliding mode control

A sliding surface is determined by

$$s = z_1^{(1)}\dot{q}. \tag{16}$$

In accordance with Eq. (16), forcing s to zero represents that $\dot{q}$ or $\bar{r}$ prompts to 0. In terms of quasi-parallel approach guideline, the line-of-sight angular velocity will be adjusted to 0 to guarantee that the interceptor hits the target [5, 6].

Theorem 1. A sliding mode control-based guidance (SMCG) scheme described by

$$a_{mq} = \left[\left(N - \frac{z_1^{(2)}}{z_1^{(3)}}\right)\left|z_2^{(1)}\right| + 2|\Delta v|\right]\dot{q} + \varepsilon sgn(\dot{q}), \tag{17}$$

where $z_2^{(1)}$, $z_1^{(2)}$, and $z_1^{(3)}$ are deduced from Eq. (11) with Eqs. (12–14), $N > 2$ is an integer, $|\Delta v|$ is obtained from Eq. (15), and ε is a_{tq}'s upper boundary, guarantees that $s = z_1^{(1)}\dot{q}$ is driven to 0.

Proof. Compute Eq. (11) with Eq. (9) and define $\bar{v} = z_2^{(1)}$ and $\bar{r} = z_1^{(1)}$.v and r are obtained as

$$r = \bar{r} + \tilde{r}, v = \bar{v} + \tilde{v}, \tag{18}$$

where derivations between estimations and real values are denoted by "$\tilde{v}$" and "$\tilde{r}$".

In terms of the deduction in Section 2, we have

$$|\tilde{r}| \leq |\Delta r|, |\tilde{v}| \leq |\Delta v|. \tag{19}$$

Define a Lyapunov function:

$$V_1 = 0.5s^2. \tag{20}$$

Since $\bar{v} < 0$, an approach scheme is defined as

$$\dot{s} = \frac{\overline{r}}{r}\{ -\left(N - 2 + \frac{z_1^{(2)}}{z_1^{(3)}} - \frac{r}{\overline{r}}\right)|\overline{v}|\dot{q} - 2\left(|\Delta v| - \tilde{v}\right)\dot{q} - \left[\varepsilon sgn(\dot{q}) - a_{tq}\right]\}. \tag{21}$$

Then,

$$\dot{V}_1 = -\frac{\overline{r}^2}{r}\left(N - 2 + \frac{z_1^{(2)}}{z_1^{(3)}} - \frac{r}{\overline{r}}\right)|\overline{v}|\dot{q}^2 - 2\frac{\overline{r}^2}{r}(|\Delta v| - \tilde{v})\dot{q}^2 - \frac{\overline{r}^2}{r}\left[\varepsilon sgn(\dot{q}) - a_{tq}\right]\dot{q}. \tag{22}$$

Equation (11) is solved with Eq. (12) or (14). Then, we get

$$0 < z_1^{(3)} \le \overline{r}. \tag{23}$$

Equation (11) is solved with Eq. (9) or (13). Because $|\tilde{r}_0| \le \delta_{r0}$, we have

$$z_1^{(2)} \ge r > 0. \tag{24}$$

Next, the following is obtained:

$$\frac{z_1^{(2)}}{z_1^{(3)}} - \frac{r}{\overline{r}} \ge 0. \tag{25}$$

Since $N > 2$, then

$$N - 2 + \frac{z_1^{(2)}}{z_1^{(3)}} - \frac{r}{\overline{r}} > 0. \tag{26}$$

Because $\varepsilon sgn(\dot{q}) - a_{tq} > 0$, $|\Delta v| - \tilde{v} > 0$, $\overline{r} > 0$, and $r > 0$, we get $\dot{V}_1 < 0$. Using Lyapunov stability theory, we can guarantee that $V_1 \to 0$. Finally $s \to 0$. Since $s = z_1^{(1)}\dot{q}$, that is, $\dot{q} \to 0$.

Remark 2. The "sgn" function in Eq. (17) is replaced by the following function to suppress the jitter:

$$a_{mq} = \left[\left(N - \frac{z_1^{(2)}}{z_1^{(3)}}\right)\left|z_2^{(1)}\right| + 2|\Delta v|\right]\dot{q} + \varepsilon\frac{\dot{q}}{|\dot{q}| + \Delta}, \tag{27}$$

where Δ is a tiny positive constant.

3.2. Improved guidance scheme based on the SMCG

ε in Eq. (27) or (17) is unchanged, which indicates that an unchanged upper boundary of a_{tq} is employed to ensure the sliding mode's subsistence. By this means, the guidance command's jitter might exist in the vicinity of the sliding mode although "sgn" is already replaced in Eq. (27). For smoothing the command, the better way is to use the adaptive approach to dynamically estimate ε [23, 24].

Proposition 1. An unchanged constant $k > 0$ exists, so that

$$\left|a_{tq}\right| \leq (N-2+k)|\overline{v}||\dot{q}|, \tag{28}$$

where upper boundaries' estimations of a_{tq} are formulated by $(N-2+k)|\overline{v}|\dot{q}$.

Theorem 2. An improved sliding mode control-based guidance (ISMCG) scheme described by

$$\begin{cases} a_{mq} = \left[\left(N - \dfrac{z_1^{(2)}}{z_1^{(3)}} + \widehat{k}\right)\left|z_2^{(1)}\right| + 2|\Delta v|\right]\dot{q}, \\ \dot{\widehat{k}} = \dfrac{1}{\gamma}\dfrac{\overline{r}^2}{r}|\overline{v}|\dot{q}^2, \end{cases} \tag{29}$$

where $z_2^{(1)}$, $z_1^{(2)}$, and $z_1^{(3)}$ are deduced from Eq. (11) with Eqs. (12–14), $N > 2$ is an integer, $|\Delta v|$ is obtained from Eq. (15), and $\gamma > 0$ is a constant, guarantees that $s = z_1^{(1)}\dot{q}$ is driven to 0.

Proof. Define $\tilde{k} = k - \widehat{k}$ and the Lyapunov function:

$$V_2 = 0.5\left(s^2 + \gamma\tilde{k}^2\right). \tag{30}$$

Then,

$$\dot{V}_2 = -\frac{\overline{r}^2}{r}\left(\frac{z_1^{(2)}}{z_1^{(3)}} - \frac{r}{\overline{r}}\right)|\overline{v}|\dot{q}^2 - 2\frac{\overline{r}^2}{r}(|\Delta v| - \tilde{v})\dot{q}^2 - \frac{\overline{r}^2}{r}\left[\left(N - 2 + \widehat{k}\right)|\overline{v}|\dot{q} - a_{tq}\right]\dot{q} + \gamma\tilde{k}\dot{\tilde{k}}. \tag{31}$$

According to Eq. (28) and $\dot{\widehat{k}} = \frac{1}{\gamma}\frac{\overline{r}^2}{r}|\overline{v}|\dot{q}^2$, we get

$$\begin{aligned} \dot{V}_2 &\leq -\frac{\overline{r}^2}{r}\left(\frac{z_1^{(2)}}{z_1^{(3)}} - \frac{r}{\overline{r}}\right)|\overline{v}|\dot{q}^2 - 2\frac{\overline{r}^2}{r}(|\Delta v| - \tilde{v})\dot{q}^2 - \frac{\overline{r}^2}{r}\left(\widehat{k} - k\right)|\overline{v}|\dot{q}^2 + \gamma\tilde{k}\dot{\tilde{k}} \\ &= -\frac{\overline{r}^2}{r}\left(\frac{z_1^{(2)}}{z_1^{(3)}} - \frac{r}{\overline{r}}\right)|\overline{v}|\dot{q}^2 - 2\frac{\overline{r}^2}{r}(|\Delta v| - \tilde{v})\dot{q}^2. \end{aligned}$$

In accordance with condition (25), because $|\Delta v| - \tilde{v} > 0$, $\overline{r} > 0$, and $r > 0$, we get $\dot{V}_2 < 0$. Using Lyapunov stability theory, we can guarantee that $V_2 \to 0$. Finally $s \to 0$. Since $s = z_1^{(1)}\dot{q}$, that is, $\dot{q} \to 0$.

Remark 3. $\overline{r}$ is employed to take the place of the real r in Eq. (29).

4. Simulations

4.1. Initial conditions

Simulations will be conducted to validate the feasibility and superiority of the proposed schemes in this part. In simulations, the maximum acceleration limit of the interceptor is 10 g for verifying the performance of the interceptor with a rather constrained maneuverability. Assume that the target is less agile than the interceptor. Control systems of them are expressed by the following first-order systems:

$$\frac{a_{ma}(s)}{a_{mq}(s)} = \frac{1}{\tau_m s + 1}, \tag{32}$$

$$\frac{a_{ta}(s)}{a_{tq}(s)} = \frac{1}{\tau_t s + 1}, \tag{33}$$

where the guidance commands are denoted by "a_{tq}" and "a_{mq}" and the responses are denoted by "a_{ta}" and "a_{ma}". $\tau_t = 0.5$ and $\tau_m = 0.2$.

Initial conditions are preset to $r_0 = 3000$ m, $v_0 = \dot{r}_0 = -350$ m/s, $q_0 = 10^{\circ}$, $\dot{q}_0 = -3\text{deg/s}$, $V_t = 500$ m/s, and $\varphi_t = 0^{\circ}$. In accordance with the Eq. (9), $\delta_{v0} = 70$ m/s and $\delta_{r0} = 300$ m are given as upper boundaries of $|\tilde{v}|$ and $|\tilde{r}|$. In Eqs. (27) and (29), $N = 3$, $\varepsilon = 8$ g, $\Delta = 0.0001$, and $\gamma = 125$. Two worst-case conditions of the initial observed relative speed $\overline{v}_0$ and distance $\overline{r}_0$ are given.

Condition 1 (C1):

$$\overline{r}_0 = r_0 - \delta_{r0}, \overline{v}_0 = v_0 - \delta_{v0}. \tag{34}$$

Condition 2 (C2):

$$\overline{r}_0 = r_0 + \delta_{r0}, \overline{v}_0 = v_0 + \delta_{v0}. \tag{35}$$

Following maneuver modes of the target, including case 1, case 2, and case 3, are used to test the performance of the proposed schemes. Assume that the interceptor is detected by the target in 2 s and then the target begins to escape.

Case 1: Square maneuver in the direction of the axis Y.

$$\begin{cases} a_{ty}(t) = 0, & t \le 2s \\ a_{ty}(t-2) = -a_{ty}(t), a_{ty}(2) = 6g. & t > 2s \end{cases} \tag{36}$$

Case 2: Sine maneuver in the direction of the axis Y.

$$\begin{cases} a_{ty}(t) = 0, & t \le 2s \\ a_{ty}(t) = 8g \cdot \sin[3(t-2)]. & t > 2s \end{cases} \tag{37}$$

Case 3: Step maneuver in the direction of the axis Y.

$$\begin{cases} a_{ty}(t) = 0, & t \le 2s \\ a_{ty}(t) = 8g. & t > 2s \end{cases} \tag{38}$$

4.2. Comparisons between the OSMG and the APNG

Compare the ISMCG and SMCG with the APNG and OSMG. The actual target normal load and relative speed are considered known in the APNG; thereby, neither Condition 1 nor Condition 2 can affect the APNG. For the OSMG, it owns a simplified formulation which has robustness to target's maneuver, and it is popular in practice. Its simplified realization for online is as follows [6].

$$a_{mq} = -3\dot{\bar{r}}_0\dot{q} + \varepsilon sgn(\dot{q}) \simeq -3\bar{v}_0\dot{q} + \varepsilon\frac{\dot{q}}{|\dot{q}| + \Delta}, \tag{39}$$

where the initial observed relative speed is denoted by "$\bar{v}_0$". ε and Δ have no difference with those of the SMCG.

The expression of the APNG is obtained as [16]

$$a_{mq} = N'|v|\dot{q} + N'\frac{a_{tq}}{2}, \tag{40}$$

where the actual target normal load and the relative speed are denoted by "a_{tq}" and "v". An optimal value of the constant N' is 3 [16].

Schemes	Case 1		Case 2		Case 3	
	Condition 1	Condition 2	Condition 1	Condition 2	Condition 1	Condition 2
APNG	0.0831	0.0831	2.7448	2.7448	0.0173	0.0173
OSMG	0.0525	0.0819	0.0129	0.1148	0.0010	0.0015
ISMCG	0.0050	0.0003	0.0010	0.0019	0.0020	0.0036
SMCG	0.0289	0.0645	0.1115	0.1253	0.0010	0.0013

Table 1. Comparisons of miss distances (m).

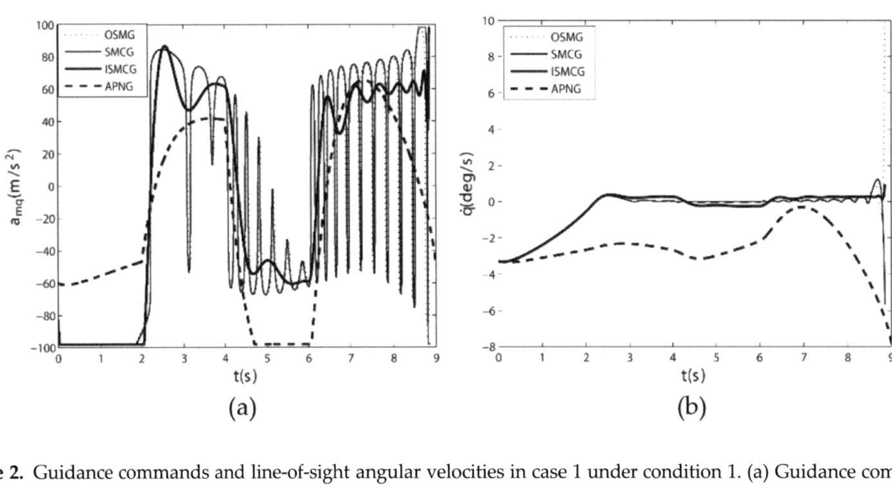

Figure 2. Guidance commands and line-of-sight angular velocities in case 1 under condition 1. (a) Guidance commands (b) Line-of-sight angular velocities.

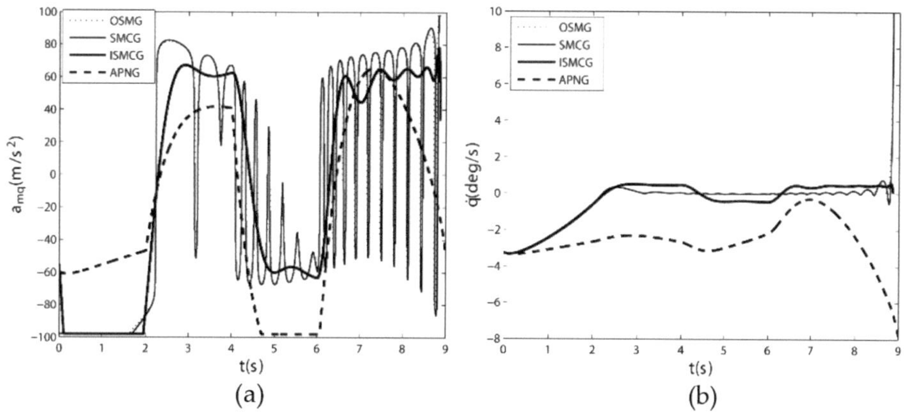

Figure 3. Guidance commands and line-of-sight angular velocities in case 1 under condition 2. (a) Guidance commands (b) Line-of-sight angular velocities.

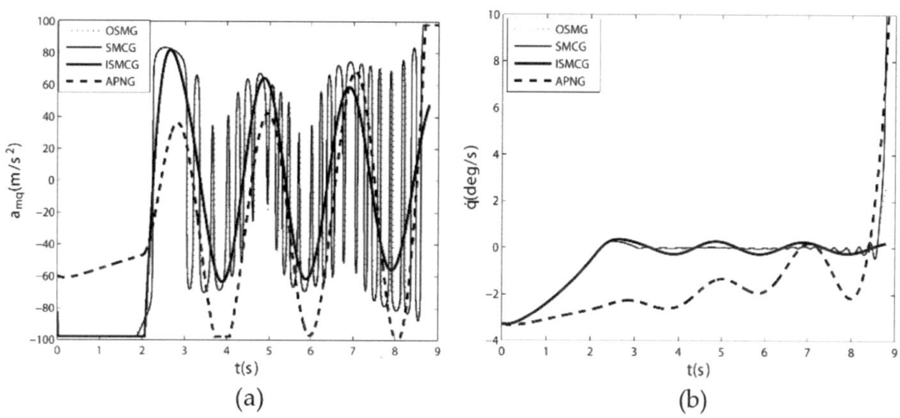

Figure 4. Guidance commands and line-of-sight angular velocities in case 2 under condition 1. (a) Guidance commands (b) Line-of-sight angular velocities.

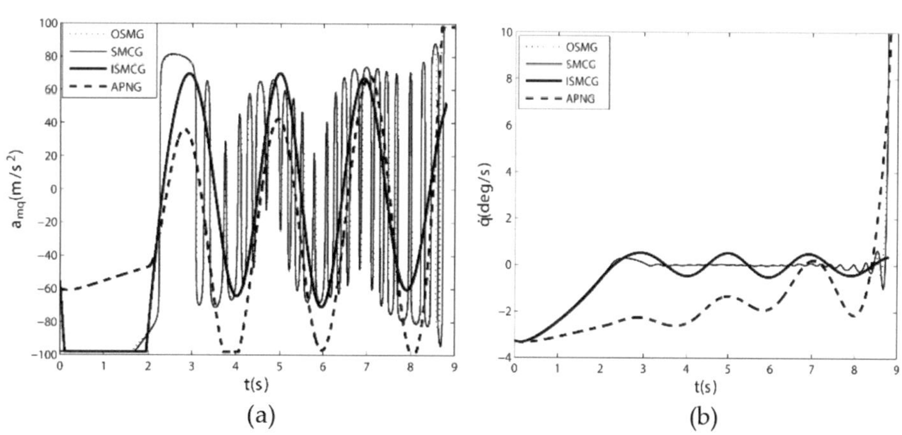

Figure 5. Guidance commands and line-of-sight angular velocities in case 2 under condition 2. (a) Guidance commands (b) Line-of-sight angular velocities.

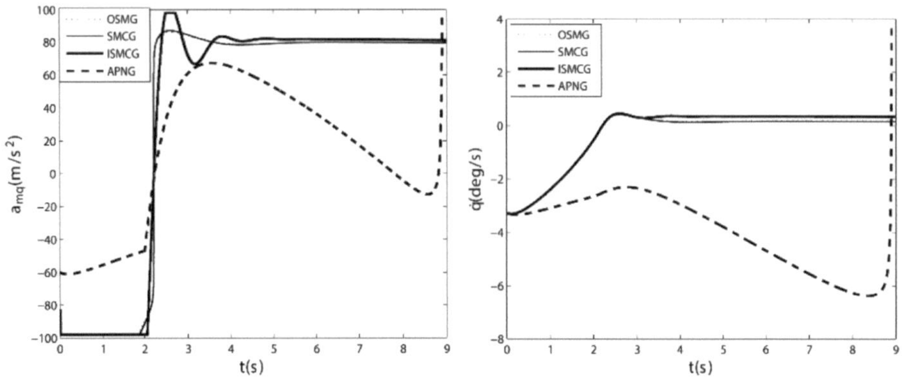

(a) (b)

Figure 6. Guidance commands and line-of-sight angular velocities in case 3 under condition 1. (a) Guidance commands (b) Line-of-sight angular velocities.

The missing distance is illustrated in **Table 1**, and the guidance commands and the line-of-sight angular velocities are shown in **Figures 2–7**. As shown in **Table 1**, these four guidance schemes all accomplish the interception task with the constraint $|a_{mq}| \leq 10$ g. In case 2, APNG's miss distances are comparatively greater, because Eq. (40) is deduced assuming a_{tq} is unchanged [17]. Nevertheless, the target might have a complicated maneuvering kind of escape. For the APNG, there is a greater miss distance to case 2, rather than to case 1 and case 3. It indicates the ANPG's limitations on intercepting unconventional maneuvering targets. **Table 1** also illustrates that the ISMCG owns the smallest miss distance in case 1 and case 2 for complicated types of target maneuvers, and the SMCG behaves like the OSMG. In case 3 for step maneuver targets, the miss distances of ISMGC, SMCG, and OSMG are small, and there is a little difference in performance of them.

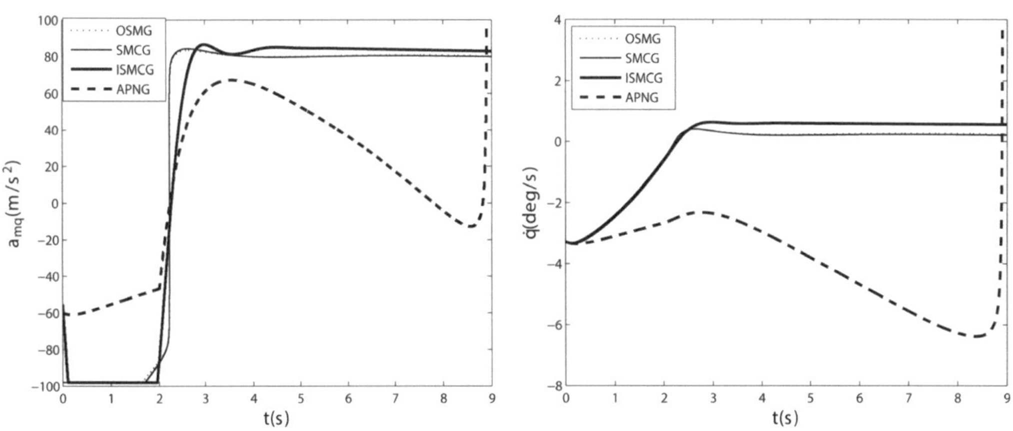

(a) (b)

Figure 7. Guidance commands and line-of-sight angular velocities in case 3 under condition 2. (a) Guidance commands (b) Line-of-sight angular velocities.

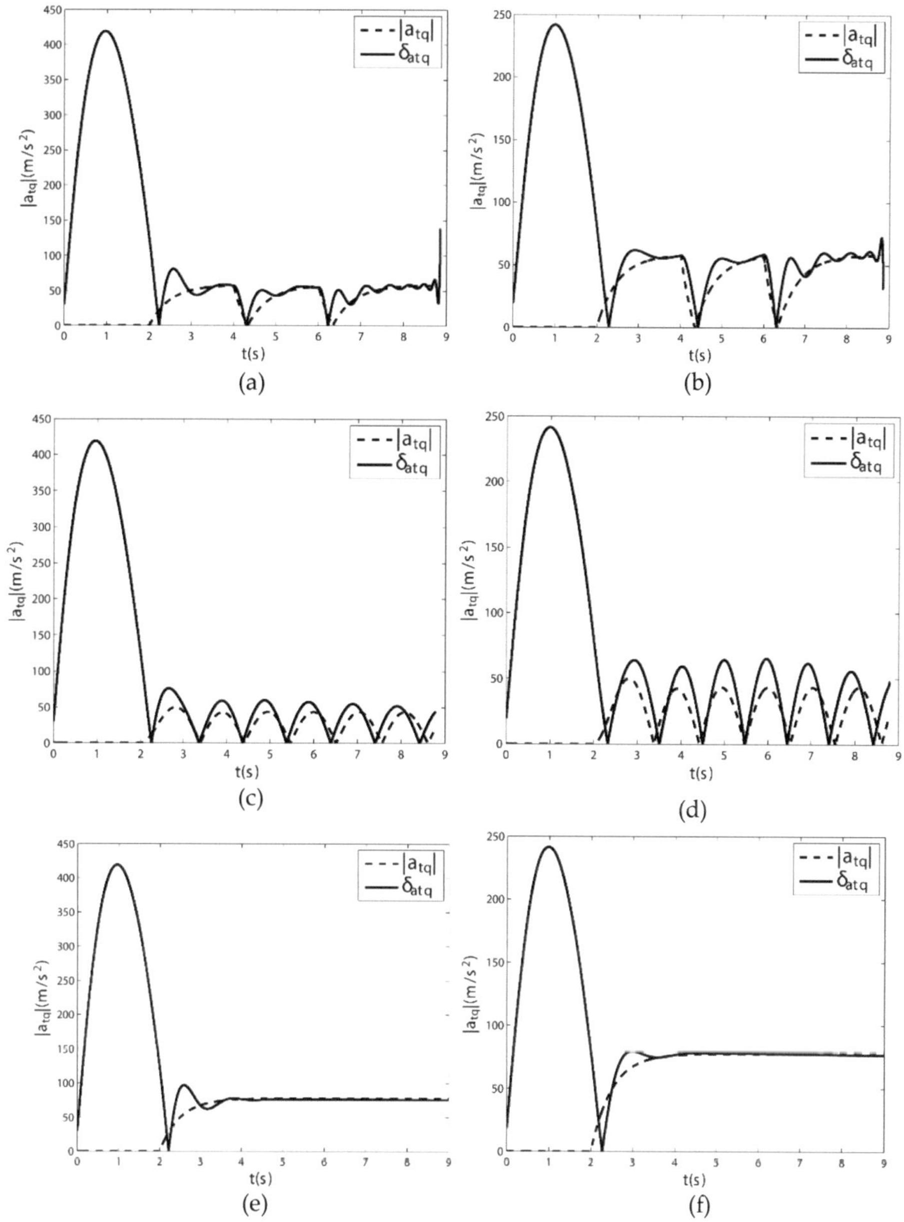

Figure 8. δ_{atq} and $|a_{tq}|$ of the ISMCG. (a) case 1 (C1); (b) case 1 (C2); (c) case 2 (C1); (d) case 2 (C2); (e) case 3 (C1); and (f) case 3 (C2).

From **Figures 2–7**, the APNG is not very appropriate to intercept complicated maneuvering targets because line-of-sight angular velocities of the APNG are greater. As a whole, the plots of the OSMG and the SMCG have little difference between each other. Their line-of-sight angular velocities are very small before the end of case 1 and case 2. Nevertheless, although continuous functions are employed to take the place of the "sgn" functions in the OSMG and the SMCG, guidance commands of the OSMCG and the SMCG all have jitters, which are detrimental to aero fins. Line-of-sight angular velocities of the ISMCG are less than those of the APNG with the actual target's acceleration. From **Figures 2(a)**, **3(a)**, **4(a)**, and **5(a)**, in case 1 and case 2, ISMCG's guidance commands are smoother than others, which are appropriate for continuous aero surfaces to track. From **Figures 2(b)**, **3(b)**, **4(b)**, and **5(b)**, because the ISMCG uses an adaptive estimation to identify a_{tq}'s upper boundary, line-of-sight angular velocities of the ISMCG are not as moderate as those of the SMCG; however, from **Table 1**, line-of-sight angular velocities in the endgame of the ISMCG are less than those of the SMCG and the OSMG in case 1 and case 2. From **Figures 6** and **7**, the guidance commands and line-of-sight angular velocities of OSMG, SMCG, and ISMCG have little differences and are superior to those of the APNG in case 3.

From **Figure 8**, it illustrates the δ_{atq} identified by the ISMCG in three cases under two conditions. Compared with $|a_{tq}|$, in the initial 2 s, δ_{atq} is larger since $\dot{q}$ and $\widehat{k}$ are larger, and then, the tracking error decreases since the ISMCG restrains the line-of-sight angular velocity. Because δ_{atq} is not the estimation of a_{tq}, tracking phases are considered and tracking errors are not concerned. Tracking phases reflect that estimations lag behind the actual target maneuver; thereby, it decides whether the compensation $\widehat{k}$ is timely and can influence the guidance precision. With tracking phases under consideration, δ_{atq} mostly tracks $|a_{tq}|$ with a tiny time delay. In fact, for the step maneuver target in case 3, from **Figure 8(e, f)**, δ_{atq} tracks $|a_{tq}|$ well. As shown in **Table 1**, small tracking phases obtain small miss distances.

5. Conclusions

In this chapter, robust guidance schemes are presented, which require states such as initial relative speed, relative distance, and error boundaries of them besides line-of-sight angular velocity. Proposed schemes' performances are validated by simulating under uncertainties for different target's maneuver modes. Two guidance schemes hit and kill maneuverable targets with fairly limited maneuverability. By comparisons with the APNG and OSMG, the ISMCG is superior, and the OSMG and SMCG perform similarly, whereas the APNG's miss distances are greater. Moreover, guidance commands of the APNG and ISMCG are smoother than those of the OSMG and SMCG for complicated maneuver modes of the target. In conclusion, ISMCG's advantage is that the guidance scheme is not required to obtain the target acceleration under uncertain conditions for different target maneuvers.

The future work concentrates on adapting the interceptor's maximum maneuverability to be a constraint condition in the proposed guidance scheme. Anti-saturation design is studied to address the control-saturation problem in Eq. (4) [18–20].

Acknowledgements

This chapter is supported by National Key R&D Program of China under Grant Nos. 2017YFD0701000 and 2016YFD0200700, National Natural Science Foundation of China under

Grant Nos. 51475472 and 61403396, and Beijing Youth Top-notch Talent Support Program under Grant No. 2017000026833ZK23.

Author details

Jian Chen[1], Yongjun Zheng[1], Yuan Ren[2]*, Yuan Tian[3], Chen Bai[4], Zhang Ren[5], Guangqi Wang[1], Nannan Du[1] and Yu Tan[1]

*Address all correspondence to: renyuan_823@aliyun.com

1 College of Engineering, China Agricultural University, Beijing, China

2 Space Engineering University, Beijing, China

3 Beijing Institute of Space Long March Vehicle, Beijing, China

4 Beijing Institute of Aerospace System Engineering, Beijing, China

5 Beihang University, Beijing, China

References

[1] Faruqi F. Differential Game Theory with Applications to Missiles and Autonomous Systems Guidance. Hoboken: John Wiley & Sons; 2017

[2] Ben-Asher J, Speyer J. Game in Aerospace: Homing Missile Guidance. Handbook of Dynamic Game Theory. Basel: Springer; 2017

[3] Khalid S, Abrar S. A low-complexity interacting multiple model filter for maneuvering target tracking. AEU-International Journal of Electronics and Communications. 2017;**73**:157-164

[4] Jauffret C, Pérez A, Pillon D. Observability: Range-only versus bearings-only target motion analysis when the observer maneuvers smoothly. IEEE Transactions on Aerospace and Electronic Systems. 2017

[5] Zhou D, Mu C, Xu W. Adaptive sliding-mode guidance of a homing missile. Journal of Guidance, Control, and Dynamics. 1999;**22**(4):589-594

[6] Zhou D, Mu C, Ling Q. and Xu W. Optimal Sliding-mode guidance of a homing-missile. In: Kamen EW, Cassandras C, editors. 38th IEEE Conference on Decision and Control; Phoenix, USA: IEEE; 1999. pp. 5131-5136

[7] Hou Z, Su M, Wang Y. and Liu L. A fuzzy optimal sliding-mode guidance for intercepting problem. In: Camisani F, editor. IFAC World Congress; Cape Town. IFAC; 2014. pp. 3401-3406

[8] Wang Z. Adaptive smooth second-order sliding mode control method with application to missile guidance. Transactions of the Institute of Measurement and Control. 2017;**39**(6): 848-860

[9] Yang P, Fang Y, Wu Y, Yong X. Finite-time convergent terminal guidance law design based on stochastic fast smooth second-order sliding mode. Optik-International Journal for Light and Electron Optics. 2016;**127**(15):6036-6049

[10] Zhang C, Wu Y. Non-singular terminal dynamic surface control based integrated guidance and control design and simulation. ISA Transactions. 2016;**63**:112-120

[11] Song H, Zhang T. Fast robust integrated guidance and control design of interceptors. IEEE Transactions on Control Systems Technology. 2016;**24**(1):349-356

[12] Ji H, Liu X, Song Z, Zhao Y. Time-varying sliding mode guidance scheme for maneuvering target interception with impact angle constraint. Journal of the Franklin Institute. 2017

[13] Hou Z, Liu L, Wang Y, Huang J, Fan H. Terminal impact angle constraint guidance with dual sliding surfaces and model-free target acceleration estimator. IEEE Transactions on Control Systems Technology. 2017;**25**(1):85-100

[14] Zhou H, Song S, Song J. Design of sliding mode guidance law with dynamic delay and impact angle constraint. International Journal of Control, Automation and Systems. 2017;**15**(1):239-247

[15] Wheeler G, Su C, Stepanenko Y. A sliding mode controller with improved adaptation laws for the upper bounds on the norm of uncertainties. Automatica. 1998;**34**(2):1657-1661

[16] Sun L, Lian P, Chang X. Capturability of retro-augmented proportional navigation guidance law against higher speed maneuvering target. In: 21st AIAA International Space Planes and Hypersonics Technologies Conference; 6–9 March 2017. Xiamen, China: American Institute of Aeronautics and Astronautics; 2017. pp. 2203/1-2203/13

[17] Cho N, Kim Y. Optimality of augmented ideal proportional navigation for maneuvering target interception. IEEE Transactions on Aerospace and Electronic Systems. 2016;**52**(2): 948-954

[18] Turner M, Sofrony J, Herrmann G. An alternative approach to anti-windup in anticipation of actuator saturation. International Journal of Robust and Nonlinear Control. 2017;**27**(6):963-980

[19] Si Y, Song S. Adaptive reaching law based three-dimensional finite-time guidance law against maneuvering targets with input saturation. Aerospace Science and Technology. 2017;**70**:198-210

[20] Li G. and Xin M. A three-dimensional anti-saturation terminal guidance law with finite-time convergence. In: 2017 American Control Conference; 24–26 May 2017. Seattle, USA: IEEE; 2017. pp. 2243-2248

Chapter 3

High-Gain Observer–Based Sliding Mode Control of Multimotor Drive Systems

Pham Tam Thanh, Dao Phuong Nam,
Tran Xuan Tinh and Luong Cong Nho

Additional information is available at the end of the chapter

http://dx.doi.org/10.5772/intechopen.71656

Abstract

Multimotor drive systems have been widely used in many modern industries. It is a nonlinear, multi-input, multi-output (MIMO) and strong-coupling complicated system, including the effect of friction, elastic, and backlash. The control law for this drive system much depends on the determining of the tension. However, it is hard to obtain this tension in practice by using a load cell or a pressure meter due to the accuracy of sensors or external disturbance. In order to solve this problem, a high-gain observer is proposed to estimate the state variables in this drive system, such as speeds and tension. An emerging proposed technique in the control law is the use of high-gain observers together with adaptive sliding mode control scheme to obtain a separation principle for the stabilization of whole system. The theory analysis and simulation results point out the good effectiveness of the proposed output feedback for the drive system.

Keywords: high-gain observer, multimotor drive systems, sliding mode control, tension, output feedback controller

1. Introduction

Multimotor drive systems have been researched by many researchers in the recent times. The control law based on neural network technique has been proposed by Yaojie Mi et al. (see [1–4], for examples). However, it is hard to find the corresponding networks as well as learning rules. Besides, the model of this system is approximately described as a linear system to use the transfer function to design the control law. Furthermore, the tracking ability or the stabilization of the whole system is not still solved under the effects of observer using neural network technique. In the multimotor drive control systems, it is necessary to obtain the belt tension to design the suitable state feedback control law. However, it is hard to measure this belt tension by using sensors, and the observer based on high-gain technique is proposed in our work.

Besides, the state feedback control design based on sliding mode control technique enables to remove efficient disturbances and uncertainties. Therefore, a high-gain observer is proposed to estimate the tension in this system and combine with the state feedback controller to obtain the output feedback control law satisfying the separation principle. The stability of whole system is obtained by the output feedback control law and verified by theory analysis and simulations.

This work is composed of 7 sections. In Section 2, the problem statements are shown and the dynamic equations of the two-motor system are described by the effect of friction, backlash, and elastic. Sections 3–5 describe the output feedback control design. Then, the high-gain observer for multimotor system is explained. Next, the sliding mode control of this system and the ability to satisfy the separation principle of output feedback controller are discussed. In Section 6, simulation results are shown. The conclusion is summarized in Section 7.

2. Problem statements

In [1], the multimotor system (in **Figure 1**) using two induction motor is described by the following dynamic Eq. (1), and the nomenclatures used in these equations are summarized in **Table 1**:

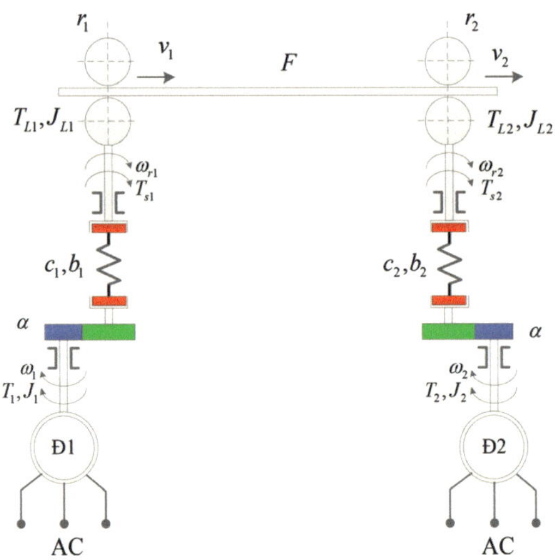

Figure 1. The two-motor drive system.

$$\begin{cases} \dfrac{dx_1'}{dt} = \dfrac{n_{p1}}{J_1}\left[\left(\mathrm{u}_1 - x_1'\right)\dfrac{n_{p1}T_{r1}}{L_{r1}}\varphi_{r1}^2 - \left(\mathrm{T}_{L1} + \mathrm{r}_1 x_3'\right)\right] \\ \dfrac{dx_2'}{dt} = \dfrac{n_{p2}}{J_2}\left[\left(\mathrm{u}_2 - x_2'\right)\dfrac{n_{p2}T_{r2}}{L_{r2}}\varphi_{r2}^2 - \left(\mathrm{T}_{L2} - \mathrm{r}_2 x_3'\right)\right] \\ \dfrac{dx_3'}{dt} = \dfrac{\mathrm{K}}{\mathrm{T}}\left(\dfrac{1}{n_{p_1}}\mathrm{r}_1\mathrm{k}_1 x_1' - \dfrac{1}{n_{p_2}}\mathrm{r}_2\mathrm{k}_2 x_2'\right) - \dfrac{x_3'}{\mathrm{T}} \end{cases} \tag{1}$$

$K = E/V$	Transfer function
E	Young's Modulus of belt
V	Expected line velocity
$T = \frac{L_0}{AV}$	Time constant of tension variation
L_0, A	Distance between racks, section area (m^2)
n_{pi}	Number of pole-pairs in the ith motor
J_1, J_2, J_{L1}, J_{L2}	Inertia moment of motors and loads (kgm^2)
T, T_L, φ_r	Motor, load torque (Nm), flux of rotor (Wb)
L_r	Self-induction of rotor (H)
$r, k, \omega_r, \omega, F$ (in Eq. (1))	Radius of roller, velocity ratio, electric angle velocity of rotor, angle velocity of stator, belt tension
$\omega_1, \omega_2, \omega_{r1}, \omega_{r2}$ (in Eqs. (2) and (3))	The angle velocity of motor and load
c_1, c_2, b_1, b_2	Stiffness and friction coefficient
$\Delta\omega_1, \Delta\omega_2$	The errors of angle speed in presence of backlash and elastic

Table 1. Dynamic parameters.

where

$x' = [x'_1, x'_2, x'_3]^T = [\omega_{r1}, \omega_{r2}, F]^T \in \mathbb{R}^3$ is state variable and $u = (u_1, u_2) = (\omega_1, \omega_2) \in \mathbb{R}^2$ is a control variable.

However, due to the effects by backlash and elastic (**Figure 1**), we extend this model to obtain the equivalent diagram (**Figure 2**) and the following dynamic Eqs. (2) and (3):

$$\begin{cases} \Delta\dot{\varphi}_1 = \dfrac{1}{T}(\omega_1 - \omega_{r1}) \\ \Delta\dot{\varphi}_2 = \dfrac{1}{T}(\omega_2 - \omega_{r2}) \\ \dot{\omega}_{r1} = J_{L1}\left[\dfrac{1}{K_{TC1}} f_{11}(\Delta\varphi_1) + K_{C1}\Delta\omega_1 f_{12}(\Delta\varphi_1) - (T_{L1} + r_1 F_{21})\right] \\ \dot{\omega}_{r2} = J_{L2}\left[\dfrac{1}{K_{TC2}} f_{21}(\Delta\varphi_2) + K_{C2}\Delta\omega_2 f_{22}(\Delta\varphi_2) - (T_{L2} + r_2 F_{12})\right] \\ \dot{F} = C_{12}\left[r_1\omega_{r1} - r_2\omega_{r2}\left(1 + \dfrac{1}{C_{12}.l}F\right)\right] \end{cases} \tag{2}$$

We denote $x_1 = \begin{bmatrix} x_{11} \\ x_{12} \end{bmatrix} = \begin{bmatrix} \Delta\varphi_1 \\ \Delta\varphi_2 \end{bmatrix} \in \mathbb{R}^2;\ x_2 = \begin{bmatrix} x_{21} \\ x_{22} \end{bmatrix} = \begin{bmatrix} \omega_{r1} \\ \omega_{r2} \end{bmatrix} \in \mathbb{R}^2;\ x_3 = \begin{bmatrix} x_{31} \\ x_{32} \end{bmatrix} = \begin{bmatrix} F_{21} \\ F_{12} \end{bmatrix} \in \mathbb{R}^2$

to obtain the following dynamic equation described in state-space representation:

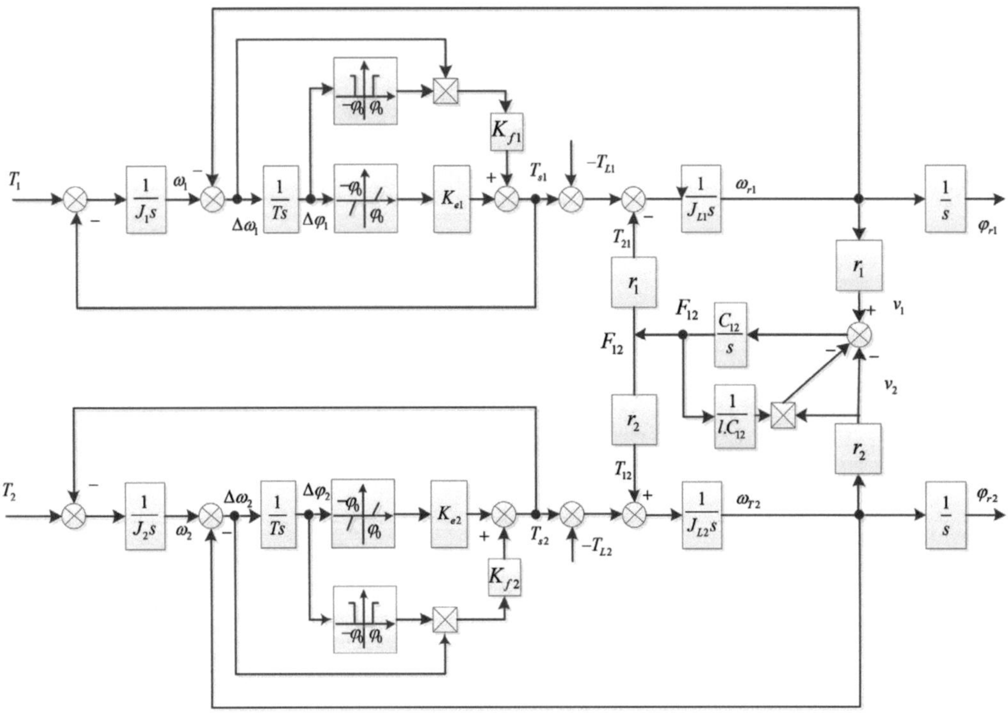

Figure 2. The equivalent diagram of the two-motor drive system.

$$\begin{cases} \dot{x}_1 = \frac{1}{T}(u - x_2) \\ \dot{x}_2 = J_L\left[\frac{1}{K_{TC}} f_1(x_1) + \mathrm{K_C}(u - x_2) f_2(x_1) - (\mathrm{T}_L + r.x_3)\right] \\ \dot{x}_3 = \mathrm{C}_{12}\left[r_1 x_{21} - r_2 x_{22}\left(1 + \frac{1}{\mathrm{C}_{12}.l} x_3\right)\right] \\ y = x_3 \end{cases} \tag{3}$$

Remark 1:

The dynamic Eqs. (2) and (3) and **Figures 1** and **2** are described by the effect of friction, backlash, and elastic and pointed out the nonlinear property of multimotor systems.

The control objective is to find the synchronous speeds $u = (\mathrm{u}_1, \mathrm{u}_2) = (\omega_1, \omega_2) \in \mathbb{R}^2$ to obtain that the desired value are tracked by tensions in the presence of friction and elastic. In order to implement this work, a new scheme is proposed to design an output feedback controller involving a high-gain observer and a sliding mode control law. Moreover, the effectiveness to satisfy the separation principle is pointed out in multimotor control system.

3. Observer design

As mentioned above, the main motivation of the work is to find an equivalent high-gain observer for the class of multimotor systems. In the following, one will present the proposed high-gain observer to estimate the tension in this system and provide a full analysis of observation error convergence.

MISO systems are described as follows:

$$\begin{cases} \frac{d}{dt}x = Ax + \gamma(x,u,y) + \varphi(u,y) \\ y = c^T x + \xi(u) \end{cases} \tag{4}$$

where $\gamma(x,u,y)$ satisfy the global Lipschitz condition $|\gamma(x,u) - \gamma(\hat{x},u)| \le \alpha|x - \hat{x}|$ and

$$A = \begin{bmatrix} 0 & 1 & 0 & \dots & 0 \\ 0 & 0 & 1 & & 0 \\ & & & \dots & \vdots \\ & & & 0 & 1 \\ 0 & 0 & & \dots & 0 \end{bmatrix}, c^T = [1, \quad 0, \quad \dots \quad 0].$$

Lemma 1 [5]: The classical high-gain observer is pointed out by the following equations:

$$\frac{d}{dt}\hat{x} = A\hat{x} + L\left(y - c^T\hat{x}\right) + \gamma(\hat{x},u) \tag{5}$$

where $L = \begin{bmatrix} h_1\varepsilon^{-1} \\ \vdots \\ h_n\varepsilon^{-n} \end{bmatrix}$ and ε is a small enough positive number and $h_n, h_{n-1}, \dots, h_1$are coefficients of a Hurwitz polynomial (6)

$$P(s) = h_n + h_{n-1}s + \dots + h_1 s^{n-1} + s^n \tag{6}$$

Remark 1:

The classical high-gain observer is the next development of Lipschitz observer with the additional contents of the coefficient ε to obtain $a < \frac{\lambda_{\min}(Q)}{2\lambda_{\max}(P)}$ without solving the LMIs problem.

However, the previous observer (5) is only suitable to systems with one output. In order to design for multi-output systems, Farza et al. develop many observers for a class of MIMO nonlinear systems [6–9]. Based on the proposed high-gain observer that is pointed out in (7) [4], we obtain the observer (8) for multimotor systems (3):

$$\dot{\widehat{x}} = f\left(u, s, \widehat{x}\right) - \begin{pmatrix} \theta C_1^q I_{n_1} \\ \theta^2 C_2^q \left[\frac{\partial f_1}{\partial x^2}(u, s, x)\right]^+ \\ \vdots \\ \theta^q C_q^q \left[\prod_{i=1}^{q-1} \frac{\partial f_k}{\partial x^{k+1}}(u, s, x)\right]^+ \end{pmatrix} \overline{C}\left(\widehat{x} - x\right) \tag{7}$$

$$\begin{cases} \dot{\widehat{x}}_1 = \frac{1}{T}(u - \widehat{x}_2) - 3\theta(\widehat{x}_3 - x_3) \\ \dot{\widehat{x}}_2 = J_L\left[\frac{1}{K_{TC}} f_1(\widehat{x}_1) + K_C(u - \widehat{x}_2) f_2(\widehat{x}_1) - (T_L + r.\widehat{x}_3)\right] + \frac{\theta^2}{T}(\widehat{x}_3 - x_3) \\ \dot{\widehat{x}}_3 = C_{12}\left[r_1\widehat{x}_{21} - r_2\widehat{x}_{22}\left(1 + \frac{1}{C_{12}.l}\widehat{x}_3\right)\right] + rJ_L\theta^3(\widehat{x}_3 - x_3) \\ y = x_3 \end{cases} \tag{8}$$

Remark 2: The convergence of observer error based on the high-gain observer (8) is pointed out in [3, 4].

4. Sliding mode control

In this section, the main work is to find a state feedback control law based on the sliding mode control technique for the class of multimotor systems.

Nonlinear systems are described as follows:

$$\frac{d}{dt}x = Ax + B(u + u_d(x, t)) \tag{9}$$

where $u_d(x, t)$ is the nonlinear term in system.
Lemma 2 [10]: The sliding mode controller is described as follows

$$u = -\left[S.Ax + \beta \text{sgn}(\sigma)\right] \tag{10}$$

based on the sliding surface:

$$\begin{aligned} &\{x : \sigma = Sx = 0\}, \\ &S = \left(B^T X^{-1} B\right)^{-1} B^T X^{-1} X \end{aligned} \tag{11}$$

with X is satisfied, the LMI problem as follows:

$$II^T\left(AX + XA^T\right)II < 0, X > 0, II = [1 \quad 0 \quad 0]^T \tag{12}$$

Remark 3. We obtain the sliding mode control for multimotor systems (2) based on Lemma 3 because it belongs to the class of systems (9).

5. Observer-integrated sliding mode control

After we obtain the output feedback control law combined between sliding mode controller and high-gain observer, the main work is to point out the ability to obtain the separation principle of the proposed solution.

Consider the nonlinear systems:

$$\begin{cases} \frac{d}{dt}x = Ax + f(x, u, t) \\ y = Cx \end{cases} \tag{13}$$

with $f(x, u, t)$ satisfying the global Lipschitz condition

$$|f(x, u, t) - f(x', u, t)| \leq a|x - x'|(\forall x, x', u) \tag{14}$$

Lemma 3 [5]: If there exists a control Lyapunov function $V(x)$ and the corresponding control input $u = r(x)$ satisfy

$$\left|\frac{\partial V}{\partial \underline{x}}\left[f(\underline{x}, u, t) - f\left(\underline{x}', u, t\right)\right]\right| \leq b\left|x - x'\right|^2, \forall x, x' > 0 \tag{15}$$

Then the output feedback control law using the observer (16) and (17) and the state feedback controller $u = r(x)$ is described as above:

$$\frac{d\widehat{x}}{dt} = A\widehat{x} + f(\widehat{x}, u, t) + L(y - C\widehat{x}) \tag{16}$$

where L is the matrix is satisfied all the real parts of eigenvalues of $(A - LC)$ that is negative and matrices P, Q satisfy the Lyapunov equation

$$(A - LC)^T P + P(A - LC) = -Q \tag{17}$$

and

$$a < \frac{\lambda_{\min}(Q) - b}{2\lambda_{\max}(P)} \tag{18}$$

Theorem 1. The whole system (**Figure 1**) is asymptotically stable by the output feedback control law with the high-gain observer (8) and the nonlinear state feedback controller (10).

Proof: Using the Lyapunov candidate function $V(x) = x^T Px$, we obtain the inequality (15) based on x being the state trajectory of multimotor system (5).

Remark 4. This result is a development from the results in [1–4], because the separation principle of output feedback controller has not been implemented in previous researches.

6. Simulation results

In this section, we consider several simulation results to demonstrate the effectiveness of the proposed output feedback control law based on the two-motor system as shown in **Table 2**. **Figures 3** and **4** show the high-performance behavior of velocity based on the proposed high-gain

n_{p1}	4
J_1	50 kgm^2
L_{r1}	0.2 H
T_{L1}	30 Nm
n_{p2}	4
J_2	55 kgm^2
L_{r2}	0.3 H
T_{L2}	25 Nm

Table 2. Multimotor system parameters.

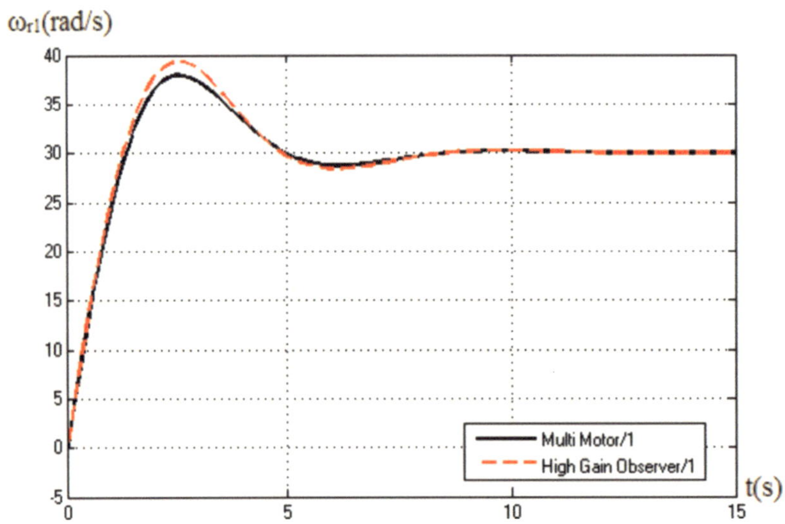

Figure 3. The velocity of motor 1 and estimation of it.

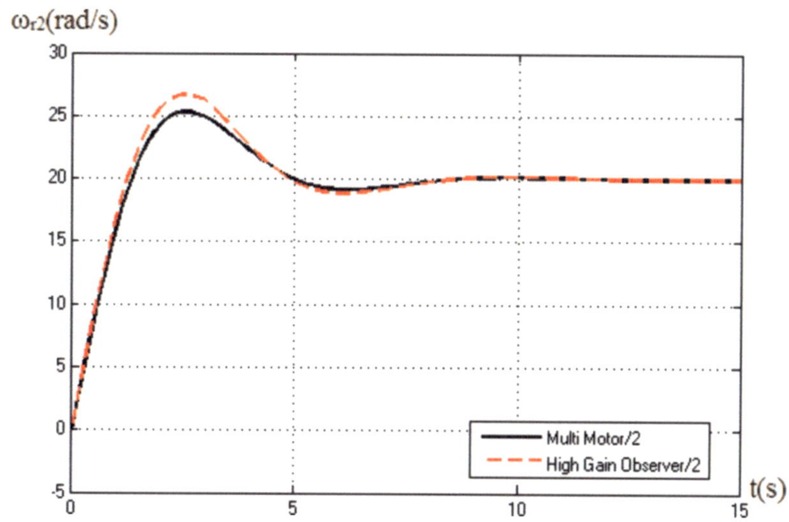

Figure 4. The velocity of motor 2 and estimation of it.

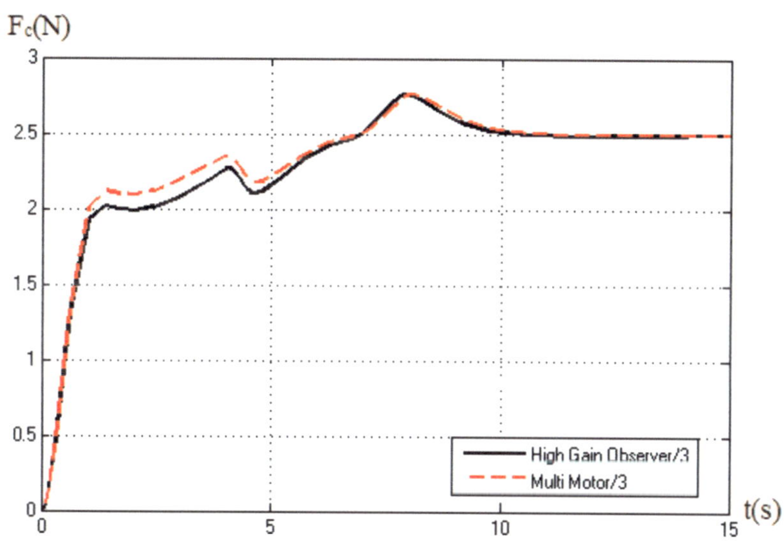

Figure 5. The tension of system and estimation of it.

observer. Moreover, we obtain the high tracking performance of tension in the presence of friction and elastic (**Figure 5**). Furthermore, **Figures 6** and **7** show the tracking performance behavior of velocity based on adaptive sliding mode control law in the presence of disturbance (**Figure 9**). **Figures 8** and **10** show the high tracking performance behavior of velocity based on adaptive sliding mode control law without disturbance.

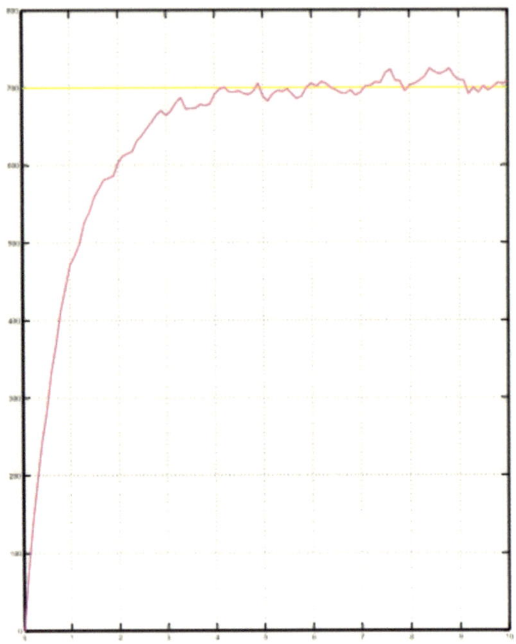

Figure 6. The behavior of the first motor's speed in the presence of disturbance.

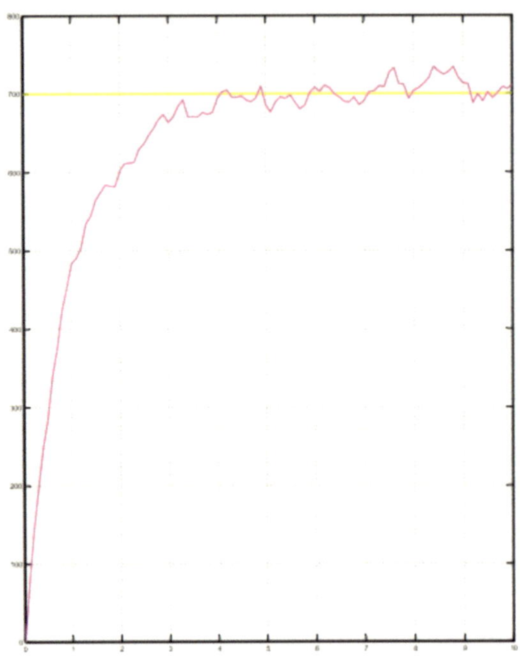

Figure 7. The behavior of the second motor's speed in the presence of disturbance.

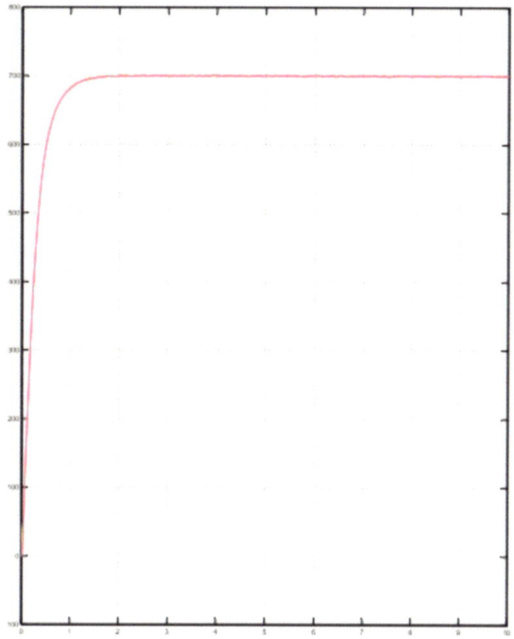

Figure 8. The behavior of the first motor's speed without disturbance.

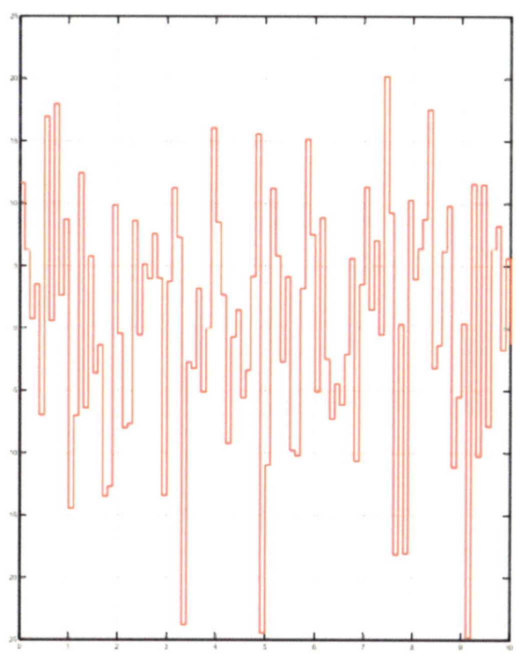

Figure 9. The behavior of the second motor's speed without disturbance.

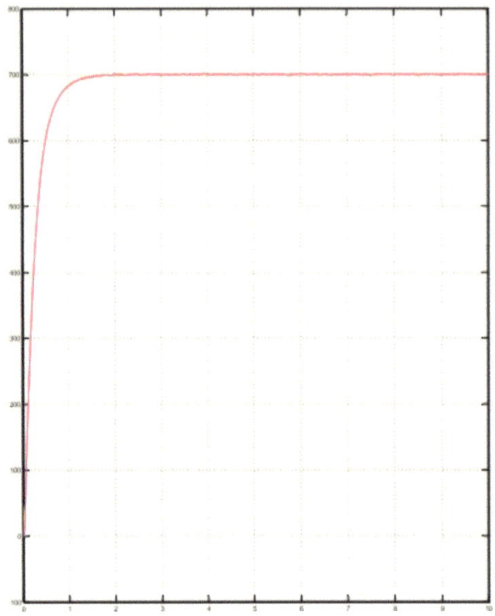

Figure 10. The system being affected by disturbance.

7. Conclusions

This chapter described an output feedback control law based on the combination between high-gain observer and sliding mode control for the two-motor system in the presence of elastic, backlash, and friction. The proposed control law allows to obtain the separation principle in the presence of friction and elastic due to the tuning of parameter in proposed high-gain observer. The effectiveness of the proposed control scheme was pointed out by theory analysis and simulation results.

Author details

Pham Tam Thanh[1]*, Dao Phuong Nam[2], Tran Xuan Tinh[2] and Luong Cong Nho[1]

*Address all correspondence to: phamtamthanh@vimaru.vn

1 Vietnam Maritime University, Vietnam

2 Le Quy Don and Hanoi University of Science and Technology, Hanoi, Vietnam

References

[1] Yaojie Mi et al. A new fuzzy adaptive combined-inversion control of two-motor drive system. In: Proceedings of the 2013 International Conference on Electrical Machines and Systems; Busan, Korea; 2013. pp. 2282-2285

[2] Jinzhao Zhang et al. An improved method for synchronous control of complex multi-motor. In: Proceedings of the IEEE International Conference on Intelligent Computing and Intelligent Systems; Shanghai, China; 2009. pp. 178-182

[3] Guohai Liu et al. Experimental research on decoupling control of multi-motor variable frequency system based on neural network generalized inverse. In: Proceedings of the IEEE International Conference on Networking, Sensing and Control; China; 2008. pp. 1476-1479

[4] Li Jinmei et al. Application of an adaptive controller with a single neuron in control of multi-motor synchronous system. In: Proceedings of the IEEE International Conference on Industrial Technology; Chengdu, China. 2008. pp. 1-6

[5] Khalil. High-gain observers in nonlinear feedback control. In: Proceedings of the International Conference on Control, Automation and System; Seoul, Korea. 2008. pp. 10-16

[6] Fazza M, M'Saad, Sekher M. A set of observer for a class of nonlinear systems. In: 16th IFACT W.C; 2005. pp. 4-8

[7] Fazza M, M'Saad, Rossignol L. Observer design for a class of mimo nonlinear systems. Automatica. 2004;**40**:135-143

[8] Said SH, Mimouni M, M'Sahli F, Farza M. High gain observer based on-line rotor and stator resistances estimation for IMs. Simulation Modelling Practice and Theory. 2011;**19**: 1518-1529

[9] Said SH, M'Sahli F, Mimouni M, Farza M. Adaptive high gain observer based output feedback predictive controller for induction motors. Computer and Electrical Engineering. 2013:1-13

[10] Nga VTT, Dong Y, Choi HH, Jung J-W. T-S fuzzy-model-based sliding-mode control for surface-mounted permanent-magnet synchronous motor considering uncertainties. IEEE Transactions on Industrial Electronics. Oct 2013;**60**(10):4281-4291

Chapter 4

Stator-Flux-Oriented Sliding Mode Control for Doubly Fed Induction Generator

Ivan Villanueva, Antonio Rosales, Pedro Ponce and Arturo Molina

Additional information is available at the end of the chapter

http://dx.doi.org/10.5772/intechopen.70714

Abstract

Doubly-fed induction generator (DFIG) is the most implemented electric machine in wind energy conversion systems (WECSs) due to reduced size converter, active and reactive power control, and economic factors. However, the power electronic stage needs an accurate controller that allows to follow the stator power regulation despite the presence of disturbances. On the other hand, sliding-mode control (SMC) offers a fast-dynamic response and provides insensitivity to matched and bounded disturbance/uncertainties, and its natural discontinuous control signals can be used for direct switching of power electronic devices. Switching frequency must be maintained inside acceptable values to avoid exceeding the maximum admissible switching frequency of semiconductors. The contribution of this chapter is a stator-flux–oriented SMC with a hysteresis band that limits the switching frequency of power electronic devices to a set value. Furthermore, the proposed SMCs ensure robustness against bounded low-voltage grid faults. Unlike other nonmodulated techniques like direct torque control (DTC), there is no necessity of modifying the controller structure for withstanding low-depth voltage dips. The controller injects negative sequence voltage/currents to compensate the unbalanced conditions. The advantages of the proposed SMC control are validated via simulations.

Keywords: doubly-fed induction generator, sliding mode control

1. Introduction

In recent years, wind energy has been established as the fastest growing energy source among renewables. From a global perspective, more than 90 countries have been involved in the installation of new renewable wind power plants [1]. In 2015, the new global installed capacity hit the record for an increase of approximately 64 GW. During 2016, more than 54 GW were installed for a total of approximately 487 GW by the end of 2016. Conservative prognoses forecast an increment of 60 GW in 2017 with a continuous annual grow of about 75 GW by 2021. With the continued improvement in wind turbines technology and ecological concerns, the wind power is now a serious competitor against heavily subsidized energy industries [1, 2].

Doubly-fed induction generator (DFIG) is a wound-rotor induction machine with voltage injection in the rotor winding. This allows a limited speed control of the electric machine, which is sufficient to implement maximum aerodynamic efficiency by maintaining the tip-speed ratio at the nominal value at most operational conditions of the turbine.

The success of DFIG in wind energy applications lies in the reduced power converter, which is typically about 25–30% of the nominal power of the electric machine; furthermore, by current injection in the rotor, it is possible to control the reactive power injected to the electric grid, which is very important for minimizing cooper loses or reactive power compensation [3]. More than 70% of the installed wind turbines use DFIG. However, this machine is very sensitive to voltage variations in the grid because the stator is directly connected to the grid [4], in contrast to other variable-speed machines that are connected through a full-size power converter.

Sliding mode control (SMC) is a nonlinear control technique that ensures finite-time convergence of the sliding surface to zero guarantying robustness against bounded disturbances and parameter variations [9]. The main disadvantage of SMC is the chattering effect (high frequency oscillations with finite amplitude) caused by unmodeled dynamics and discretization [9]. On the other hand, power electronics are controlled by means of the injection of discontinuous signals matching with the discontinuous nature of conventional SMC; therefore, conventional SMC can be used for direct switching of power electronics on DFIG applications, avoiding modulation. SMC has been successfully implemented in DFIG control and tested under unbalanced conditions and harmonics [11, 15–18]. However, the proposals given in [16, 17] require modulation, and the tested faults are moderate since these do not represent a brusque variation in the stator voltage. The SMC presented in [15–18] works under unbalanced conditions but implementation of the SMC regarding the commutation of the power electronics is not discussed.

In this chapter, it is presented a SMC with the following advantages: (a) do not required modulation; (b) do not require modifications of the controller structure to withstand stator voltage perturbances (compared with classical control approach as direct-torque-control [DTC], see [5]); and (c) the controller can regulate torque and reactive power even under unbalanced conditions, which is equivalent to negative current regulation. Therefore, the proposed SMC offers a very simple alternative that requires neither symmetrical decomposition nor pulse width modulation and is not affected by parameter variation. Furthermore, the DFIG system with SMC is characterized in the frequency domain for estimating the commutation frequency of power electronics. A maximum switching frequency value is ensured by means of the addition of hysteresis in the sign function. The hysteresis value is computed applying the Tsypkin's method, a theoretically exact technique to analyze nonlinear systems [12, 14]. Due to the nature of the power converter, the voltage gain seen by the controller is variable in time, which makes difficult to maintain the switching frequency constant. However, it is possible to compute a maximum switching frequency value (minimum hysteresis value), which provides a commutation frequency inside of the acceptable values given in the datasheet of the power electronics.

The chapter presents the dynamic modeling of the electric machine and the wind turbine, the controller design, frequency domain characterization, and finally, fault-ride through capability is evaluated analyzing the regulated current from a sequence components point of view.

2. Aerodynamic model of wind turbine

Horizontal axis wind turbines are used to extract mechanical power from the wind resource based on the lifting force of the rotor blades. The mechanical power is a function of the kinematic energy of the wind and the power coefficient:

$$P_{wind} = \frac{1}{2} C_P(\Lambda, \gamma)\rho A V^3 \tag{1}$$

The power coefficient is a function of the tip speed ratio and the pitch angle. Since very complex aerodynamic analyses are required to characterize turbine blades, the power coefficient is usually approximated using mathematical expressions such as [7]:

$$C_P = c_1(c_2\kappa - c_3\gamma - c_4\gamma^{c_5} - c_6)e^{-c_7\kappa} \tag{2}$$

where $\kappa = \frac{1}{\Lambda + 0.06\,\gamma} - \frac{0.035}{1+\gamma^3}$ and c_1 to c_7 are coefficients dependent on the blades geometry.

The tip speed ratio is the relationship between the tangential speed of the blades tip and the wind speed if it is expressed using the DFIG mechanical speed and the gearbox ratio we obtain:

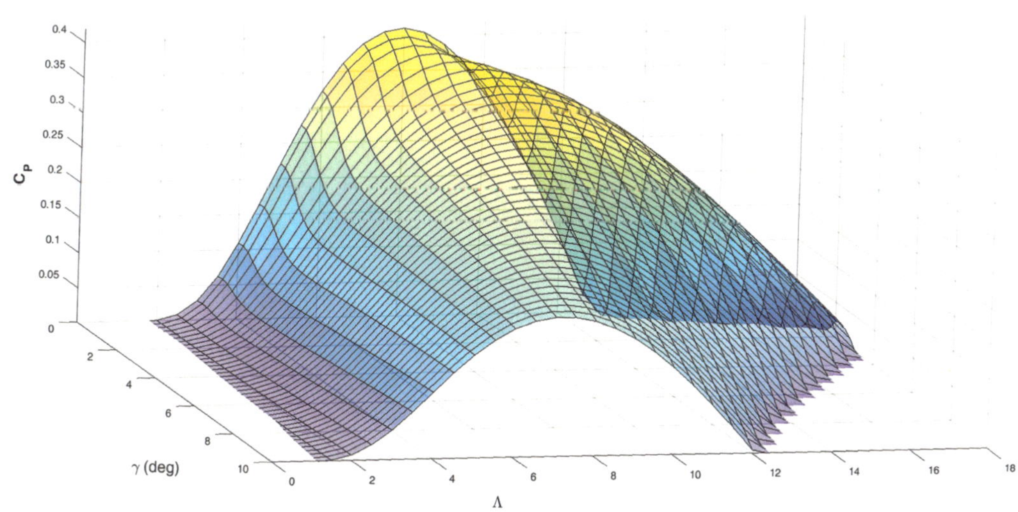

Figure 1. Power coefficient as a function of tip speed ratio and pitch angle.

$$\Lambda = \frac{\omega_m r}{\eta V} \tag{3}$$

Pitch angle is normally used for aerodynamically reduce power extraction when the wind speed is above the nominal value. For normal operation, it is maintained constant, while the rotor speed is controlled by the DFIG to maintain the tip speed ratio constant, for the blades model shown in **Figure 1**, the nominal pitch angle is zero and the nominal tip speed ratio is 8 for a maximum power coefficient of approximately 0.41. The parameters used to generate the displayed function are $c_1=0.5; c_2=116; c_3=0.4; c_4=0; c_6=5; c_7=21$:

3. Dynamic modeling of the DFIG

The dynamic equivalent circuit of DFIG can be expressed in an arbitrary reference frame rotating at a speed equal to ω [6, 8]:

$$v_{sd} = \frac{d\lambda_{sd}}{dt} - \omega\lambda_{sq} + R_s i_{sd} \tag{4}$$

$$v_{sq} = \frac{d\lambda_{sq}}{dt} + \omega\lambda_{sd} + R_s i_{sq} \tag{5}$$

$$v_{rd} = \frac{d\lambda_{rd}}{dt} - (\omega - \omega_m)\lambda_{rq} + R_r i_{rd} \tag{6}$$

$$v_{rq} = \frac{d\lambda_{rq}}{dt} + (\omega - \omega_m)\lambda_{rd} + R_r i_{rq} \tag{7}$$

$$\lambda_{sd} = L_s i_{sd} + L_m i_{rd} \tag{8}$$

$$\lambda_{sq} = L_s i_{sq} + L_m i_{rq} \tag{9}$$

$$\lambda_{rd} = L_r i_{rd} + L_m i_{sd} \tag{10}$$

$$\lambda_{rq} = L_r i_{rq} + L_m i_{sq} \tag{11}$$

The electromagnetic torque can be expressed as an interaction between rotor current and stator magnetic flux:

$$T_{em} = \frac{3PL_m}{2L_s}\left(i_{rd}\lambda_{sq} - i_{rq}\lambda_{sd}\right) \tag{12}$$

A common expression of the stator reactive power in terms of the stator voltage and rotor current is:

$$Q_s = \frac{3L_m}{2L_s}\left(v_{sd}i_{rq} - v_{sq}i_{rd}\right) + \frac{3}{2L_s}\left(\lambda_{sd}v_{sq} - \lambda_{sq}v_{sd}\right) \tag{13}$$

The second term of Eq. (13) is the reactive power required to magnetize the electric machine.

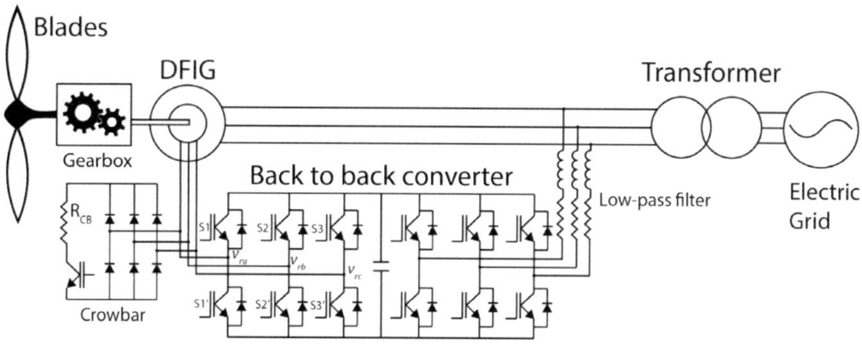

Figure 2. Basic scheme of DFIG-based wind turbine.

The basic scheme of a wind turbine is shown in **Figure 2**. A back-to-back power converter is necessary to send the required voltage to the rotor. In the grid side, it is also necessary a converter since the power flow must be bidirectional, from the electric machine to the grid at super-synchronous operation of the machine and from the electric grid to the machine at sub-synchronous operation. Grid side converter is usually controlled by cascade controller, which maintains DC-link voltage constant by regulating the grid current. The reactive power can also be controlled; hence, the wind turbine operates at unitary power factor, or it can inject reactive power to the grid similarly to an electrically excited synchronous generator. For purposes of this work, let us consider that the DC-link voltage is maintained constant by the grid-side controller.

3.1. Rotor-side power converter model

DFIG rotors are typically connected in star with the neutral connection isolated, the voltage if measured from phase a to the neutral point of the rotor star will be a combination of the phase voltages, which can be summarized using the following matrix equation:

$$\begin{bmatrix} v_{an} \\ v_{bn} \\ v_{cn} \end{bmatrix} = \frac{V_{DC}}{3} \begin{bmatrix} 2 & -1 & -1 \\ -1 & 2 & -1 \\ -1 & -1 & 2 \end{bmatrix} \overbrace{\begin{bmatrix} S_1 \\ S_2 \\ S_3 \end{bmatrix}}^{S_{123}} \quad (14)$$

Each component of the vector $S_{1,2,3}$ has only two valid states, 0 or 1. Then, each phase voltage has six different possible values: $\pm\frac{2}{3}V_{DC}$, $\pm\frac{1}{3}V_{DC}$, 0. The rotor side converter can be analyzed as a discontinuous sign function with variable gain.

4. Sliding mode controller design

It is well known from vector control that orienting the machine model presented in Section 3 in the stator flux reference frame is an effective way for decoupling active power (or torque) and reactive power control by means of rotor current regulation. In normal operation, the stator voltage vector will lead the stator flux by approximately 90 degrees neglecting voltage drop due to the stator resistance. From Eqs. (12) and (13) oriented at the stator flux direction, i.e., $\lambda_{sq}=0$ and $v_{sd}=0$, we obtain a decoupled control system: $T_{em}=-\frac{3PL_m}{2L_s}\lambda_{sd}\boldsymbol{i_{rq}}$ and $Q_s=-\frac{3L_m}{2L_s}v_{sq}\boldsymbol{i_{rd}}+\frac{3}{2L_s}\lambda_{sd}v_{sq}$. However, under unbalanced conditions, the phase shift between voltage and flux will not be constant; therefore, robust control is necessary to withstand this perturbation.

From Eq. (8), the direct axis component of stator current can be expressed as:

$$i_{sd}=\frac{\lambda_{sd}-L_m i_{rd}}{L_s} \tag{15}$$

Substituting Eq. (15) in Eq. (10):

$$\lambda_{rd}=L_r' i_{rd}+\frac{L_m}{L_s}\lambda_{sd} \tag{16}$$

where $L_r'=L_r-\frac{L_m^2}{L_s}$.

The quadrature component of rotor flux can be obtained in a similar way:

$$\lambda_{rq}=L_r' i_{rq}+\frac{L_m}{L_s}\lambda_{sq} \tag{17}$$

Substituting Eqs. (16) and (17) in (6) and (7) and solving for $\frac{di_{rd}}{dt}$ and $\frac{di_{rq}}{dt}$:

$$\frac{di_{rd}}{dt}=\frac{1}{L_r'}\left(v_{rd}-R_r i_{rd}-\frac{L_m}{L_s}\frac{d\lambda_{sd}}{dt}+\omega_r L_r' i_{rq}+\frac{L_m}{L_s}\omega_r\lambda_{sq}\right) \tag{18}$$

$$\frac{di_{rq}}{dt}=\frac{1}{L_r'}\left(v_{rq}-R_r i_{rq}-\frac{L_m}{L_s}\frac{d\lambda_{sq}}{dt}-\omega_r L_r' i_{rd}-\frac{L_m}{L_s}\omega_r\lambda_{sd}\right) \tag{19}$$

where $\omega_r=\omega_s-\omega_m$. Since the reference frame selected rotate at a synchronous speed, ω_r is equivalent to the rotor current angular speed and the slip angular frequency.

In **Figure 3**, the equivalent system for current regulation is shown. Under normal operation, the stator flux induces a voltage in the quadrature axis loop. The induced voltage is proportional to the slip and affects only the active power (or torque) regulation loop. Under a grid fault, the stator flux is directly affected and can be analyzed as another perturbation affecting the current regulation. Analyzing the flux in positive, negative, and natural fluxes, we can see the influence of the induced voltage (perturbation) to the current regulation loop. The positive sequence flux has an induced voltage proportional to $s\omega_s$, the slip is less than 0.3 for this type of machine; therefore, the induced voltage is low. The natural flux does not rotate; therefore, the induced voltage is proportional to the mechanical speed of the machine, and the negative flux rotates opposite to the reference frame orientation, the induced voltage is very large, proportional to near twice the synchronous speed $(2-s)\omega_s$ [8]. If the grid fault is very large, no control strategy could withstand this kind of perturbation, since the induced voltage may be larger than the rotor voltage. For that reason, external protection devices are required, and the simplest one is the crowbar.

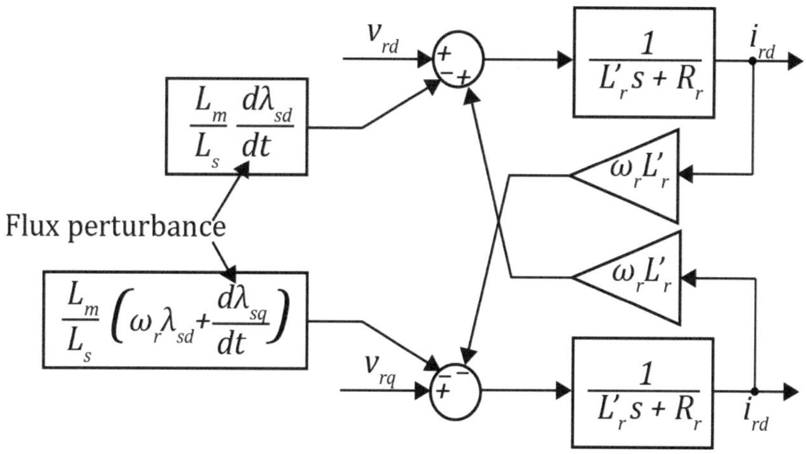

Figure 3. Equivalent current plant including stator flux perturbation.

SMC is robust against bounded disturbances and parameter variations; hence, it is a good alternative to control perturbed plants as the DFIG model shown in **Figure 3**. As we want to control the torque and reactive power, the following sliding surfaces are selected:

$$\sigma_{T_{em}} = T_{em} - T^*_{em}; \qquad \sigma_{Q_s} = Q_s - Q^*_s \tag{20}$$

where T^*_{em} and Q^*_s are the desired (reference) value of torque and reactive power, considering that the reference values are much slower than the dynamics of the system and that the stator flux is perfectly aligned with the reference frame d axis:

$$\dot{\sigma}_{T_{em}} = \dot{T}_{em} = -\frac{3PL_m}{2L_s}\left(\dot{i}_{rq}\lambda_{sd} + i_{rq}\dot{\lambda}_{sd}\right) \tag{21}$$

On the other hand, the voltage vector may have a d component due to voltage perturbation:

$$\dot{\sigma}_{Q_s} = \dot{Q}_s = \frac{3L_m}{2L_s}\left(\dot{v}_{sd} i_{rq} + v_{sd}\dot{i}_{rq} - \dot{v}_{sq} i_{rd} - v_{sq}\dot{i}_{rd}\right) + \frac{3}{2L_s}\left(\dot{\lambda}_{sd} v_{sq} + \lambda_{sd}\dot{v}_{sq}\right) \tag{22}$$

Substituting Eqs. (18) and (19) in (22) and (23):

$$\dot{\sigma}_{T_{em}} = \overbrace{kP\lambda_{sd}\left(R_r i_{rq} + \frac{L_m}{L_s}\omega_r\lambda_{sd} + \omega_r L'_r i_{rd}\right)}^{F_{Tem}} \overbrace{-kP\lambda_{sd} v_{rq}}^{d_{Tem}} \overbrace{-kPL'_r i_{rq}\dot{\lambda}_{sd}}^{P_{Tem}} \tag{23}$$

$$\begin{aligned} \dot{\sigma}_{Q_s} &= \overbrace{kv_{sq}\left(R_r i_{rq} - \omega_r L'_r i_{rq}\right)}^{F_{Q_s}} \overbrace{-kv_{sq} v_{rd}}^{d_{Q_s}} \\ &\underbrace{+kv_{sq}\left[\frac{L_m}{L_s}\dot{\lambda}_{sd} + \frac{L'_r}{v_{sq}}\left(\dot{v}_{sd} i_{rq} + v_{sd}\dot{i}_{rq} - \dot{v}_{sq} i_{rd}\right)\right] + \frac{3}{2L_s}\left(\dot{\lambda}_{sd} v_{sq} + \lambda_{sd}\dot{v}_{sq}\right)}_{P_{Q_s}} \end{aligned} \tag{24}$$

where $k = \frac{3L_m}{2L_s L_r'}$, expressing Eqs. (23) and (24) in matrix form:

$$\begin{bmatrix} \dot{\sigma}_{T_{em}} \\ \dot{\sigma}_{Q_s} \end{bmatrix} = \begin{bmatrix} F_{T_{em}} \\ F_{Q_s} \end{bmatrix} - k \begin{bmatrix} 0 & P\lambda_{sd} \\ v_{sq} & 0 \end{bmatrix} \begin{bmatrix} v_{rd} \\ v_{rq} \end{bmatrix} + \begin{bmatrix} P_{T_{em}} \\ P_{Q_s} \end{bmatrix} \tag{25}$$

Under normal conditions, all the terms dependent on $v_{sd}, \dot{v}_{sd}, \dot{v}_{sq}$ and $\dot{\lambda}_{sd}$ are equal to zero. Under abnormal conditions those terms can be analyzed as perturbations $P_{T_{em}}$ and P_{Q_s}. The control signals v_{rd} and v_{rq} appears in the first derivate of the sliding surface, thus the relative degree of the control system is one. Then, finite time convergence to the sliding surface and robustness against bounded disturbance/uncertainties can be achieved using the control signal:

$$\begin{bmatrix} v_{rd} \\ v_{rq} \end{bmatrix} = \begin{bmatrix} -M_d \operatorname{sgn}\left(\sigma_{Q_s}\right) \\ -M_q \operatorname{sgn}\left(\sigma_{T_{em}}\right) \end{bmatrix} \tag{26}$$

where $M_d, M_q > 0$. A detailed description of M_d and M_q computation is given in section 4.1.

Remark: A discontinuous function is intentionally selected because the nature of the rotor-side power converter is discontinuous as well; therefore, the control signal can be easily used directly from the controller algorithm to the power converter without modulation. On the contrary, any continuous control; e.g. saturation, sigmoid function, etc. employed to smooth the discontinuous control in Eq. (26) must be modulated to be implementable in a power converter; therefore, chattering will be present no matter the control strategy used.

The desired voltage need to be send through the rotor-side power converter. The voltage seen from the stator flux reference frame is obtained using park transform:

$$\begin{bmatrix} v_{rd} \\ v_{rq} \end{bmatrix} = \overbrace{\begin{bmatrix} \cos\theta_d & -\sin\theta_d \\ \sin\theta_d & \cos\theta_d \end{bmatrix} \frac{2}{3} \begin{bmatrix} 1 & -1/2 & -1/2 \\ 0 & \sqrt{3}/2 & -\sqrt{3}/2 \end{bmatrix}}^{T} \begin{bmatrix} v_{an} \\ v_{bn} \\ v_{cn} \end{bmatrix} \tag{27}$$

where the reference frame angle is $\theta_d = \theta_m - \theta_{sf}$.

Using Eqs. (14) (25) and (27), it is possible to establish a relationship between dq and abc quantities:

$$\begin{bmatrix} \sigma_{T_{em}} \\ \sigma_{Q_s} \end{bmatrix} = \overbrace{-k \begin{bmatrix} 0 & P\lambda_{sd} \\ v_{sq} & 0 \end{bmatrix} T \frac{V_{DC}}{3} \begin{bmatrix} 2 & -1 & -1 \\ -1 & 2 & -1 \\ -1 & -1 & 2 \end{bmatrix}}^{D} \begin{bmatrix} \sigma_a \\ \sigma_b \\ \sigma_c \end{bmatrix} \tag{28}$$

Since the matrix D is not square, the Moore-Penrose pseudo-inverse is used:

$$\overbrace{\begin{bmatrix} \sigma_a \\ \sigma_b \\ \sigma_c \end{bmatrix}}^{\sigma_{abc}} = D^{+} \begin{bmatrix} \sigma_{T_{em}} \\ \sigma_{Q_s} \end{bmatrix} \tag{29}$$

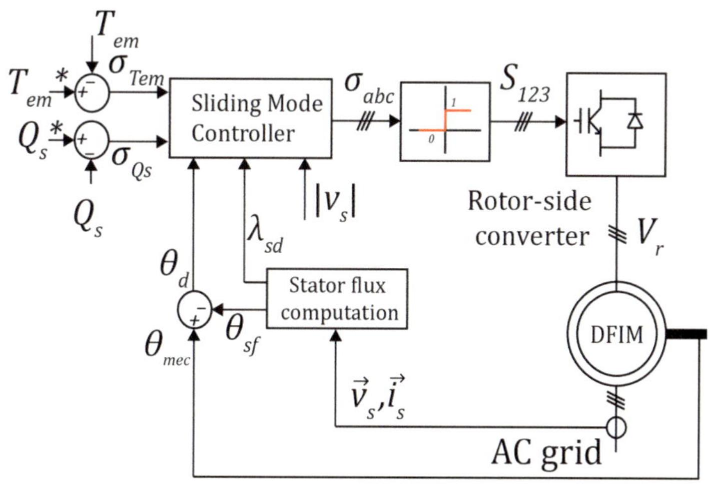

Figure 4. Basic scheme of SMC.

where:

$$D^{+} = D^{T}\left(DD^{T}\right)^{-1} = -\frac{3}{kV_{DC}}\begin{bmatrix} \frac{1}{P\lambda_{sd}}\sin\theta_d & \frac{1}{v_{sq}}\cos\theta_d \\ \frac{1}{P\lambda_{sd}}\sin\left(\theta_d + 2\pi/3\right) & \frac{1}{v_{sq}}\cos\left(\theta_d + 2\pi/3\right) \\ \frac{1}{P\lambda_{sd}}\sin\left(\theta_d - 2\pi/3\right) & \frac{1}{v_{sq}}\cos\left(\theta_d - 2\pi/3\right) \end{bmatrix}$$

The constant term $\frac{3}{kV_{DC}}$, which contains parameters of the electric machine, can be removed from the transformation matrix, since the controller will only evaluate the sign of σ_{abc}, thus the controller is robust against parametric uncertainties; under normal conditions, the stator flux direct component (λ_{sd}) and stator voltage quadrature component (v_{sq}) are constant. In **Figure 4** is shown the basic scheme of the presented controller. The output of Eq. (29) is the equivalent *abc* sliding surface, then a function similar to sign is used to evaluate the switch state (0 means lower leg activated and 1 means upper leg activated). The resulting control system does not require modulation since the switching state is determined directly from the control system.

4.1. Sliding mode existence condition

For a relative degree one system

$$\dot{\sigma} = F(x) + d(x)u \qquad (x \in \mathbb{R}^n) \tag{30}$$

With a scalar control

$$u = -M\,\text{sgn}\,(\sigma) \tag{31}$$

The condition for satisfying the existence of the sliding mode is [9]:

$$d(x)M > |F(x)| \tag{32}$$

The stator-flux–oriented SMC has a very similar form of system of Eq. (30). Then, existence of SMC is ensured if the following conditions are met:

$$M_{q,\,min} > |(F_{T_{em}} + P_{T_{em}})/d_{T_{em}}|; \circledcirc M_{d,\,min} > |(F_{Q_s} + P_{Q_s})/d_{Q_s}| \tag{33}$$

Therefore, choosing M_d and M_q using Eq. (33) ensures finite convergence to the sliding variable and insensitivity to bounded disturbance/uncertainties. The only requirement to compute the gains M_d and M_q is the knowledge of the bounds of the system and the disturbance/uncertainties, $|F_{Q_s,\,T_{em}}|$, $|d_{Q_s,\,T_{em}}|$ and $|P_{Q_s,\,T_{em}}|$. Note that real implementation of control gains in Eqs. (33) depends of DC-link voltage V_{DC}, which is variable in practice, and it must be ensured that V_{DC} is regulated correctly to ensure a robust performance of the rotor side of the DFIG system.

4.2. Switching frequency limitation

In **Figure 4** it is shown the scheme of an ideal sliding mode controller; however, it requires infinite switching frequency, which is not possible in real physical systems, therefore, the most common solution for this issue is to include a hysteresis loop to the ideal sign function [10] since hysteresis makes the switching frequency finite. Note that sigmoid functions and saturation can be implemented to reduce switching frequency as well as attenuate chattering. However, these functions are continuous and/or contain linear parts requiring modulation for its application to power electronics, while a sign function with hysteresis can be injected without modulation.

A widely used method to determine limit cycles and the oscillation frequency is the sinusoidal describing function (DF), which can be used for segmented nonlinear system composed by a linear system and a nonlinear part [12], that is the case of a linear plant controlled by a relay-based control system (see **Figure 5**). A requirement for applying DF is that the linear system $L(s)$ must have a low-pass filter behavior. Furthermore, only the first harmonic is considered in the analysis. However, for relative degree-one systems as the one presented in this chapter, this technique is not suitable [12]. DF will predict no oscillations because it ignores the contribution of the harmonics. On the other hand, Tsypkin's method is an exact method in which one can

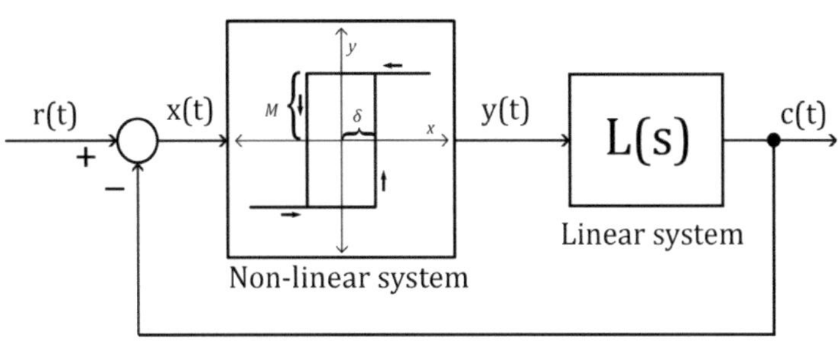

Figure 5. Relay-based control of a linear system.

select the number of harmonics to be considered, and it can be applied to estimate oscillations in relative degree one systems. Therefore, the DFIG system with SMC presented in this chapter is analyzed in the frequency domain using Tsypkin's method.

Expressing Eqs. (4)–(11) in matrix form:

$$\dot{x} = Ax + Bu; \qquad y = Cx \tag{34}$$

where:

$$x = \begin{bmatrix} i_{sd} & i_{sq} & i_{rd} & i_{rq} \end{bmatrix}^T; u = \begin{bmatrix} v_{sd} & v_{sq} & v_{rd} & v_{rq} \end{bmatrix}^T;$$

$$A = \frac{1}{L_r'L_s}\begin{bmatrix} -R_sL_r & \omega_m L_m^2 + \omega_s L_r' L_s & R_r L_m & \omega_m L_m L_r \\ -\omega_m L_m^2 - \omega_s L_r' L_s & -R_s L_r & -\omega_m L_m L_r & R_r L_m \\ R_s L_m & -\omega_m L_s L_m & -R_r L_s & -\omega_m L_r L_s + \omega_s L_r' L_s \\ \omega_m L_s L_m & R_S L_m & \omega_m L_r L_s - \omega_s L_r' L_s & -R_r L_s \end{bmatrix};$$

$$B = \frac{1}{L_r'L_s}\begin{bmatrix} L_r & 0 & -L_m & 0 \\ 0 & L_r & 0 & -L_r \\ -L_m & 0 & L_s & 0 \\ 0 & -L_m & 0 & L_s \end{bmatrix}; C = \begin{bmatrix} 0 & 0 & 1 & 0 \\ 0 & 0 & 0 & 1 \end{bmatrix}.$$

The system presents a nonlinearity due to the mechanical speed ω_m, which is dependent on the electromagnetic torque and the mechanical equation; however, since the time scales of the electrical quantities is much smaller than the time scale of the mechanical system, we are going to study system of Eq. (34) as a set of linear systems varying the mechanical speed in a range of ±30% about the synchronous speed. Other supposition is that the system is fully decoupled, that is, v_{rd} controls i_{rd} and v_{rq} controls i_{rq}. So, we can obtain both transfer functions from the diagonal elements of the following matrix equation:

$$G(s) = C(sI - A)^{-1}B + D \tag{35}$$

where I is a 4 x 4 identity matrix. The diagonal transfer functions will be relative degree one systems with four poles and three zeros:

$$L(s) = \frac{I_{rd}(s)}{V_{rd}(s)} = \frac{I_{rq}(s)}{V_{rq}(s)} = \frac{a_3 s^3 + a_2 s^2 + a_1 s + a_0}{d_4 s^4 + d_3 s^3 + d_2 s^2 + d_1 s + d_0} \tag{36}$$

Tsypkin's locus is defined as [12]:

$$T(j\omega) = \sum_{n\ odd}^{\infty} \mathrm{Re}[L_1(jn\omega)] + j\sum_{n\ odd}^{\infty} \frac{1}{n}\mathrm{Im}[L_1(jn\omega)] \tag{37}$$

The conditions required to predict a limit cycle oscillating at an angular frequency ω_0 are [12]:

$$\text{Im}[T(j\omega_0)] = \frac{\pi}{4}\left[L(\infty) - \frac{\delta}{M}\right] \tag{38}$$

$$\text{Re}[T(j\omega_0)] < \frac{\pi}{4\omega_0}\lim_{s\to\infty}[sL_1(s)] \tag{39}$$

In **Figure 6** it is shown the graphical solution for a DFIG with the characteristics shown in **Table 1**, the mechanical speed has minor influence at high frequencies; therefore, it is valid to consider the DFIG as a linear system (note that the number of pole pairs is not considered in (34); therefore, the mechanical speed reported is the one of an equivalent two pole machine). Supposing that the power inverter has a maximum switching frequency of $f_{max} = 4000$ Hz $\rightarrow$ $\omega_0 = 8000\pi\frac{\text{rad}}{\text{s}}$, from Eq. (38), we can calculate the hysteresis width. We know that the maximum gain the system can have is $M_{max} = \frac{2}{3}V_{dc}$, then the hysteresis loop that will ensure a maximum switching frequency of 4000 Hz is:

$$\delta = \frac{(0.3094)(4)}{\pi} * \frac{2}{3}600 = 157.57\ A$$

However, the control system was designed to direct control torque and reactive power, so we need to calculate the hysteresis width of those quantities, which can be easily done since the control system is decoupled, and the current direct and quadrature components are directly related with reactive power and torque, respectively.

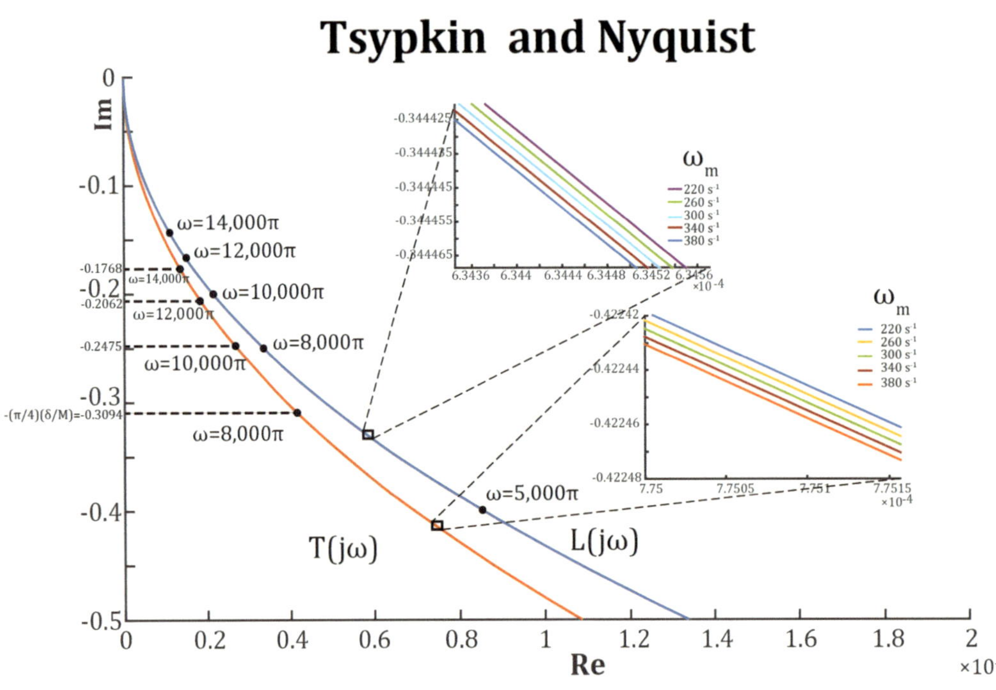

Figure 6. Graphical solution using Tsypkin's method.

Nominal power = 2 MW
Voltage = 690 V / 50 Hz
DC-link voltage = 1200 V
Stator/rotor turns ratio = 1/2
Mutual inductance (L_m) = 2.5 mH
Stator inductance (L_s) = 2.58 mH
Rotor inductance (L_r) = 2.58 mH
Stator resistance (R_s) = 2.6 mΩ
Rotor resistance (R_r) = 2.9 mΩl
Pole pairs (P) = 2

Table 1. Machine parameters used in simulation.

$$\delta_{Q_s} = \frac{3L_m v_{sq}}{2L_s}\delta; \qquad \delta_{T_{em}} = \frac{3PL_m\lambda_{sd}}{2L_s}\delta \tag{40}$$

Moreover, as can be seen from **Figure 6** and Eqs. (38) and (39), the addition of hysteresis can be used to attenuated chattering since the hysteresis width is directly proportional to amplitude of chattering while it is inversely proportional to the frequency.

4.3. Step-by-step design

The design of the SMC proposed in this section can be summarized in the next step-by-step algorithm:

1. Select the sliding mode surfaces for torque and reactive power using Eq. (20).

2. Choose the gains, M_d and M_q, of the controller in Eq. (26) using the conditions given in Eq. (33) to guarantying robustness.

3. Compute a hysteresis value for the torque and reactive power controllers (sign functions) using Eqs. (40) to ensure an acceptable switching frequency in the power electronics.

Then, following the step-by-step process described in this subsection, a robust SMC to regulate torque and reactive power in DFIG systems can be designed. Furthermore, practical implementation is considered in the design since a method to compute a hysteresis value, limiting undesired high frequency commutation in power electronics, is provided.

5. Complex power and torque under unbalanced conditions

The active and reactive power can be obtained from electrical quantities seen from a stationary reference frame:

$$P_s = \frac{3}{2}\mathrm{Re}\left(\vec{v_s}\,\overline{\vec{i_s}}\right) = \frac{3}{2}\left(v_{s\alpha}i_{s\alpha} + v_{s\beta}i_{s\beta}\right) \tag{41}$$

$$Q_s = \frac{3}{2}\mathrm{Im}\left(\vec{v_s}\,\overline{\vec{i_s}}\right) = \frac{3}{2}\left(v_{s\beta}i_{s\alpha} - v_{s\alpha}i_{s\beta}\right) \tag{42}$$

where the operator $\overline{\vec{x}}$ is the complex conjugate.

In case of unbalanced conditions, the symmetrical components methods can be used for simplifying analysis, since zero sequence does not produce complex power, only positive and negative sequences are analyzed:

$$\vec{v_s} = \vec{v}_{s_1} + \vec{v}_{s_2} = v_{s\alpha 1} + v_{s\alpha 2} + j\left(v_{s\beta 1} + v_{s\beta 2}\right) \tag{43}$$

$$\vec{i_s} = \vec{i}_{s_1} + \vec{i}_{s_2} = i_{s\alpha 1} + i_{s\alpha 2} + j\left(i_{s\beta 1} + i_{s\beta 2}\right) \tag{44}$$

Substituting (43) and (44) in Eqs. (41) and (42) yields:

$$P_s = \overbrace{\frac{3}{2}\left(v_{s\alpha 1}i_{s\alpha 1} + v_{s\beta 1}i_{s\beta 1}\right)}^{P_{s11}} + \overbrace{\frac{3}{2}\left(v_{s\alpha 1}i_{s\alpha 2} + v_{s\beta 1}i_{s\beta 2}\right)}^{P_{s12}} + \overbrace{\frac{3}{2}\left(v_{s\alpha 2}i_{s\alpha 1} + v_{s\beta 2}i_{s\beta 1}\right)}^{P_{s21}} + \overbrace{\frac{3}{2}\left(v_{s\alpha 2}i_{s\alpha 2} + v_{s\beta 2}i_{s\beta 2}\right)}^{P_{s22}} \tag{45}$$

$$Q_s = \overbrace{\frac{3}{2}\left(v_{s\beta 1}i_{s\alpha 1} - v_{s\alpha 1}i_{s\beta 1}\right)}^{Q_{s11}} + \overbrace{\frac{3}{2}\left(v_{s\beta 1}i_{s\alpha 2} - v_{s\alpha 1}i_{s\beta 2}\right)}^{Q_{s12}} + \overbrace{\frac{3}{2}\left(v_{s\beta 2}i_{s\alpha 1} - v_{s\alpha 2}i_{s\beta 1}\right)}^{Q_{s21}} + \overbrace{\frac{3}{2}\left(v_{s\beta 2}i_{s\alpha 2} - v_{s\alpha 2}i_{s\beta 2}\right)}^{Q_{s22}} \tag{46}$$

On the other hand, electromagnetic torque can be obtained using the well-known equation:

$$T_{em} = \frac{3P}{2}\mathrm{Im}\left(\vec{\lambda}_s\,\overline{\vec{i_s}}\right) \tag{47}$$

Using the symmetrical components theory, an unbalance condition can be modeled with invariant positive and negative sequence components; therefore, at steady state, the derivate term of Eqs. (4) and Eq. (5) are zero leading to the following positive and negative stator flux components:

$$\vec{\lambda}_{s1} = \frac{\vec{v}_{s1} - R_s\vec{i}_{s1}}{j\omega_s}; \qquad \vec{\lambda}_{s2} = \frac{\vec{v}_{s2} - R_s\vec{i}_{s2}}{-j\omega_s} \tag{48}$$

Substituting Eqs. (43) and (44) and (48) in (47):

$$T_{em} = \frac{3P}{2\omega_s}\mathrm{Re}\left[\vec{v}_{s1}\overline{\vec{i}_{s1}} + \vec{v}_{s1}\overline{\vec{i}_{s2}} - \vec{v}_{s2}\overline{\vec{i}_{s1}} - \vec{v}_{s2}\vec{i}_{s2} - R_s\left(\left|\vec{i}_{s1}\right|^2 - \left|\vec{i}_{s2}\right|^2\right)\right] \tag{49}$$

Comparing Eq. (50) with Eq. (45), it is easy to see that the same terms appear in both equations:

$$T_{em} = \frac{P}{\omega_s}\left[P_{s11} + P_{s12} - P_{s21} - P_{s22} + \frac{3R_s}{2}\left(\left|\vec{i}_{s2}\right|^2 - \left|\vec{i}_{s1}\right|^2\right)\right] \tag{50}$$

The terms P_{s12} and P_{s21} are the cause of oscillation in torque and power when an unbalanced dip occurs. Since the condition for canceling torque oscillations ($P_{s12}=P_{s21}$) is opposite to the condition for canceling active power oscillations ($P_{s12}= -P_{s21}$), it is not possible to cancel both at the same time. It is preferable to cancel torque oscillations; otherwise the mechanical components may be severely damaged. On the other hand, to cancel reactive power oscillations, the following condition must be met $Q_{s12}= -Q_{s21}$.

6. Simulation results

To test the controller, a DFIG was simulated using the parameters displayed in **Table 1**. The blades model surface is shown in **Figure 1** with gearbox ratio ($\eta=85.8$) rotor radius ($r=40$ m) and air density ($\rho=1.25$ kg\m^3). The nominal wind speed of 12 m/s, the nominal tip speed ratio $\Lambda_{opt}=7.9533$ and the maximum power coefficient is $C_{p,\max}=0.4109$. The pitch controller is ideal chopping the extracted aerodynamic power to the nominal power (2 MW) for wind speed above the nominal value. Considering that the maximum switching frequency of the converter is 7000 Hz; from **Figure 6** the desired hysteresis width is $\delta=90.04A$; therefore, $\delta_{Q_s}=128\times 10^3$ and $\delta_{T_{em}}=811$.

The following electromagnetic torque reference is used for maintaining the tip-speed ratio at the optimal value without measuring wind speed:

$$T_{ref} = -\frac{\frac{1}{2}\pi\rho R^5 C_{p,\max}}{\Lambda_{opt}^2\eta^3}\omega_m^2 \tag{51}$$

On the other hand, the reactive power reference is maintained at zero.

The wind speed profile is shown in **Figure 7** II, the wind speed was taken from real measurements reported by the Department of Wind Energy, Technical University of Denmark [19] with values oscillating in all the operational range of the wind speed.

The power extracted by the blades is shown in **Figure 7** III, during the high wind speed periods, the ideal pitch controller maintains the extracted power at the nominal value of 2 MW, while the mechanical speed is controlled during the rest of the time to optimize power extraction as shown in **Figure 7** IV. The rotor converter nominal power limits the operational speed of the wind turbine $P_r\approx sP_s$.

The references are followed even under unbalanced grid conditions, a two-phase voltage dip is simulated at the terminals of the electric machine, the voltage dip is 20% of the nominal value. In **Figure 7** I, it is displayed the detail of the voltage dip. The voltage dip starts at 95 seconds and ends at 98 seconds of the simulation, the time axis in **Figure 7** I is chopped from 95.05 to

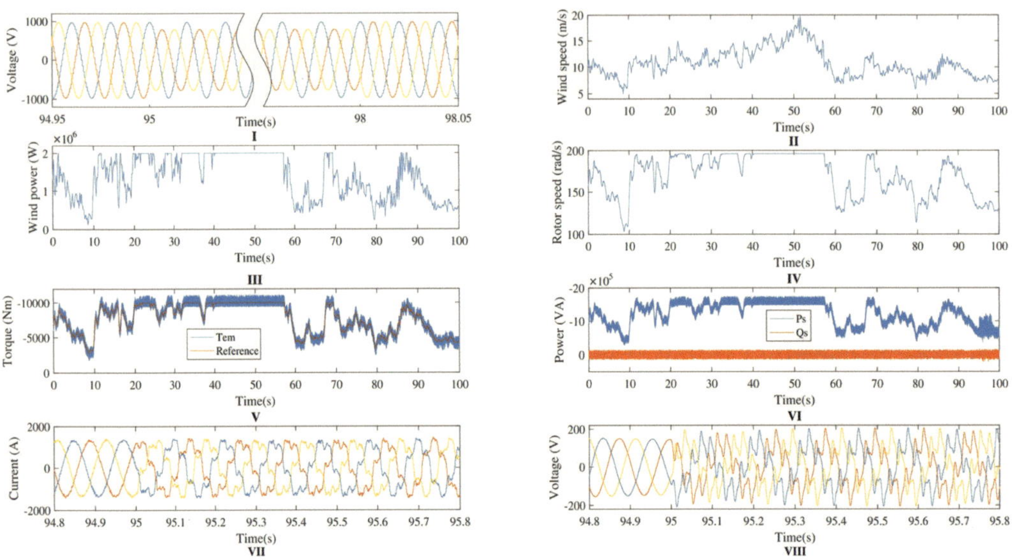

Figure 7. Simulation results: I stator voltage, II wind speed profile, III extracted aerodynamic wind power, IV rotor speed, V electromagnetic torque, VI stator active and reactive power, VII rotor current, VIII rotor filtered voltage.

97.95 s, in order to better see the stator voltage waveform during the fault. As demonstrated in Section 5, it is not possible to maintain both torque and reactive power constant during unbalanced conditions; therefore, the stator reactive power is affected (see **Figure 7** VI) since the controller forces the electromagnetic torque to be constant. As a result, the controller injects a negative sequence current to the rotor in order to cancel torque oscillations out (see **Figure 7** VII), it is worth mention that other control strategies as vector control requires a double control loop for controlling positive and negative sequence separated [13], which not only complicates the current regulation algorithm but also requires sequence decoupling of rotor current, a very complex issue that is avoided by using a robust control strategy.

Finally, the filtered rotor voltage is displayed in **Figure 7** VIII. The controller does not require modulation and automatically injects negative sequence voltage to regulate torque and reactive power under unbalanced conditions.

7. Conclusions

A stator-flux–oriented sliding mode control, which regulates torque and reactive power in DFIG, is presented. The controller is not dependent on electric machine parameters and do not require modulation, injecting the desired voltage vector directly to a two-level power converter. Despite the proposed SMC controller has a variable switching frequency, which is not desired in practical applications, the switching frequency is limited by a hysteresis loop in the torque and reactive power controllers. The hysteresis value is calculated by means of the frequency domain characterization of the DFIG system with SMC, via Tsypkin's method.

Then, the safe operation of the power converter regarding commutation frequency is ensured. Furthermore, the proposed SMC is capable to reject stator flux variations providing a desired operation even under unbalanced grid conditions. Compared with classical control techniques as DTC, the presented SMC does not require the modification of the control loop. The equations for compute the control gains that ensure robustness and existence of SMC are given. Simulations validating the advantages of the proposed SMC are shown.

Nomenclature

v	Voltage [V]
i	Current [A]
λ	Magnetic flux [Wb]
s, r	Stator, rotor sub index
d, q	Direct, quadrature reference frame axis component
α, β	Alpha, beta reference frame axis component
L_m	Magnetic inductance [H]
L_s	Stator inductance [H]
L_r	Rotor inductance [H]
P	Number of pole pairs.
P_s, Q_s	Stator active, reactive power [W, VA]
T_{em}	Electromagnetic torque [N·m]
V_{DC}	DC-link voltage [V]
ω_s	Synchronous speed [rad/s]
θ_{sf}	Stator flux angular position [rad]
ω_m	Rotor speed [rad/s]
θ_m	Rotor mechanical angle [rad]
ω_r	Rotor electrical speed [rad/s]
s	Slip
P_{wind}	Aerodynamic power [W]
C_P	Power coefficient
γ	Pitch angle [deg]
Λ	Tip-speed ratio

ρ	Air density [kg/m^3]
A	Turbine transverse area [m^2]
r	Turbine radius [m]
V	Wind speed [m/s]
η	Gearbox ratio
s	Complex frequency variable

Author details

Ivan Villanueva*, Antonio Rosales, Pedro Ponce and Arturo Molina

*Address all correspondence to: ivan.villanueva@itesm.mx

Tecnologico de Monterrey, Mexico

References

[1] GWEC. Global Wind Report 2016 [Internet]. 2017. Available from: http://www.gwec.net/publications/global-wind-report-2/global-wind-report-2016/# [Accessed: 07/17/2017]

[2] European Commission. JRC Wind Energy Status Report: 2016 Edition [Internet]. March 2017. Available from: https://ec.europa.eu/jrc/en/news/jrc-wind-energy-status-report-2016-edition [Accessed: 07/17/2017]

[3] Tapia A, Tapia G, Ostaloza JX, Saenz JR. Modeling and control of a wind turbine driven doubly fed induction generator. IEEE Transactions on Energy Conversion. 2003;**18**(2):194-204

[4] Xu L, Wang Y. Dynamic modeling and control of DFIG-based wind turbine under unbalanced network conditions. IEEE Transactions on Power Systems. 2007;**22**(1):314-323

[5] Abad G, Rodriguez MA, Poza J, Canales M. Direct torque control of doubly fed induction machine-based wind turbines under voltage dips and without crowbar protection. IEEE Transactions on Energy Conversion. 2012;**25**(2):586-588

[6] Ong CM. Dynamic Simulation of Electric Machinery Using Matlab/Simulink. Prentice Hall: Upper Saddle River, NJ; 1998. 626 p

[7] Lubosny Z. Wind Turbine Operation in Electric Power Systems. Springer; 2003. 262 p

[8] Abad G, Lopez J, Rodriguez M, Marroyo L, Iwanski G. Doubly Fed Induction Machine: Modeling and Control for Wind Energy Generation. John Wiley & Sons: Hoboken, NJ; 2011. 625 p

[9] Utkin V, Guldner J, Shi J. Sliding Mode Control in Electro-Mechanical Systems. 2nd ed. CRC Press; Boca Raton, Florida, 2009. p. 503

[10] Venkataramanan R. Sliding Mode Control of Power Converters [Dissertation]. California Institute of Technology; Pasadena, California, 1986

[11] Villanueva I, Rosales A, Ponce P, Molina A. Stator-flux-oriented sliding mode controller for DFIG with variable hysteresis loop for limiting switch frequency of rotor-side power converter. In: IEEE, editor. Industrial Technology (ICIT), 2017 IEEE International Conference; 2017

[12] Gelb A, Vander Velde WE. Multiple-Input Describing Functions and Nonlinear System Design. New York: McGraw-Hill; 1968. 655 p

[13] Xu L. Coordinated control of DFIG's rotor and grid side converters during network unbalance. IEEE Transactions on Power Electronics. 2008;**23**(3):1041-1049

[14] Gupta G, Gosh A. Frequency-domain characterization of sliding mode control of an inverter used in DSTATCOM application. IEEE Transactions on Circuits and Systems. 2006;**53**(3):662-672

[15] Martinez MI, Tapia G, Susperregui A, Camblong H. Sliding mode control for DFIG rotor and grid side converters under unbalanced and harmonically distorted grid voltage. IEEE Transactions on Energy Conversions. 2012;**27**(2):328-339

[16] Beltran B, Benbouzid M, Ahmed-Ali T. High-order sliding mode control of a DFIG-based wind turbine for power maximization and grid fault tolerance. In: Electric Machines and Drives Conference; IEEE; 2009. p. 183-189

[17] Benbouzid M, Beltran B, Amirat Y, Yao G, Han J, Mangel H. Second order sliding mode control of DFIG-based wind turbines fault ride through capability enhancement. ISA Transactions. 2014;**53**(3):827-833

[18] Martinez M, Susperregui A, Tapia G, Xu L. Sliding mode control of a wind turbine driven double-fed induction generator under non-ideal grid voltages. IET Renewable Power Generation. 2013;**7**(4):370-379

[19] Hansen KS, Larsen GC. Database of Wind Characteristics [Internet]. Available from: http://www.winddata.com/ [Accessed: January 2016]

Section 2

Robust Control Systems with Model Reference Adaptive Approach

Chapter 5

Robust Adaptive Output Tracking for Quadrotor Helicopters

Keyvan Mohammadi and Andrea L'Afflitto

Additional information is available at the end of the chapter

http://dx.doi.org/10.5772/intechopen.70723

Abstract

Quadrotor helicopters are drawing considerable attention both for their mobility and their potential to perform multiple tasks in complete autonomy. Moreover, the numerous limitations characterizing these aircraft, such as their underactuation, make quadrotors ideal testbeds for innovative theoretical approaches to the problem of controlling autonomous mechanical systems. In this chapter, we propose a robust model reference adaptive control architecture and design an autopilot for quadrotors, which guarantees satisfactory output tracking despite uncertainties in the vehicle's mass, matrix of inertia, and location of the center of mass. The feasibility of our results is supported by a detailed analysis of the quadrotor's equations of motion. Specifically, considering the vehicle's equations of motion as a time-varying nonlinear dynamical system and avoiding the common assumption that the vehicle's Euler angles are small at all times, we prove that the proposed autopilot guarantees satisfactory output tracking and verifies sufficient conditions for a weak form of controllability of the closed-loop system known as strong accessibility. A numerical example illustrates the applicability of the theoretical results presented and clearly shows how the proposed autopilot outperforms in strong wind conditions autopilots designed using a commonly employed proportional-derivative control law and a conventional model reference adaptive control law.

Keywords: robust model reference adaptive control, *e*-modification, quadrotors, autopilot design, output tracking

1. Introduction

Quadrotor helicopters, also known as "quadrotors," are currently employed in diverse scenarios, which range from search and rescue missions to infrastructure inspection, precision agriculture, and wildlife monitoring ([1, Ch. 1], [2, 3]). Employing quadrotors in enclosed industrial environments or in proximity of untrained personnel is still considered as a challenge for the high reliability required to these aircraft. Additional complexity in the use of quadrotors for commercial applications, such as parcel delivery, is that users demand satisfactory trajectory following capabilities without tuning the controller's gains prior to each mission, whenever the payload is changed.

Autopilots for commercial-off-the-shelf quadrotors are currently designed assuming that the vehicle's and the payload's inertial properties are known and constant in time. Moreover, it is assumed that the propulsion system is able to deliver maximum thrust whenever needed. These assumptions considerably simplify the design of control algorithms for quadrotor helicopters, but also undermine these vehicles' reliability in challenging work conditions, such as in case the propulsion system is partly damaged or the payload is not rigidly attached to the vehicle. For instance, the authors in [4] show that if the payload's mass and matrix of inertia vary in time, then autopilots for quadrotors designed using classical control techniques, such as the proportional-derivative control, are inadequate to guarantee satisfactory trajectory tracking.

In recent years, numerous authors, such as Bouadi et al. [5]; Dydek et al. [6]; Jafarnejadsani et al. [7]; Loukianov [8]; Mohammadi & Shahri [9]; Zheng et al. [10], to name a few, employed nonlinear robust control techniques, such as sliding mode control, model reference adaptive control (MRAC), adaptive sliding mode control, and $\mathcal{L}_1$ adaptive control, to design autopilots for quadrotors that are able to account for inaccurate modeling assumptions and compensate failures in the propulsion system. These autopilots are generally designed assuming perfect knowledge of the location of the quadrotor's center of mass, supposing that the vehicle's Euler angles are small at all times, and neglecting the inertial counter-torque. Furthermore, in several cases also the aerodynamic force and the corresponding moment are omitted. Because of these simplifying assumptions, these autopilots are inadequate for aircraft performing aggressive maneuvers, flying in adverse weather conditions, and transporting payloads not rigidly connected to the vehicle's frame [11]. The vehicle's guidance system is usually delegated to avoiding obstacles detected by proximity sensors and cameras installed aboard. For details, see the recent works by Faust et al. [12]; Gao & Shen [13]; Lin & Saripalli [14].

In the first part of this chapter, we present the equations of motion of quadrotors and analyze those properties needed to design effective nonlinear robust controls that enable output tracking. Specifically, we present the equations of motion of quadrotors without assuming *a priori* that the Euler angles are small and without neglecting the inertial counter-torque and the gyroscopic effect. Since the inertial counter-torque cannot be expressed as an algebraic function of the quadrotor's state and control vectors, we account for this effect as an unmatched time-varying disturbance on the vehicle's dynamics and hence, we consider the equations of motion of a quadrotor as a nonlinear time-varying dynamical system. Successively, we verify for the first time sufficient conditions for the strong accessibility of quadrotors' altitude and rotational dynamics; strong accessibility [15] is a weak form of controllability for nonlinear time-varying dynamical systems. As a result of this analysis, we show that a conservative control law for quadrotors must prevent rotations of a $\pm\pi/2$ angle about either of the two horizontal axes of the body reference frame; otherwise, the vehicle may be uncontrollable.

In the second part of this chapter, we present a robust autopilot for quadrotors, which is based on a version of the *e*-modification of the MRAC architecture [16]. This autopilot is characterized by numerous unique features. For instance, we assume that the quadrotor's inertial properties, such as mass, moment of inertia, and location of the center of mass are *unknown*. Moreover, we assume that the quadrotor's reference frame is centered at an arbitrary point, which does not necessarily coincide with the vehicle's center of mass. In addition, we suppose that the coefficients characterizing the aerodynamic force and moment are *unknown*. In order to reduce the rotational kinematic and dynamic equations to a form that is suitable for MRAC, we employ an output-feedback linearization approach so that the controlled rotational dynamics is captured by a linear dynamical system, whose virtual input is designed using the proposed robust MRAC architecture.

We design the autopilot's outer loop so that it regulates the vehicle's position in the inertial frame and the inner loop so that it regulates the vehicle's attitude. In conventional autopilots for quadrotors, the outer loop regulates the vehicle's position in the horizontal plane and the inner loop controls the quadrotor's altitude and orientation. As in conventional architectures, our outer loop defines the reference pitch and roll angles for the inner loop to track. However, our control architecture allows us to verify *a priori* that the reference pitch and roll angles meet the sufficient conditions for strong accessibility of the quadrotor's altitude and rotational dynamics. Conventional autopilots' outer loop may generate large reference pitch and roll angles that are not guaranteed to lay in the vehicle's reachable set.

The conventional MRAC architecture ([17], Ch. 9) is designed to regulate time-invariant dynamical systems and, for this reason, autopilots for quadrotors based on the classical MRAC are unable to account for time-varying terms in the vehicle's dynamics, such as the inertial counter-torque. Moreover, autopilots for quadrotors based on the classical MRAC architecture are robust to both matched and parametric uncertainties, but not unmatched uncertainties, such as aerodynamic forces. The autopilots presented in this chapter, instead, are robust to unmatched uncertainties as well and account for the fact that quadrotors are inherently time-varying dynamical systems.

A numerical example illustrates our theoretical framework by designing a control law that allows a quadrotor to follow a circular trajectory, although the vehicle's inertial properties are unknown, one of the motors is suddenly turned off, the payload is dropped over the course of the mission, and the wind blows at 16 m/s, which is usually considered as a prohibitive velocity for conventional quadrotors. In this example, we clearly show how the proposed robust control algorithm outperforms both the classical proportional-derivative (PD) control and the conventional MRAC. Indeed, it is shown that quadrotors implementing autopilots based on the PD framework crash as soon as one motor is turned off. Moreover, we verify that quadrotors implementing autopilots based on the classical MRAC framework [6] are unable to fly in the presence of wind gusts faster than 6 m/s. Lastly, we show that, to fly in a wind blowing at 16 m/s, our autopilot requires a control effort that is smaller than the one required by a conventional MRAC-based autopilot to fly in a 6 m/s wind.

2. Notation and definitions

In this section, we establish the notation and the definitions used in this chapter. Let $\mathbb{R}$ denote the set of real numbers, $\mathbb{R}^n$ the set of $n \times 1$ real column vectors, and $\mathbb{R}^{n \times m}$ the set of $n \times m$ real matrices. We write $\mathbf{1}_n$ for the $n \times n$ identity matrix, $0_{n \times m}$ for the zero $n \times m$ matrix, and A^{T} for the transpose of the matrix $A \in \mathbb{R}^{n \times m}$. Given $a = [a_1, a_2, a_3]^{\mathrm{T}} \in \mathbb{R}^3$, $a^{\times} \triangleq \begin{bmatrix} 0 & -a_3 & a_2 \\ a_3 & 0 & -a_1 \\ -a_2 & a_1 & 0 \end{bmatrix}$ denotes the *cross product operator*. We write $\|\cdot\|$ for the Euclidean vector norm and $\|\cdot\|_{\mathrm{F}}$ for the Frobenius matrix norm, that is, given $B \in \mathbb{R}^{n \times m}$, $\|B\|_{\mathrm{F}} \triangleq \left[\mathrm{tr}\left(BB^{\mathrm{T}}\right)\right]^{\frac{1}{2}}$. The Fréchet derivative of the continuously differentiable function $V: \mathbb{R}^n \to \mathbb{R}$ at x is denoted by $V'(x) \triangleq \partial V(x)/\partial x$.

Definition 2.1 ([18], Def. 6.8). *The Lie derivative of the continuously differentiable function* $V: \mathbb{R}^n \to \mathbb{R}$ *along the vector field* $f: \mathbb{R}^n \to \mathbb{R}^n$ *is defined as*

$$L_f V(x) \triangleq V'(x) f(x), \qquad x \in \mathbb{R}^n. \tag{1}$$

The *zeroth-order* and the *higher-order Lie derivatives* are, respectively, defined as

$$L_f^0 V(x) \triangleq V(x), \qquad L_f^k V(x) \triangleq L_f\left(L_f^{k-1} V(x)\right), \qquad x \in \mathbb{R}^n, \qquad k \geq 1. \tag{2}$$

Given the continuously differentiable functions $f, g: \mathbb{R}^n \to \mathbb{R}^n$, the *Lie bracket* of $f(\cdot)$ and $g(\cdot)$ is defined as

$$\mathrm{ad}_f g(x) \triangleq \frac{\partial g(x)}{\partial x} f(x) - \frac{\partial f(x)}{\partial x} g(x), \qquad x \in \mathbb{R}^n. \tag{3}$$

To recall the definition of uniform ultimate boundedness, consider the nonlinear time-varying dynamical system

$$\dot{x}(t) = f(t, x(t)), \qquad x(t_0) = x_0, \qquad t \geq t_0, \tag{4}$$

where $x(t) \in \mathbb{R}^n$, $t \geq t_0$, $f: [t_0, \infty) \times \mathbb{R}^n \to \mathbb{R}^n$ is jointly continuous in its arguments, $f(t, \cdot)$ is locally Lipschitz continuous in x uniformly in t for all t in compact subsets of $t \in [t_0, \infty)$, and $0 = f(t, 0)$, $t \geq t_0$.

Definition 2.2 ([19], Def. 4.6). *The nonlinear time-varying dynamical system* (4) *is uniformly ultimately bounded* with ultimate bound $b > 0$ if there exists $c > 0$ independent of t_0 and for every $a \in (0, c)$, there exists $T = T(a, c) \geq 0$, independent of t_0, such that if $\|x_0\| \leq a$, *then* $\|x(t)\| \leq b$, $t \geq t_0 + T$.

3. Robust MRAC for output tracking

In order to enable robust output tracking, in this section we present a robust nonlinear control law that is based on the e-modification of the conventional model reference adaptive control [16]. This control law guarantees that after a finite-time transient, the plant's measured output tracks a given reference signal within some bounded error despite model uncertainties and external disturbances. In practice, the proposed controller guarantees uniform ultimate boundedness of the output tracking error.

Consider the nonlinear time-varying plant and the plant sensors' dynamics

$$\dot{x}_{\mathrm{p}}(t) = A_{\mathrm{p}}x_{\mathrm{p}}(t) + B_{\mathrm{p}}\Lambda\left[u(t) + \Theta^{\mathrm{T}}\Phi\left(x_{\mathrm{p}}(t)\right)\right] + \hat{\xi}(t), \quad x_{\mathrm{p}}(t_0) = x_{\mathrm{p},0}, \quad t \geq t_0, \tag{5}$$

$$\dot{y}(t) = \varepsilon C_{\mathrm{p}}x_{\mathrm{p}}(t) - \varepsilon y(t), \qquad y(t_0) = C_{\mathrm{p}}x_{\mathrm{p},0}, \tag{6}$$

where $x_{\mathrm{p}}(t) \in \mathcal{D}_{\mathrm{p}} \subseteq \mathbb{R}^{n_{\mathrm{p}}}$, $t \geq t_0$, denotes the *plant's trajectory*, $0_{n_{\mathrm{p}}\times 1} \in \mathcal{D}_{\mathrm{p}}$, $u(t) \in \mathbb{R}^m$ denotes the *control input*, $y(t) \in \mathbb{R}^m$ denotes the *measured output*, $\varepsilon > 0$, $A_{\mathrm{p}} \in \mathbb{R}^{n_{\mathrm{p}} \times n_{\mathrm{p}}}$ is *unknown*, $B_{\mathrm{p}} \in \mathbb{R}^{n_{\mathrm{p}} \times m}$, $C_{\mathrm{p}} \in \mathbb{R}^{m \times n_{\mathrm{p}}}$, $\Lambda \in \mathbb{R}^{m \times m}$ is diagonal, positive-definite, and *unknown*, $\Theta \in \mathbb{R}^{N \times m}$ is *unknown*, the *regressor vector* $\Phi : \mathbb{R}^{n_{\mathrm{p}}} \to \mathbb{R}^N$ is Lipschitz continuous in its argument, and $\hat{\xi} : [t_0, \infty) \to \mathbb{R}^{n_{\mathrm{p}}}$ is continuous in its argument and *unknown*. We assume that $\|\hat{\xi}(t)\| \leq \hat{\xi}_{max}$, $t \geq t_0$, and Λ is such that the pair $(A_{\mathrm{p}}, B_{\mathrm{p}}\Lambda)$ is controllable and $\Lambda_{min}\mathbf{1}_m \leq \Lambda$, for some $\Lambda_{min} > 0$. Both Λ and $\Theta^{\mathrm{T}}\Phi(x_{\mathrm{p}})$, $x_{\mathrm{p}} \in \mathcal{D}_{\mathrm{p}}$, capture the plant's matched and parametric uncertainties, such as malfunctions in the control system; the term $\hat{\xi}(\cdot)$ captures the plant's unmatched uncertainties, such as external disturbances.

Eq. (6) models the plant sensors as linear dynamical systems, whose uncontrolled dynamics is exponentially stable and characterized by the parameter $\varepsilon > 0$ ([20], Ch. 2). Given the *reference signal* $y_{\mathrm{cmd}} : [t_0, \infty) \to \mathbb{R}^m$, which is continuous with its first derivative, define $y_{\mathrm{cmd},2}(t) \triangleq \dot{y}_{\mathrm{cmd}}(t)$, $t \geq t_0$, and assume that both $y_{\mathrm{cmd}}(\cdot)$ and $y_{\mathrm{cmd},2}(\cdot)$ are bounded, that is, $\|y_{\mathrm{cmd}}(t)\| \leq y_{\max,1}$, $t \geq t_0$, and $\|y_{\mathrm{cmd},2}(t)\| \leq y_{\max,2}$, for some $y_{\max,1}, y_{\max,2} > 0$.

The following theorem provides a robust MRAC for the nonlinear time-varying dynamical system (5) and (6) such that the measured output $y(\cdot)$ is able to eventually track the reference signal $y_{\mathrm{cmd}}(\cdot)$ with bounded error, that is, there exist $b > 0$ and $c > 0$ independent of t_0, and for every $a \in (0, c)$, there exists a finite-time $T = T(a, c) \geq 0$, independent of t_0, such that if $\|y(t_0) - y_{\mathrm{cmd}}(t_0)\| \leq a$, then

$$\|y(t) - y_{\mathrm{cmd}}(t)\| \leq b, \qquad t \geq t_0 + T. \tag{7}$$

For the statement of this result, let $n \triangleq n_{\mathrm{p}} + m$ and $x(t) \triangleq \left[x_{\mathrm{p}}^{\mathrm{T}}(t), \left[y(t) \quad y_{\mathrm{cmd}}(t)\right]^{\mathrm{T}}\right]^{\mathrm{T}} \in \mathbb{R}^n$, $t \geq t_0$, note that (5) and (6) are equivalent to

$$\dot{x}(t) = Ax(t) + B\Lambda\left[u(t) + \Theta^{\mathrm{T}}\Phi\left(x_{\mathrm{p}}(t)\right)\right] + \xi(t), \qquad x(t_0) = \begin{bmatrix} x_{\mathrm{p},0} \\ C_{\mathrm{p}}x_{\mathrm{p},0} - y_{\mathrm{cmd}}(t_0) \end{bmatrix}, \qquad t \geq t_0, \tag{8}$$

where $x(t) \in \mathcal{D} \subseteq \mathbb{R}^n$, $\mathcal{D} \triangleq \mathcal{D}_{\mathrm{p}} \times \mathbb{R}^m$, $A \triangleq \begin{bmatrix} A_{\mathrm{p}} & 0_{n_{\mathrm{p}}\times m} \\ \varepsilon C_{\mathrm{p}} & -\varepsilon \mathbf{1}_m \end{bmatrix}$, $B \triangleq \begin{bmatrix} B_{\mathrm{p}} \\ 0_{m\times m} \end{bmatrix}$, $B_1 \triangleq \begin{bmatrix} 0_{n_{\mathrm{p}}\times m} \\ -\mathbf{1}_m \end{bmatrix}$, and $\xi(t) \triangleq B_1\left[y_{\mathrm{cmd},2}(t) + \varepsilon y_{\mathrm{cmd}}(t)\right] + \begin{bmatrix} \mathbf{1}_{n_{\mathrm{p}}} \\ 0_{m\times n_{\mathrm{p}}} \end{bmatrix}\hat{\xi}(t)$, and consider the *reference dynamical model*

$$\dot{x}_{ref}(t) = A_{ref}x_{ref}(t) + B_{ref}y_{cmd}(t), \qquad x_{ref}(t_0) = \begin{bmatrix} x_{p,0} \\ C_p x_{p,0} - y_{cmd}(t_0) \end{bmatrix}, \qquad t \geq t_0, \tag{9}$$

where $A_{ref} = \begin{bmatrix} A_{ref,1} & 0_{n_p \times m} \\ 0_{m \times n_p} & A_{ref,2} \end{bmatrix}$, $A_{ref,1} \in \mathbb{R}^{n_p \times n_p}$ is Hurwitz, $A_{ref,2} \in \mathbb{R}^{m \times m}$ is Hurwitz, and $B_{ref} \in \mathbb{R}^{n \times m}$ is such that the pair (A_{ref}, B_{ref}) is controllable.

Theorem 3.1 *Consider the nonlinear time-varying dynamical system given by* Eqs. (5) *and* (6), *the augmented dynamical system* (8), *and the linear time-invariant reference dynamical model* (9). *Define* $e(t) \triangleq x(t) - x_{ref}(t)$, $t \geq t_0$, *and let*

$$\gamma(t, x_p, x) = \hat{K}_x^T(t)x + \hat{K}_{cmd}^T(t)y_{cmd}(t) - \hat{\Theta}^T(t)\Phi(x_p), \qquad (t, x_p, x) \in [t_0, \infty) \times \mathcal{D}_p \times \mathcal{D}, \tag{10}$$

where

$$\dot{\hat{K}}_x(t) = -\Gamma_x\left[x(t)e^T(t)PB + \sigma_1\|B^T Pe(t)\|\hat{K}_x(t)\right], \qquad \hat{K}_x(t_0) = 0_{n\times m}, \qquad t \geq t_0, \tag{11}$$

$$\dot{\hat{K}}_{cmd}(t) = -\Gamma_{cmd}\left[y_{cmd}(t)e^T(t)PB + \sigma_2\|B^T Pe(t)\|\hat{K}_{cmd}(t)\right], \qquad \hat{K}_{cmd}(t_0) = 0_{m\times m}, \tag{12}$$

$$\dot{\hat{\Theta}}(t) = \Gamma_\Theta\left[\Phi(x_p(t))e^T(t)PB - \sigma_3\|B^T Pe(t)\|\hat{\Theta}(t)\right], \qquad \hat{\Theta}(t_0) = 0_{N\times m}, \tag{13}$$

the learning gain matrices $\Gamma_x \in \mathbb{R}^{n \times n}$, $\Gamma_{cmd} \in \mathbb{R}^{m \times m}$, and $\Gamma_\Theta \in \mathbb{R}^{N \times N}$ are symmetric positive-definite, $P \in \mathbb{R}^{n \times n}$ is the symmetric positive-definite solution of the Lyapunov equation

$$0 = A_{ref}^T P + PA_{ref} + Q, \tag{14}$$

$Q \in \mathbb{R}^{n \times n}$ is symmetric positive-definite, and $\sigma_1, \sigma_2, \sigma_3 > 0$. If there exists $K_x \in \mathbb{R}^{n \times m}$ and $K_{cmd} \in \mathbb{R}^{m \times m}$ such that

$$A_{ref} = A + B\Lambda K_x^T, \tag{15}$$

$$B_{ref} = B\Lambda K_{cmd}^T, \tag{16}$$

then the nonlinear time-varying dynamical system (8) with $u(t) = \gamma(t, x_p(t), x(t))$, $t \geq t_0$, is uniformly ultimately bounded and (7) is verified.

Proof: Let $\Delta K_x \triangleq \hat{K}_x - K_x$, $\Delta K_{cmd} \triangleq \hat{K}_{cmd} - K_{cmd}$, and $\Delta\Theta \triangleq \hat{\Theta} - \Theta$ and consider the Lyapunov function candidate

$$V(e, \Delta K_x, \Delta K_{cmd}, \Delta\Theta) = e^T Pe + tr\left(\left[\Delta K_x^T \Gamma_x^T \Delta K_x + \Delta K_{cmd}^T \Gamma_{cmd}^T \Delta K_{cmd} + \Delta\Theta^T \Gamma_\Theta^T \Delta\Theta\right]\Lambda\right),$$

$$(e, \Delta K_x, \Delta K_{cmd}, \Delta\Theta) \in \mathbb{R}^n \times \mathbb{R}^{n\times m} \times \mathbb{R}^{m\times m} \times \mathbb{R}^{N\times m}, \tag{17}$$

where $tr(\cdot)$ denotes the *trace operator*. The error dynamics is given by

$$\dot{e}(t) = A_{\text{ref}}e(t) + B\Lambda\left[\Delta K_x^{\text{T}}x(t) + \Delta K_{\text{cmd}}^{\text{T}}y_{\text{cmd}}(t) - \Delta\Theta^{\text{T}}\Phi(x_{\text{p}}(t))\right] + \xi(t), \quad e(t_0) = 0, \qquad t \geq t_0, \tag{18}$$

and using the same arguments as in ([17], pp. 324-325), one can prove that

$$\dot{V}(e, \Delta K_x, \Delta K_{\text{cmd}}, \Delta\Theta) < 0 \qquad (e, \Delta K_x, \Delta K_{\text{cmd}}, \Delta\Theta) \in \Omega, \tag{19}$$

for some compact set $\Omega \subset \mathbb{R}^n \times \mathbb{R}^{n\times m} \times \mathbb{R}^{m\times m} \times \mathbb{R}^{N\times m}$. Thus, it follows from Theorem 4.18 of Khalil [19] that the nonlinear dynamical system given by (18) and (11)–(13) is uniformly ultimately bounded.

Next, let $x_{\text{ref}}(t) = \left[x_{\text{ref},1}^{\text{T}}(t), x_{\text{ref},2}^{\text{T}}(t)\right]^{\text{T}}$, $t \geq t_0$, verify (9), where $x_{\text{ref},1}(t) \in \mathbb{R}^{n_{\text{p}}}$ and $x_{\text{ref},2}(t) \in \mathbb{R}^m$. It follows from the uniform ultimate boundedness of (18) that

$$\left\| y(t) - y_{cmd}(t) - x_{\text{ref},2}(t) \right\| \leq \hat{b}, \qquad t \geq T + t_0, \tag{20}$$

for some $\hat{b} > 0$ and $T \geq 0$, which are independent of t_0. Moreover, since A_{ref} is block-diagonal and Hurwitz, $B_1 = [0_{m\times n_{\text{p}}}, -1_m]^{\text{T}}$, and $y_{\text{cmd}}(\cdot)$ is bounded, it follows from (9) that $x_{\text{ref},2}(\cdot)$ is uniformly bounded ([18], 245), that is, $\|x_{\text{ref},2}(t)\| \leq b_2$, $t \geq t_0$, for some $b_2 \geq 0$ independent of t_0. Thus, it follows from (20) that (7) is verified with $b = \hat{b} + b_2$. □

It is important to notice that although the matrix A_{p}, which characterizes the plant's uncontrolled linearized dynamics, is unknown and hence, the augmented matrix A is unknown, the structure of the matrix A_{p} is usually known. Thus, in problems of practical interest it is generally possible to verify the *matching conditions* (15) and (16), although the matrix A is unknown ([17], Ch. 9).

Remark 3.1 If the adaptive gains $\hat{K}_x(\cdot)$, $\hat{K}_{\text{cmd}}(\cdot)$, and $\hat{\Theta}(\cdot)$ verify (11)–(13), respectively, with $\sigma_1 = \sigma_2 = \sigma_3 = 0$, then the control law (10) reduces to the conventional model reference adaptive control law for the augmented dynamical system (8) ([17], p. 298). However, conventional MRAC does not guarantee uniform ultimate boundedness of the closed-loop system in the presence of the matched uncertainty $\hat{\xi}(\cdot)$ ([17], pp. 317-319).

4. Modeling assumptions on quadrotors' dynamics

Let $\mathbb{I} = \{O; X, Y, Z\}$ denote an orthonormal reference frame fixed with the Earth and centered at some point O, and let $\mathbb{J} = \{A; x(t), y(t), z(t)\}$, $t \geq t_0$, denote an orthonormal reference frame fixed with the quadrotor and centered at some point A, which is arbitrarily chosen. The axes of the reference frames $\mathbb{I}$ and $\mathbb{J}$ form two orthonormal bases of $\mathbb{R}^3$ and if a vector $a \in \mathbb{R}^3$ is expressed in $\mathbb{I}$, then this vector is denoted by $a^{\mathbb{I}}$. Alternatively, if $a \in \mathbb{R}^3$ is expressed in $\mathbb{J}$, then no superscript is used. In this chapter, we consider the reference frame $\mathbb{I}$ as an inertial reference frame; quadrotors move at subsonic velocities and are usually operated at altitudes considerably lower than 10 kilometers and hence, the error induced by this modeling assumption is negligible ([21], Ch. 5). The Z axis is chosen so that the quadrotor's weight is given by $F_{\text{g}}^{\mathbb{I}} = m_{\text{Q}}gZ$, where $m_{\text{Q}} > 0$ denotes the vehicle's mass and g denotes the gravitational acceleration; the X and Y axes are chosen arbitrarily. The axis $z(\cdot)$ points down and the axis $x(\cdot)$ is aligned to one of the quadrotor's arms; see **Figure 1**.

The attitude of the reference frame $\mathbb{J}$ with respect to the reference frame $\mathbb{I}$ is captured by the roll, pitch, and yaw angles using a 3-2-1 rotation sequence ([22], Ch. 1). In particular, we denote by $\psi:[t_0,\infty)\to[0,2\pi)$ the yaw angle and $\phi,\theta:[t_0,\infty)\to\left(-\frac{\pi}{2},\frac{\pi}{2}\right)$ the roll and pitch angles, respectively. The angular velocity of $\mathbb{J}$ with respect to $\mathbb{I}$ is denoted by $\omega:[t_0,\infty)\to\mathbb{R}^3$ ([22], Def. 1.9). The position of the point A with respect to the origin O of the inertial reference frame $\mathbb{I}$ is denoted by $r_A:[t_0,\infty)\to\mathbb{R}^3$ and the velocity of A with respect to $\mathbb{I}$ is denoted by $v_A:[t_0,\infty)\to\mathbb{R}^3$.

The position of the quadrotor's center of mass C with respect to the reference point A is denoted by $r_C\in\mathbb{R}^3$. The matrix of inertia of the quadrotor, excluding its propellers, with respect to A is denoted by $I\in\mathbb{R}^{3\times3}$ and the matrix of inertia of each propeller with respect to A is denoted by $I_P\in\mathbb{R}^{3\times3}$. The spin rate of the ith propeller is denoted by $\Omega_{P,i}:[t_0,\infty)\to\mathbb{R}$, $i=1,\ldots,4$. In this chapter, we model the quadrotor's frame as a rigid body and propellers as thin disks. Moreover, we assume that the vehicle's inertial properties, such as the mass m_Q, the inertia matrix I, and the location of the center of mass r_C, are constant, but *unknown*. The quadrotor's estimated mass is denoted by $\widehat{m}_Q>0$ and the quadrotor's estimated matrix of inertia with respect to A is given by the symmetric, positive-definite matrix $\widehat{I}\in\mathbb{R}^{3\times3}$.

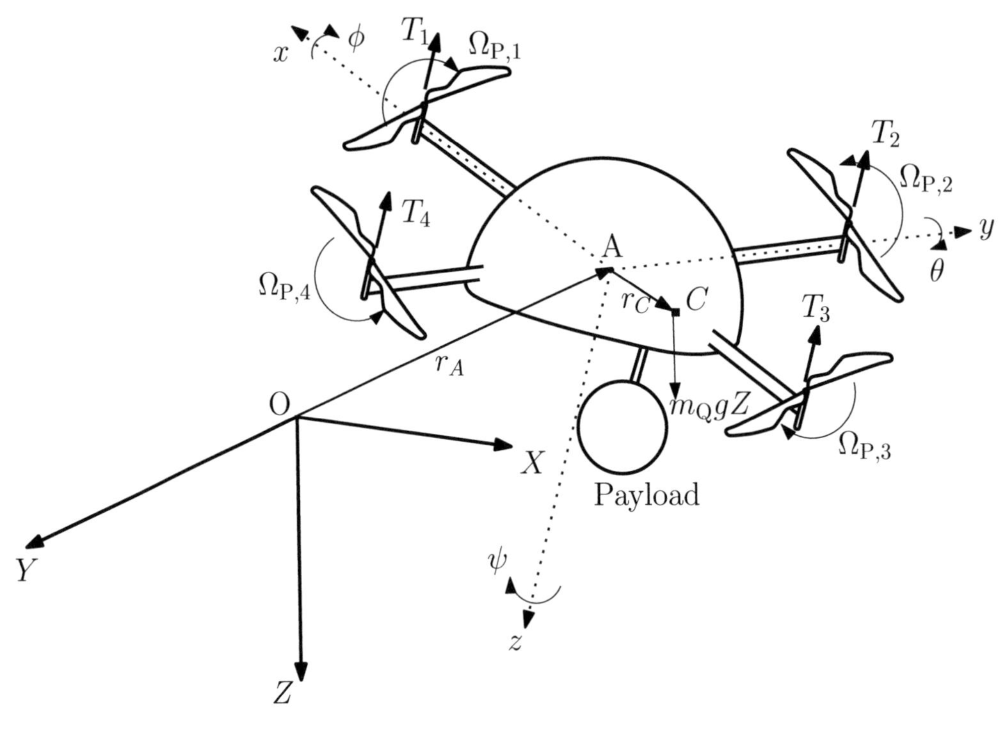

Figure 1. Schematic representation of a quadrotor helicopter.

5. Quadrotors' equations of motion

In this section, we present the equations of motion of quadrotors. Specifically, a quadrotor's *translational kinematic equation* is given by ([22], Ex. 1.12)

$$\dot{r}_A^{\mathbb{I}}(t) = R(\phi(t), \theta(t), \psi(t))v_A(t), \qquad r_A^{\mathbb{I}}(t_0) = r_{A,0}^{\mathbb{I}}, \qquad t \geq t_0, \tag{21}$$

where

$$R(\phi,\theta,\psi) \triangleq \begin{bmatrix} \cos\psi & -\sin\psi & 0 \\ \sin\psi & \cos\psi & 0 \\ 0 & 0 & 1 \end{bmatrix} \begin{bmatrix} \cos\theta & 0 & \sin\theta \\ 0 & 1 & 0 \\ -\sin\theta & 0 & \cos\theta \end{bmatrix} \begin{bmatrix} 1 & 0 & 0 \\ 0 & \cos\phi & -\sin\phi \\ 0 & \sin\phi & \cos\phi \end{bmatrix},$$

$$(\phi,\theta,\psi) \in \left(-\frac{\pi}{2},\frac{\pi}{2}\right) \times \left(-\frac{\pi}{2},\frac{\pi}{2}\right) \times [0, 2\pi),$$

and the *rotational kinematic equation* is given by ([22], Th. 1.7)

$$\begin{bmatrix} \dot{\phi}(t) \\ \dot{\theta}(t) \\ \dot{\psi}(t) \end{bmatrix} = \Gamma(\phi(t), \theta(t))\omega(t), \qquad \begin{bmatrix} \phi(t_0) \\ \theta(t_0) \\ \psi(t_0) \end{bmatrix} = \begin{bmatrix} \phi_0 \\ \theta_0 \\ \psi_0 \end{bmatrix}, \tag{22}$$

where

$$\Gamma(\phi,\theta) \triangleq \begin{bmatrix} 1 & \sin\phi\tan\theta & \cos\phi\tan\theta \\ 0 & \cos\phi & -\sin\phi \\ 0 & \sin\phi\sec\theta & \cos\phi\sec\theta \end{bmatrix}, \qquad (\phi,\theta) \in \left(-\frac{\pi}{2},\frac{\pi}{2}\right) \times \left(-\frac{\pi}{2},\frac{\pi}{2}\right).$$

Under the modeling assumptions outlined in Section 4, a quadrotor's *translational dynamic equation* is given by [4]

$$\begin{aligned} F_{\mathrm{T}}(t) + F_{\mathrm{g}}(\phi(t), \theta(t)) + F(v_{\mathrm{A}}(t)) = \\ m_{\mathrm{Q}}[\dot{v}_A(t) + \omega^\times(t)v_A(t) + \dot{\omega}^\times(t)r_C + \omega^\times(t)\omega^\times(t)r_C], \quad v_A(t_0) = v_{A,0}, \quad t \geq t_0, \end{aligned} \tag{23}$$

where $F_{\mathrm{T}}(t) = [0, 0, u_1(t)]^{\mathrm{T}}$ denotes the *thrust force*, that is, the force produced by the propellers that allows a quadrotor to hover,

$$F_{\mathrm{g}}(\phi,\theta) = m_{\mathrm{Q}}g[-\sin\theta, \cos\theta\sin\phi, \cos\theta\cos\phi]^{\mathrm{T}}, \qquad (\phi,\theta) \in \left(-\frac{\pi}{2},\frac{\pi}{2}\right) \times \left(-\frac{\pi}{2},\frac{\pi}{2}\right), \tag{24}$$

denotes the *quadrotor's weight*, and $F : \mathbb{R}^3 \to \mathbb{R}^3$ denotes the aerodynamic force acting on the quadrotor [23]. The *rotational dynamic equation* of a quadrotor is given by [4]

$$M_{\mathrm{T}}(t) + M_{\mathrm{g}}(\phi(t), \theta(t)) + M(\omega(t)) = m_{\mathrm{Q}} r_{\mathrm{C}}^{\times}[\dot{v}_A(t) + \omega^{\times}(t) v_A(t)] + I\dot{\omega}(t) + \omega^{\times}(t) I \omega(t)$$
$$+ I_{\mathrm{P}} \sum_{i=1}^{4} \left[0, 0, \dot{\Omega}_{\mathrm{P},i}(t)\right]^{\mathrm{T}} + \omega^{\times}(t) I_{\mathrm{P}} \sum_{i=1}^{4} [0, 0, \Omega_{\mathrm{P},i}(t)]^{\mathrm{T}}, \quad \omega(t_0) = \omega_0, \quad t \geq t_0, \tag{25}$$

where $M_T(t) = [u_2(t), u_3(t), u_4(t)]^{\mathrm{T}}$ denotes the *moment of the forces induced by the propellers*, $M_{\mathrm{g}}(\phi, \theta) \triangleq r_{\mathrm{C}}^{\times} F_{\mathrm{g}}(\phi, \theta)$, $(\phi, \theta) \in \left(-\frac{\pi}{2}, \frac{\pi}{2}\right) \times \left(-\frac{\pi}{2}, \frac{\pi}{2}\right)$, denotes the moment of the quadrotor's weight with respect to A, and $M : \mathbb{R}^3 \to \mathbb{R}^3$ denotes the moment of the aerodynamic force with respect to A. The terms $I_{\mathrm{P}} \sum_{i=1}^{4} \left[0, 0, \dot{\Omega}_{P_i}(t)\right]^{\mathrm{T}}$, $t \geq t_0$, and $\omega^{\times}(t) I_{\mathrm{P}} \sum_{i=1}^{4} \left[0, 0, \Omega_{P_i}(t)\right]^{\mathrm{T}}$ in (25) are known as *inertial counter-torque* and *gyroscopic effect*, respectively. In this chapter, we refer to (21)–(23) and (25) as the *equations of motion of a quadrotor helicopter*.

We model the aerodynamic force and the moment of the aerodynamic force as

$$F(v_A) = -\|v_A\| K_F v_A, \qquad v_A \in \mathbb{R}^3, \tag{26}$$

$$M(\omega) = -\|\omega\| K_M \omega, \qquad \omega \in \mathbb{R}^3, \tag{27}$$

where $K_F, K_M \in \mathbb{R}^{3 \times 3}$ are diagonal, positive-definite, and *unknown*; for details, refer to [23]. The aerodynamic force (26) is expressed in the reference frame $\mathbb{J}$. The next result allows expressing $F(\cdot)$ in the reference frame $\mathbb{I}$.

Proposition 5.1 *Consider the translational kinematic equation* (21) *and let* (26) *capture the aerodynamic forces acting on a quadrotor. It holds that*

$$F^{\mathbb{I}}(v_A) = -\|\dot{r}_A^{\mathbb{I}}\| R(\phi, \theta, \psi) K_F R^{\mathrm{T}}(\phi, \theta, \psi) \dot{r}_A^{\mathbb{I}}, \ (\dot{r}_A, \phi, \theta, \psi) \in \mathbb{R}^3 \times \left(-\frac{\pi}{2}, \frac{\pi}{2}\right) \times \left(-\frac{\pi}{2}, \frac{\pi}{2}\right) \times [0, 2\pi). \tag{28}$$

Proof: It follows from (26) that

$$F^{\mathbb{I}}(v_A) = -\|v_A\| [K_F v_A]^{\mathbb{I}} = -\|v_A\| R(\phi, \theta, \psi) K_F v_A \tag{29}$$

for all $(v_A, \phi, \theta, \psi) \in \mathbb{R}^3 \times \left(-\frac{\pi}{2}, \frac{\pi}{2}\right) \times \left(-\frac{\pi}{2}, \frac{\pi}{2}\right) \times [0, 2\pi)$, and it follows from (21) that

$$F^{\mathbb{I}}(v_A) = -\|v_A\| R(\phi, \theta, \psi) K_F R^{-1}(\phi, \theta, \psi) \dot{r}_A^{\mathbb{I}}. \tag{30}$$

Eq. (28) now follows from (30), since $R(\cdot, \cdot, \cdot)$ is an orthogonal matrix and hence, per definition, $R^{-1}(\phi, \theta, \psi) = R^{\mathrm{T}}(\phi, \theta, \psi)$, $(\phi, \theta, \psi) \in \left(-\frac{\pi}{2}, \frac{\pi}{2}\right) \times \left(-\frac{\pi}{2}, \frac{\pi}{2}\right) \times [0, 2\pi)$, ([22], Def. A.13) and $\|v_A\| = \left\|R^{\mathrm{T}}(\phi, \theta, \psi) \dot{r}_A\right\| = \|\dot{r}_A\|$ ([24], p. 132). □

Eq. (26) captures the aerodynamic drag acting on a quadrotor in absence of wind. If the wind velocity $v_W^{\mathbb{I}} : [t_0, \infty) \to \mathbb{R}^3$ is not identically equal to zero, then it follows from (28) that the aerodynamic force is given by

$$F^{\mathbb{I}}(t,v_A) = -\left\|v_{\rm w}^{\mathbb{I}}(t) - \dot{r}_A^{\mathbb{I}}\right\| R(\phi,\theta,\psi) K_F R^{\rm T}(\phi,\theta,\psi)\left[v_{\rm w}^{\mathbb{I}}(t) - \dot{r}_A^{\mathbb{I}}\right],$$
$$(t,\dot{r}_A,\phi,\theta,\psi) \in [t_0,\infty) \times \mathbb{R}^3 \times \left(-\frac{\pi}{2},\frac{\pi}{2}\right) \times \left(-\frac{\pi}{2},\frac{\pi}{2}\right) \times [0,2\pi). \quad (31)$$

It is reasonable to assume that the wind velocity does not affect the moment of the aerodynamic force (27).

A quadrotor's *control vector* is given by $u(t) = [u_1(t), u_2(t), u_3(t), u_4(t)]^{\rm T}$, $t \geq t_0$, that is, the third component of the thrust force $F_{\rm T}(\cdot)$ and the moment of the forces induced by the propellers $M_{\rm T}(\cdot)$. One can verify that ([1], Ch. 2)

$$\begin{bmatrix} u_1(t) \\ u_2(t) \\ u_3(t) \\ u_4(t) \end{bmatrix} = \begin{bmatrix} 1 & 1 & 1 & 1 \\ 0 & l & 0 & -l \\ l & 0 & -l & 0 \\ -c_{\rm T} & c_{\rm T} & -c_{\rm T} & c_{\rm T} \end{bmatrix} \begin{bmatrix} T_1(t) \\ T_2(t) \\ T_3(t) \\ T_4(t) \end{bmatrix}, \quad t \geq t_0, \quad (32)$$

where $T_i : [t_0, \infty) \to \mathbb{R}$, $i = 1, \ldots, 4$, denotes the component of the force produced by the ith propeller along the $-z(\cdot)$ axis of the reference frame $\mathbb{J}$, $l > 0$ denotes the length of each propeller's arm, and $c_{\rm T} > 0$ denotes each propeller's drag coefficient.

Remark 5.1 The state vector for the equations of motion of a quadrotor (21)–(23) and (25) is defined as $\left[r_A^T, v_A^T, \phi, \theta, \psi, \omega^T\right]^T \in \mathbb{R}^{12}$ and the inertial counter-torque $I_P \sum_{i=1}^4 \left[0, 0, \dot{\Omega}_{P_i}(t)\right]^T$, $t \geq t_0$, can be explicitly related through algebraic expressions to neither the state vector x nor the control input u. Thus, the inertial counter-torque must be considered as a time-varying term in a quadrotor's rotational dynamic equations. Furthermore, it is common practice not to relate the gyroscopic effect $\omega^\times(t) I_P \sum_{i=1}^4 \left[0, 0, \Omega_{P_i}(t)\right]^T$, $t \geq t_0$, with the control input u through (32) ([1], Ch. 2). Hence, also the gyroscopic effect must be accounted for as a time-varying term in a quadrotor's equations of motion. For these reasons, (21)–(23) and (25) are a nonlinear *time-varying* dynamical system.

6. Proposed control system for quadrotors

In this section, we outline a control strategy for quadrotors and verify that this strategy does not defy the vehicle's limits given by its controllability and underactuation.

6.1. Proposed control strategy

The configuration of a quadrotor, whose frame is modeled as a rigid body, is uniquely identified by the position in the inertial space of the reference point A, that is, $r_A^{\mathbb{I}}(t) = [r_X(t), r_Y(t), r_Z(t)]^{\rm T}$, $t \geq t_0$, and the Euler angles $\phi(t)$, $\theta(t)$, and $\psi(t)$. Observing the equations of motion of a quadrotor (21)–(23) and (25), one can show that the four control inputs $u_1(\cdot), \ldots, u_4(\cdot)$ are unable to instantaneously and simultaneously accelerate the six independent generalized coordinates

$r_X(\cdot)$, $r_Y(\cdot)$, $r_Z(\cdot)$, $\phi(\cdot)$, $\theta(\cdot)$, and $\psi(\cdot)$, and hence quadrotors are underactuated mechanical systems ([25], Def. 2.9). However, it follows from (21)–(23) and (25) that the control inputs $u_1(\cdot)$, …, $u_4(\cdot)$ are able to instantaneously and simultaneously accelerate the independent generalized coordinates $r_Z(\cdot)$, $\phi(\cdot)$, $\theta(\cdot)$, and $\psi(\cdot)$, which uniquely capture the vehicle's altitude and orientation dynamics.

In practical applications, quadrotors are employed to transport detection devices, such as antennas or cameras, that must be taken to some specific location and pointed in some given direction. For this reason, one usually needs to regulate a quadrotor's position $r_A^{\mathbb{I}}(\cdot)$ and yaw angle $\psi(\cdot)$. To meet this goal despite quadrotors' underactuation, we apply the following *control strategy.* Let $[r_{X,\mathrm{ref}}(t), r_{Y,\mathrm{ref}}(t), r_{Z,\mathrm{ref}}(t)]^{\mathrm{T}} \in \mathbb{R}^3$, $t \geq t_0$, denote the quadrotor's *reference trajectory,* let $\psi_{ref}(t) \in [0, 2\pi)$ denote the quadrotor's *reference yaw angle,* and assume that $r_{X,\mathrm{ref}}(\cdot)$, $r_{Y,\mathrm{ref}}(\cdot)$, $r_{Z,\mathrm{ref}}(\cdot)$, and $\psi_{\mathrm{ref}}(\cdot)$, are continuous with their first two derivatives and bounded with their first derivatives. It follows from Example 1.4 of [22] that (21) and (23) are equivalent to

$$\ddot{r}_A^{\mathbb{I}}(t) = \begin{bmatrix} u_X(t) \\ u_Y(t) \\ u_Z(t) \end{bmatrix} + m_Q^{-1} F^{\mathbb{I}}(v_A(t), \omega(t)) - \ddot{r}_C^{\mathbb{I}}(t) + \mu^{\mathbb{I}}(t), \quad \begin{bmatrix} r_A^{\mathbb{I}}(t_0) \\ v_A^{\mathbb{I}}(t_0) \end{bmatrix} = \begin{bmatrix} r_{A,0}^{\mathbb{I}} \\ v_{A,0}^{\mathbb{I}} \end{bmatrix}, \quad t \geq t_0, \tag{33}$$

where [26]

$$\mu^{\mathbb{I}}(t) \triangleq m_Q^{-1} u_1(t) \left[R(\phi(t), \theta(t), \psi(t)) - R\left(\phi_{ref}(t), \theta_{ref}(t), \psi_{ref}(t)\right) \right] Z, \tag{34}$$

$$\phi_{ref}(t) \triangleq \sin^{-1} \frac{u_X(t) \sin \psi_{ref}(t) - u_Y(t) \cos \psi_{ref}(t)}{\sqrt{u_X^2(t) + u_Y^2(t) + u_Z^2(t)}}, \tag{35}$$

$$\theta_{ref}(t) \triangleq \tan^{-1} \frac{u_X(t) \cos \psi_{ref}(t) + u_Y(t) \sin \psi_{ref}(t)}{u_Z(t)}. \tag{36}$$

Thus, a feedback control law for the *virtual* control input $[u_X(\cdot), u_Y(\cdot), u_Z(\cdot)]^{\mathrm{T}}$ is designed so that, after a finite-time transient, $r_A(\cdot)$ tracks $[r_{X,\mathrm{ref}}(\cdot), r_{Y,\mathrm{ref}}(\cdot), r_{Z,\mathrm{ref}}(\cdot)]^{\mathrm{T}}$ with bounded error. Furthermore, a feedback control law for the control input $[u_2(\cdot), u_3(\cdot), u_4(\cdot)]^{\mathrm{T}}$ is designed so that, after a finite-time transient, $[\phi(\cdot), \theta(\cdot), \psi(\cdot)]^{\mathrm{T}}$ tracks $[\phi_{\mathrm{ref}}(\cdot), \theta_{\mathrm{ref}}(\cdot), \psi_{\mathrm{ref}}(\cdot)]^{\mathrm{T}}$ with bounded error. Since the quadrotor's mass m_Q is unknown, we compute the component of the quadrotor's thrust along the $z(\cdot)$ axis of the reference frame $\mathbb{J}$ as

$$u_1(t) = \widehat{m}_Q \sqrt{u_X^2(t) + u_Y^2(t) + u_Z^2(t)}, \qquad t \geq t_0. \tag{37}$$

Figure 2 provides a schematic representation of the proposed control strategy.

Eqs. (35) and (36) constrain the nonlinear dynamical system given by (33), (22), and (25) and enforce its underactuation. Note that (36) is well-defined, since $u_Z(t) \neq 0$, $t \geq t_0$, is a necessary condition for a quadrotor to fly, and (35) is well-defined since $u_1(t) \neq 0$, $t \geq t_0$, is a necessary condition to fly and

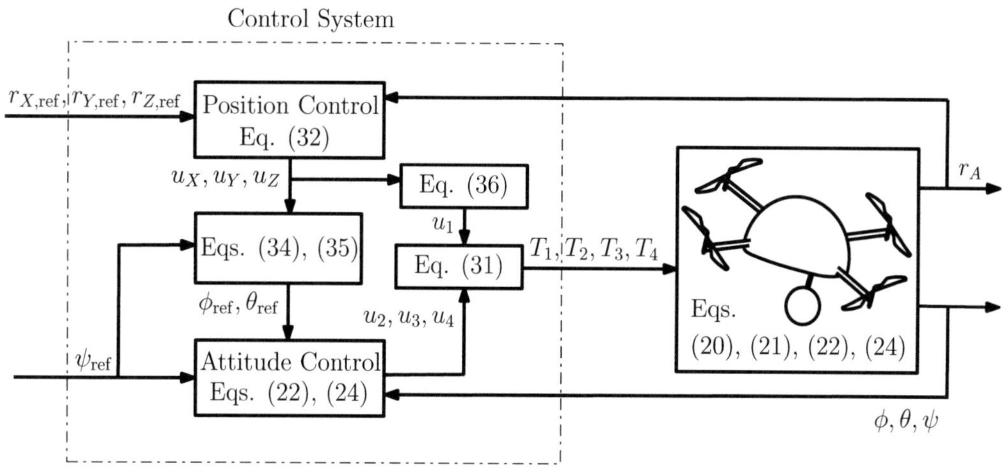

Figure 2. Proposed control scheme for a quadrotor.

$$\frac{u_X(t)\sin\psi_{ref}(t) - u_Y(t)\cos\psi_{ref}(t)}{\sqrt{u_X^2(t)+u_Y^2(t)+u_Z^2(t)}} = \sqrt{\frac{u_X^2(t)+u_Y^2(t)}{u_X^2(t)+u_Y^2(t)+u_Z^2(t)}}\cos\left(\psi_{ref}(t)+\alpha(t)\right) \le 1,$$

where $\alpha(t) \triangleq \tan^{-1}(u_Y(t)/u_X(t))$. Alternative control strategies, which rely on the assumption that the roll and pitch angles are small, are provided by Islam et al. [27]; Kotarski et al. [28]; Liu & Hedrick [29].

6.2. Strong accessibility of quadrotors

In this section, we prove that quadrotors are not controllable whenever the $z(\cdot)$ axis is parallel to the horizontal plane. Specifically, it is well-known that $\theta(t) \in \left(-\frac{\pi}{2}, \frac{\pi}{2}\right)$, $t \ge t_0$, is a necessary condition for a 3-2-1 rotation sequence to uniquely identify a quadrotor's orientation in space and guarantee finiteness of the yaw rate for finite angular velocities ([22], p. 19). In the following, we verify for the first time the conditions for quadrotors' *strong accessibility* [15], which is a weaker form of controllability for nonlinear dynamical systems [30–32], and prove that if $\phi(t^*) = \left|\frac{\pi}{2}\right|$ for some $t^* \ge t_0$, then the quadrotor may not be controllable, that is, there may not exist a continuous control input that is able to regulate the vehicle's altitude and orientation dynamics at $t = t^*$.

To the authors' best knowledge, the controllability of quadrotors' altitude and orientation dynamics has been studied considering simplified models, which assume that the vehicle's pitch and roll angles are small at all times [5, 33]. Moreover, existing results on the controllability of quadrotors neglect the fact that, as discussed in Remark 5.1, these vehicles are time-varying dynamical systems and rely on sufficient conditions for the controllability of time-invariant dynamical systems [30, 34].

In the following, we recall the notions of reachable set and strong accessibility for the nonlinear time-varying dynamical system

$$\dot{x}(t) = f(t, x(t)) + G(x(t))u(t), \qquad x(t_0) = x_0, \qquad t \geq t_0, \tag{38}$$

where $x(t) \in \mathbb{R}^n$, $t \geq t_0$, $u(t) \in \mathbb{R}^m$ is continuous, and both $f: [t_0, \infty) \times \mathbb{R}^n \to \mathbb{R}^n$ and $G: \mathbb{R}^n \to \mathbb{R}^{n \times m}$ are continuously differentiable.

Definition 6.1 ([15]). Consider the nonlinear time-varying dynamical system (38), let $\mathcal{M}$ be a real analytic manifold of dimension n, and let $y \in \mathcal{M}$ and $t_1, t_2 \geq t_0$. *The reachable set* $\mathcal{R}(y, t_1, t_2)$ *of* (38) *from* (y, t_1) *at* t_2 is the set of all states that can be reached at time t_2 by following the solution of (38) with initial condition y, initial time t_1, and some continuous control input $u(\cdot)$. The nonlinear time-varying dynamical system (38) is *strongly accessible at* $y \in \mathcal{M}$ *at time* t_1 *if* $\mathcal{R}(y, t_1, t_2)$ has a non-empty interior in $\mathcal{M}$ for every $t_2 > t_1$. The nonlinear time-varying dynamical system (38) is *strongly accessible* on $\mathcal{M}$ if it is strongly accessible at every $y \in \mathcal{M}$ and every $t_1 \geq t_0$.

In practice, Definition 6.1 states that if the nonlinear time-varying dynamical system (38) is strongly accessible on $\mathcal{M}$, then for every point in the reachable set of (38), there exists a continuous control input such that the system's trajectory is contained both in the reachable set and the manifold $\mathcal{M}$ at all times. The next theorem provides sufficient conditions for the strong accessibility of the nonlinear dynamical system (38). For the statement of this result, consider the augmented time-invariant dynamical system

$$\dot{\tilde{x}}(t) = \tilde{f}(\tilde{x}(t)) + \tilde{G}(\tilde{x}(t))u(t), \qquad \tilde{x}(0) = \left[x_0^{\mathrm{T}}, t_0\right]^{\mathrm{T}}, \qquad t \geq 0, \tag{39}$$

where $\tilde{x} \triangleq \left[x^{\mathrm{T}}, t\right]^{\mathrm{T}}$, $\tilde{f}(\tilde{x}) \triangleq \left[f^{\mathrm{T}}(x), 1\right]^{\mathrm{T}}$, $\tilde{G}(\tilde{x}) \triangleq \left[G^{\mathrm{T}}(x), 0_{n \times 1}\right]^{\mathrm{T}}$, and recall that the *controllability matrix* of the augmented time-invariant dynamical system (39) is defined as [15]

$$\mathcal{C}(\tilde{x}) \triangleq \left[\tilde{g}_1(\tilde{x}), \ldots, \tilde{g}_m(\tilde{x}), \mathrm{ad}_{\tilde{f}}\tilde{g}_1(\tilde{x}), \ldots, \mathrm{ad}_{\tilde{f}}\tilde{g}_m(\tilde{x})\right], \qquad \tilde{x} \in \mathbb{R}^n \times [t_0, \infty), \tag{40}$$

where $\tilde{G}(\tilde{x}) = [\tilde{g}_1(\tilde{x}), \ldots \tilde{g}_m(\tilde{x})]$.

Theorem 6.1 ([15]). *Consider the nonlinear dynamical system* (38). *If* $\mathrm{rank}\, \mathcal{C}(\tilde{x}) = n$ *for all* $\tilde{x} \in \mathcal{M} \times [t_0, \infty)$, *then* (38) *is strongly accessible.*

It follows from (21)–(23) and (25) that a quadrotor's altitude and orientation are captured by (38) with $n = 8$, $m = 4$, $x = \left[r_Z, \phi, \theta, \psi, \dot{r}_Z, \omega^{\mathrm{T}}\right]^{\mathrm{T}}$, $f: [t_0, \infty) \times \mathcal{D} \times \mathbb{R}^4 \to \mathbb{R}^4$, $\mathcal{D} = [0, \infty) \times \left(-\frac{\pi}{2}, \frac{\pi}{2}\right) \times \left(-\frac{\pi}{2}, \frac{\pi}{2}\right) \times [0, 2\pi) \times \mathbb{R} \times \mathbb{R}^3$, and

$$G(x) = m_{\mathrm{Q}}^{-1} \begin{bmatrix} 0_{4\times1} & 0_{4\times3} \\ \cos\phi\cos\theta & 0_{1\times3} \\ 0_{3\times1} & m_{\mathrm{Q}}\Gamma(\phi, \theta) I^{-1} \end{bmatrix}; \tag{41}$$

the explicit expression for $f(\cdot, \cdot)$ is omitted for brevity. In this case, the controllability matrix $\mathcal{C}(\cdot)$ of the fully actuated, augmented time-invariant dynamical system (39) is such that

$$\det C(\tilde{x}) = \frac{\cos\phi}{m_Q^2 \det^2 I'}, \qquad \tilde{x} \in \mathbb{R}^8 \times [t_0, \infty), \tag{42}$$

and it follows from Theorem 6.1 that $\phi(t) \in \left(-\frac{\pi}{2}, \frac{\pi}{2}\right)$, $t \geq t_0$, is a sufficient condition to guarantee that a quadrotor's altitude $r_Z(t)$ and orientation $[\phi(t), \theta(t), \psi(t)]^{\mathrm{T}}$ can be regulated by some continuous control input $u(t)$, while $\left[\dot{r}_Z(t), \dot{\phi}(t), \dot{\theta}(t), \dot{\psi}(t)\right]^{\mathrm{T}}$ remain bounded at all times; in practice, to preserve controllability, a conservative control law for quadrotors must prevent rotations of an angle of $\pm\pi/2$ about the $x(\cdot)$ axis of the reference frame $\mathbb{J}$. Note that it follows from (35) that $\phi_{\mathrm{ref}}(t) \in \left(-\frac{\pi}{2}, \frac{\pi}{2}\right)$, $t \geq t_0$, and hence the reference roll angle verifies sufficient conditions for strong accessibility of quadrotors' altitude and orientation dynamics.

7. Nonlinear robust control of quadrotors

In this section, we apply the results presented in Sections 3–6 to design control laws so that a quadrotor can follow a given trajectory with bounded error. Specifically, we design a control law for $u(\cdot)$ so that a quadrotor can track both the given reference trajectory $[r_{X,\mathrm{ref}}(t), r_{Y,\mathrm{ref}}(t), r_{Z,\mathrm{ref}}(t)]^{\mathrm{T}}$, $t \geq t_0$, and the reference yaw angle $\psi_{\mathrm{ref}}(t)$. In practice, we design control laws both for the virtual control input $[u_X(t), u_Y(t), u_Z(t)]^{\mathrm{T}}$, $t \geq t_0$, and the moment of the propellers' thrust $[u_2(t), u_3(t), u_4(t)]^{\mathrm{T}}$, so that a quadrotor tracks $[r_{X,\mathrm{ref}}(t), r_{Y,\mathrm{ref}}(t), r_{Z,\mathrm{ref}}(t)]^{\mathrm{T}}$, the reference roll angle (35), the reference pitch angle (36), and the reference yaw angle $\psi_{\mathrm{ref}}(t)$.

It follows form (33) that if the aerodynamic force is modeled as in (31), then a quadrotor's translational kinematic and dynamic equations are given by

$$\dot{x}_{\mathrm{p,P}}(t) = A_{\mathrm{p,P}} x_{\mathrm{p,P}}(t) + B_{\mathrm{p,P}} \Lambda_{\mathrm{P}} \left(\begin{bmatrix} u_X(t) \\ u_Y(t) \\ u_Z(t) \end{bmatrix} + \Theta_{\mathrm{P}}^{\mathrm{T}} \Phi(x_{\mathrm{p,P}}(t)) \right) + \widehat{\xi}_{\mathrm{P}}(t),$$

$$x_{\mathrm{p,P}}(t_0) = \left[r_{A,0}^{\mathrm{T}}, \left(v_{A,0}^{\mathbb{I}}\right)^{\mathrm{T}} \right]^{\mathrm{T}}, t \geq t_0, \tag{43}$$

$$\dot{y}_{\mathrm{P}}(t) = \varepsilon C_{\mathrm{p,P}} x_{\mathrm{p,P}}(t) - \varepsilon y_{\mathrm{P}}(t), \qquad y_{\mathrm{P}}(t_0) = C_{\mathrm{p,P}} x_{\mathrm{p,P}}(t_0), \tag{44}$$

where $x_{\mathrm{p,P}}(t) = \left[\left(r_A^{\mathbb{I}}(t)\right)^{\mathrm{T}}, \left(\dot{r}_A^{\mathbb{I}}(t)\right)^{\mathrm{T}} \right]^{\mathrm{T}}$, $t \geq t_0$, $A_{\mathrm{p,P}} = \begin{bmatrix} 0_{3\times3} & \mathbf{1}_3 \\ 0_{3\times3} & 0_{3\times3} \end{bmatrix}$, $B_{\mathrm{p,P}} = \begin{bmatrix} 0_{3\times3} \\ \mathbf{1}_3 \end{bmatrix}$, $\Lambda_{\mathrm{P}} = m_Q^{-1}\mathbf{1}_3$, $C_{\mathrm{p,P}} = [\mathbf{1}_3, 0_{3\times3}]$, $\widehat{\xi}_{\mathrm{P}}(t) \in \mathbb{R}^6$, $[\mathbf{1}_3, 0_{3\times3}]\widehat{\xi}_{\mathrm{P}}(t) = 0_{3\times3}$, and

$$[0_{3\times3}, \mathbf{1}_3]\widehat{\xi}_{\mathrm{P}}(t) = m_Q^{-1} \left\| v_W^{\mathbb{I}}(t) - \dot{r}_A^{\mathbb{I}}(t) \right\| B_{\mathrm{p,P}} R(\phi(t), \theta(t), \psi(t))$$

$$\cdot K_F R^{\mathrm{T}}(\phi(t), \theta(t), \psi(t)) \left[v_W^{\mathbb{I}}(t) - \dot{r}_A^{\mathbb{I}}(t) \right] + \mu^{\mathbb{I}}(t) - \ddot{r}_C^{\mathbb{I}}(t). \tag{45}$$

Although $\Theta_{\rm P}^{\rm T}\Phi(x_{\rm p,P}(\cdot))$ follows neither from (33) nor (28), this nonlinear term has been introduced to account for failures of the control system; in this section, we assume that

$$\Phi(z) = \tanh z, \qquad z \in \mathbb{R}^n, \tag{46}$$

which is globally Lipschitz continuous. Since any quadrotor's velocity and acceleration are bounded, it follows from (45) that also the unmatched uncertainty $\widehat{\xi}_{\rm P}(\cdot)$ is bounded. Eq. (44) captures the plant sensor's dynamics ([20], Ch. 2).

It follows from (22) and (25) that a quadrotor's rotational dynamics is captured by

$$\begin{bmatrix}\dot\phi(t)\\ \dot\theta(t)\\ \dot\psi(t)\\ \dot\omega(t)\end{bmatrix} = f(\phi(t),\theta(t),\psi(t),\omega(t)) + \begin{bmatrix}0_{3\times 3}\\ \widehat{I}^{-1}\end{bmatrix}\begin{bmatrix}u_2(t)\\ u_3(t)\\ u_4(t)\end{bmatrix} + \widehat{\xi}_{\rm A}(t),$$

$$[\phi(t_0),\theta(t_0),\psi(t_0),\omega^{\rm T}(t_0)]^{\rm T} = [\phi_0,\theta_0,\psi_0,\omega_0^{\rm T}]^{\rm T}, \qquad t \geq t_0, \tag{47}$$

where $f(\phi,\theta,\psi,\omega) = \left[\omega^{\rm T}\Gamma^{\rm T}(\phi,\theta),\left(-\widehat{I}^{-1}\omega^{\times}\widehat{I}\omega\right)^{\rm T}\right]^{\rm T}$, $\widehat{\xi}_{\rm A}(t) \in \mathbb{R}^6$, $[\mathbf{1}_3, 0_{3\times 3}]\widehat{\xi}_{\rm A}(t) = 0_3$,

$$[0_{3\times 3},\mathbf{1}_3]\widehat{\xi}_{\rm A}(t) = I^{-1}r_{\rm C}^{\times}\left[F_{\rm g}(\phi(t),\theta(t)) - m_{\rm Q}\ddot{r}_A^{\rm I}(t)\right] + I^{-1}M(\omega(t)) + \left[\widehat{I}^{-1}\omega^{\times}(t)\widehat{I} - I^{-1}\omega^{\times}(t)I\right]\omega(t) +$$
$$\left(I-\widehat{I}\right)[u_2(t),u_3(t),u_4(t)]^{\rm T} - I^{-1}I_{\rm P}\sum_{i=1}^{4}\begin{bmatrix}0\\0\\ \dot\Omega_{{\rm P},i}(t)\end{bmatrix} - I^{-1}\omega^{\times}(t)I_{\rm P}\sum_{i=1}^{4}\begin{bmatrix}0\\0\\ \Omega_{{\rm P},i}(t)\end{bmatrix}, \tag{48}$$

$F_{\rm g}(\cdot,\cdot)$ is given by (24), $r_{\rm A}(\cdot)$ verifies (43), and $M(\cdot)$ is given by (27). Let $x_{\rm p,A} = [\phi,\dot\phi,\theta,\dot\theta,\psi,\dot\psi]^{\rm T}$, $\eta(x_{\rm p,A}) = [\dot\phi,\dot\theta,\dot\psi]^{\rm T}$, $\beta(x_{\rm p,A}) = \left[L_f^2\phi, L_f^2\theta, L_f^2\psi,\right]^{\rm T}$, and $v \in \mathbb{R}^3$; the explicit expression of $\beta(\cdot)$ is omitted for brevity. By proceeding as in Example 6.3 of [35], one can prove that the nonlinear dynamical system (47) is feedback linearizable ([31], Ch. 5). Specifically, (47) with

$$[u_2,u_3,u_4]^{\rm T} = \widehat{I}\Gamma^{-1}(\phi,\theta)\left[\eta(x_{\rm p,A}) - \beta(x_{\rm p,A}) + v\right],$$

$$(x_{\rm p,A}, v) \in \left(-\frac{\pi}{2},\frac{\pi}{2}\right) \times \mathbb{R} \times \left(-\frac{\pi}{2},\frac{\pi}{2}\right) \times \mathbb{R} \times [0,2\pi) \times \mathbb{R} \times \mathbb{R}^3, \tag{49}$$

is equivalent to

$$\dot{x}_{\rm p,A}(t) = A_{\rm p,A}x_{\rm p,A}(t) + B_{\rm p,A}\Lambda_{\rm A}\left[v(t) + \Theta_{\rm A}^{\rm T}\Phi(x_{\rm p,A}(t))\right] + \widehat{\xi}_{\rm A}(t),$$
$$x_{\rm p,A}(t_0) = \left[\phi_0,\dot\phi_0,\theta_0,\dot\theta_0,\psi_0,\dot\psi_0\right]^{\rm T}, \qquad t \geq t_0, \tag{50}$$

$$\dot{y}_{\rm A}(t) = \varepsilon C_{\rm p,A} x_{\rm p,A}(t) - \varepsilon y_{\rm A}(t), \qquad y_{\rm A}(t_0) = C_{\rm p,A} x_{\rm p,A}(t_0), \tag{51}$$

where

$$A_{\rm p,A} = \begin{bmatrix} 0 & 1 & 0 & 0 & 0 & 0 \\ 0 & 1 & 0 & 0 & 0 & 0 \\ 0 & 0 & 0 & 1 & 0 & 0 \\ 0 & 0 & 0 & 1 & 0 & 0 \\ 0 & 0 & 0 & 0 & 0 & 1 \\ 0 & 0 & 0 & 0 & 0 & 1 \end{bmatrix}, \quad B_{\rm p,A} = \begin{bmatrix} 0 & 0 & 0 \\ 1 & 0 & 0 \\ 0 & 0 & 0 \\ 0 & 1 & 0 \\ 0 & 0 & 0 \\ 0 & 0 & 1 \end{bmatrix}, \quad C_{\rm p,A} = \begin{bmatrix} 1 & 0 & 0 \\ 0 & 0 & 0 \\ 0 & 1 & 0 \\ 0 & 0 & 0 \\ 0 & 0 & 1 \\ 0 & 0 & 0 \end{bmatrix}^{\rm T}, \tag{52}$$

$\left[\dot{\phi}_0, \dot{\theta}_0, \dot{\psi}_0\right]^{\rm T} = \Gamma(\phi_0, \theta_0)\omega_0$, $\Lambda \in \mathbb{R}^{3\times 3}$ is diagonal positive-definite, and $\Phi(\cdot)$ given by (46). Although $\Lambda_{\rm A} = \mathbf{1}_3$ [35], we assume that $\Lambda_{\rm A}$ is *unknown* and accounts for failures of the propulsion system and erroneous modeling assumptions. Similarly, the term $\Theta_{\rm A}^{\rm T}\Phi(x_{\rm p,A}(\cdot))$ has been introduced to capture matched uncertainties. Since any quadrotor's angular velocity, angular acceleration, and propeller's spin rate are bounded, it follows from (48) that also the unmatched uncertainty $\hat{\xi}_{\rm A}(\cdot)$ is bounded. Eq. (51) captures the plant sensor's dynamics ([20], Ch. 2).

The next theorem provides feedback control laws both for $[u_X(\cdot), u_Y(\cdot), u_Z(\cdot)]^{\rm T}$ and $[u_2(\cdot), u_3(\cdot), u_4(\cdot)]^{\rm T}$ so that the measured output signal $y_{\rm P}(\cdot)$ tracks the reference signal

$$y_{\rm cmd,P}(t) = [r_{X,\rm ref}(t), r_{Y,\rm ref}(t), r_{Z,\rm ref}(t)]^{\rm T}, \qquad t \geq t_0, \tag{53}$$

and the measured output signal $y_{\rm A}(\cdot)$ tracks the reference signal

$$y_{\rm cmd,A}(t) = \left[\phi_{\rm ref}(t), \theta_{\rm ref}(t), \psi_{\rm ref}(t)\right]^{\rm T}, \tag{54}$$

where $\phi_{\rm ref}(\cdot)$ and $\theta_{\rm ref}(\cdot)$ are given by (35) and (36), respectively, with some bounded error despite model uncertainties, external disturbances, and failures of the propulsion system. For the statement of this result, consider both the nonlinear dynamical system given by (43) and (44) and the nonlinear dynamical system given by (50) and (51), and note that these systems are equivalent to (5) and (6) with $x_{\rm p} = \left[x_{\rm p,P}^{\rm T}, x_{\rm p,A}^{\rm T}\right]^{\rm T}$, $\mathcal{D}_{\rm p} = \mathbb{R}^3 \times \mathbb{R}^3 \times \left(-\frac{\pi}{2}, \frac{\pi}{2}\right) \times \mathbb{R} \times \left(-\frac{\pi}{2}, \frac{\pi}{2}\right) \times \mathbb{R} \times [0, 2\pi) \times \mathbb{R}$, $u = [u_X, u_Y, u_Z, v^{\rm T}]^{\rm T}$, $y_{\rm cmd} = \left[y_{\rm cmd,P}^{\rm T}, y_{\rm cmd,A}^{\rm T}\right]^{\rm T}$, $y_{\rm cmd,2} = \left[y_{\rm cmd,2,P}^{\rm T}, y_{\rm cmd,2,A}^{\rm T}\right]^{\rm T}$, $n_{\rm p} = 12$, $m = 6$, $\hat{\xi} = \left[\hat{\xi}_{\rm P}^{\rm T}, \hat{\xi}_{\rm A}^{\rm T}\right]^{\rm T}$, and

$$A_{\rm p} = \begin{bmatrix} A_{\rm p,P} & 0_{6\times 6} \\ 0_{6\times 6} & A_{\rm p,A} \end{bmatrix}, \quad B_{\rm p} = \begin{bmatrix} B_{\rm p,P} & 0_{6\times 3} \\ 0_{6\times 3} & B_{\rm p,A} \end{bmatrix}, \quad C_{\rm p} = \begin{bmatrix} C_{\rm p,P} & 0_{3\times 6} \\ 0_{3\times 6} & C_{\rm p,A} \end{bmatrix},$$

$$\Lambda = \begin{bmatrix} \Lambda_{\rm P} & 0_{3\times 3} \\ 0_{3\times 3} & \Lambda_{\rm A} \end{bmatrix}, \quad \Theta = \begin{bmatrix} \Theta_{\rm P} & 0_{6\times 3} \\ 0_{6\times 3} & \Theta_{\rm A} \end{bmatrix}, \quad \Phi(x_{\rm p}) = \begin{bmatrix} \Phi(x_{\rm p,P}) \\ \Phi(x_{\rm p,A}) \end{bmatrix}.$$

Theorem 7.1 *Consider the nonlinear dynamical system given by* (43) *and* (44)*, the nonlinear dynamical system given by* (50) *and* (51)*, the reference signals* (53) *and* (54)*, the augmented dynamical system* (8)*, the reference dynamical model* (9)*, the feedback control law* $\gamma(\cdot,\cdot,\cdot)$ *given by* (10)*, and the adaptation laws* (11)–(13)*. If there exist* $K_x \in \mathbb{R}^{18\times 6}$ *and* $K_{\rm cmd} \in \mathbb{R}^{6\times 6}$ *such that* (15) *and* (16) *and satisfied, then* (8) *with* $u=\gamma(t,x_{\rm p},x)$ *is uniformly ultimately bounded. Furthermore, there exist* $b>0$ *and* $c>0$ *independent of* t_0*, and for every* $a\in(0,c)$*, there exists a finite-time* $T=T(a,c)\geq 0$*, independent of* t_0*, such that if* $\|y_{\rm P}(t_0)-y_{\rm cmd,P}(t_0)\|\leq a$ *and* $\|y_{\rm A}(t_0)-y_{\rm cmd,A}(t_0)\|\leq a$*, then*

$$\|y_{\rm P}(t)-y_{\rm cmd,P}(t)\|\leq b, \qquad t\geq t_0+T, \tag{55}$$

$$\|y_{\rm A}(t)-y_{\rm cmd,A}(t)\|\leq b. \tag{56}$$

Lastly, the thrust force generated by the quadrotor's propellers is such that

$$u_1(t)=\hat{m}_{\rm Q}\left\|\gamma_{\rm P}\left(t,x_{\rm p}(t),x(t)\right)\right\|, \qquad t\geq t_0, \tag{57}$$

and the moment of the thrust force generated by the quadrotor's propellers is given by

$$\begin{bmatrix} u_2(t)\\ u_3(t)\\ u_4(t)\end{bmatrix}=\hat{I}\Gamma^{-1}(\phi(t),\theta(t))\left[\eta\left(x_{\rm p,A}(t)\right)-\beta\left(x_{\rm p,A}(t)\right)+\gamma_{\rm A}\left(t,x_{\rm p}(t),x(t)\right)\right], \tag{58}$$

where $\gamma(t,x_{\rm p},x)=\left[\gamma_{\rm P}^{\rm T}(t,x_{\rm p},x),\gamma_{\rm A}^{\rm T}(t,x_{\rm p},x)\right]^{\rm T}$, $(t,x_{\rm p},x)\in[t_0,\infty)\times\mathbb{R}^{12}\times\mathbb{R}^{18}$, $\gamma_{\rm P}(t,x_{\rm p},x)\in\mathbb{R}^3$, and $\gamma_{\rm A}(t,x_{\rm p},x)\in\mathbb{R}^3$.

Proof: Uniform ultimate boundedness of (8) with $u=\gamma(t,x_{\rm p},x)$ is a direct consequence of Theorem 3.1. Thus, both the nonlinear dynamical system given by (43) and (44) with $[u_X,u_Y,u_Z]^{\rm T}=\gamma_{\rm P}(t,x_{\rm p},x)$, $(t,x_{\rm p},x)\in[t_0,\infty)\times\mathcal{D}_{\rm p}\times\mathcal{D}$, and the nonlinear dynamical system given by (50) and (51) with $[u_2,u_3,u_4]^{\rm T}=\gamma_{\rm A}(t,x_{\rm p},x)$ are uniformly ultimately bounded. Consequently, it follows from Definition 2.2 that there exist $b>0$ and $c>0$ independent of t_0, and for every $a\in(0,c)$, there exists a finite-time $T=T(a,c)\geq 0$, independent of t_0, such that if $\|y_{\rm P}(t_0)-y_{\rm cmd,P}(t_0)\|\leq a$ and $\|y_{\rm A}(t_0)-y_{\rm cmd,A}(t_0)\|\leq a$, then (55) and (56) are satisfied. Lastly, (57) directly follows from (37), and (58) directly follows from (49).

8. Illustrative numerical example

In this section, we provide a numerical example to illustrate both the applicability and the advantages of the theoretical results presented in this chapter. Specifically, we design a nonlinear robust control algorithm that allows a quadrotor helicopter to follow a circular trajectory, although the vehicle's inertial properties are unknown, one of the motors is suddenly turned off, the payload is dropped over the course of the mission, and the wind blows at strong velocity.

Consider a quadrotor of mass $m_Q = 1$ kg and matrix of inertia $I = \mathbf{1}_3$ kg m^2, let the propellers be characterized by the matrix of inertia $I_p = \begin{bmatrix} 0.025 & 0 & 0 \\ 0 & 0.025 & 0 \\ 0 & 0 & 0.05 \end{bmatrix}$ kg m^2, and let that the sensor's dynamics be characterized by $\varepsilon = 10$. We assume that the vehicle's mass and matrix of inertia are unknown and estimated to be $\widehat{m}_Q = 1.25$ kg and $\widehat{I} = 0.8 \cdot \mathbf{1}_3$ kg m^2, respectively. Moreover, we assume that the aerodynamic force (31) and the aerodynamic moment (27) are characterized by $K_F = K_M = 0.01 \cdot \mathbf{1}_3$, which we assume unknown, and the wind velocity is given by $v_W^{\mathbb{I}}(t) = [16, 0, 0]^T$ m/s, $t \geq t_0$; it is worthwhile to note that this wind speed is considered as excessive for quadrotors equipped with conventional autopilots.

Figure 3 shows the quadrotor's trajectory obtained applying the control laws (57) and (58) to track a circular path of radius 0.3 m at an altitude of 0.75 m despite the fact that the quadrotor's payload of 0.5 kg is dropped at $t \geq 40$ s and one of the motors is turned off at $t = 90$ s. These results have been obtained by setting $\sigma_1 = \sigma_2 = \sigma_3 = 2$, $\Gamma_{cmd} = 100 \cdot \mathbf{1}_6$, and Γ_x and Γ_Θ as block-diagonal matrices, whose non-zero blocks are $\Gamma_{x,(1,1)} = 1000 \cdot \mathbf{1}_9$, $\Gamma_{x,(2,2)} = 2000 \cdot \mathbf{1}_9$, $\Gamma_{\Theta,(1,1)} = 200 \cdot \mathbf{1}_9$, abd $\Gamma_{\Theta,(2,2)} = 1600 \cdot \mathbf{1}_9$.

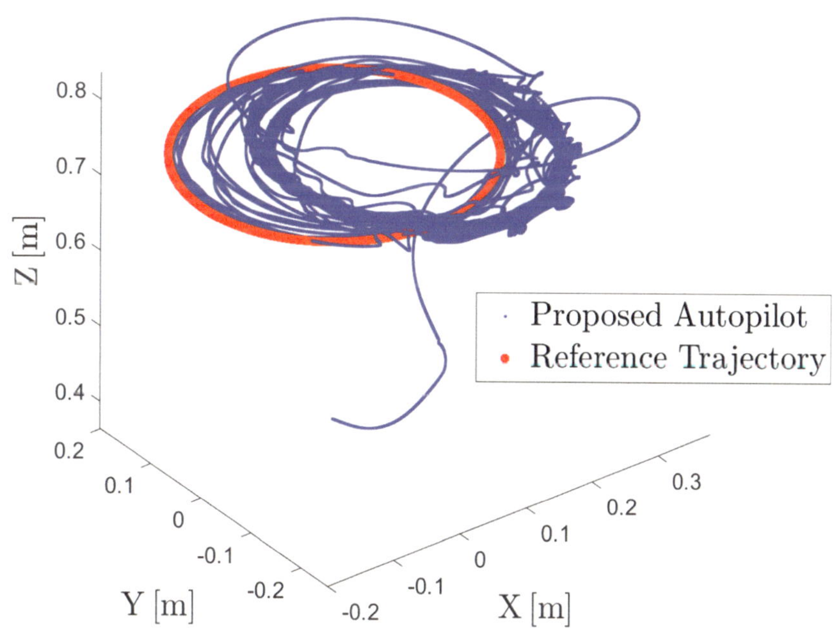

Figure 3. Reference trajectory and trajectory followed by the quadrotor implementing the proposed control algorithm. The vehicle is disturbed by some wind constantly blowing at 16 m/s.

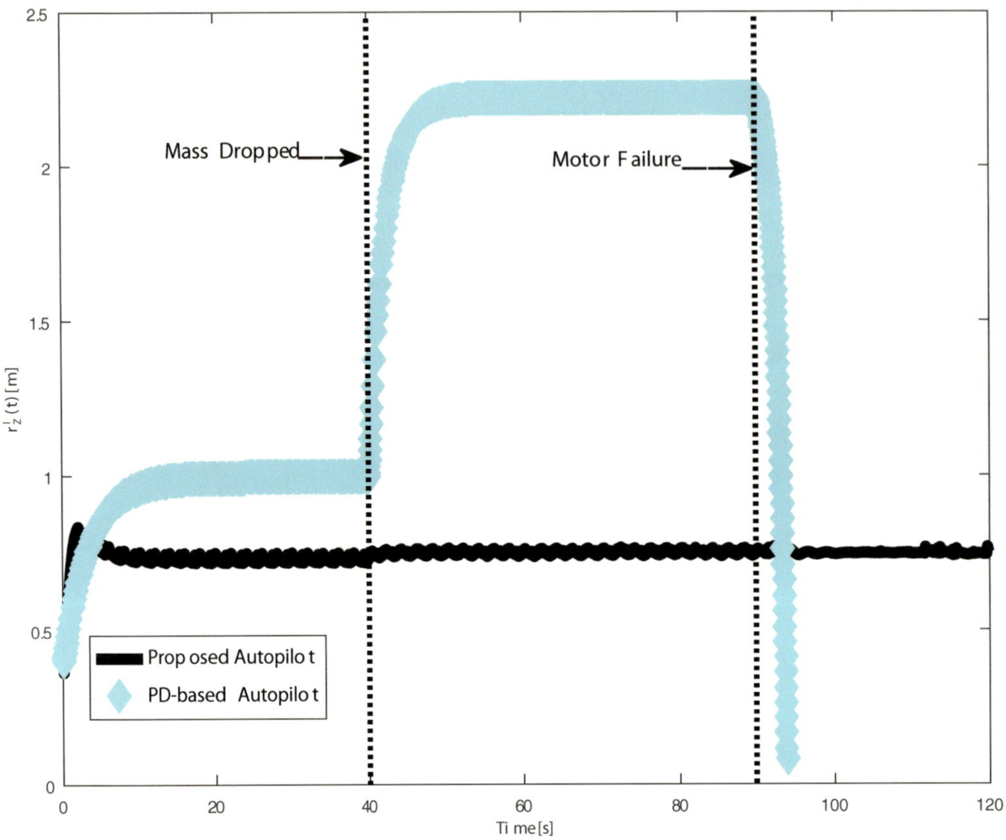

Figure 4. Altitude of a quadrotor implementing the proposed control algorithm and altitude of an identical quadrotor implementing an autopilot based on the classical PD control.

Figure 4 shows both the quadrotor's altitude as function of time and the altitude of an identical quadrotor implementing an autopilot based on the classical PD framework [36] and flying in absence of wind. It is clear how the quadrotor implementing our control algorithm is able to fly at the desired altitude despite the fact that the payload is dropped at $t=40$ s and a motor is turned off at $t=90$ s. The quadrotor implementing the PD algorithm is unable to reach the desired altitude because of the large error in the vehicle's mass' estimate. Moreover, this quadrotor reaches a considerably higher altitude after the payload is dropped and crashes after one of the propellers is turned off.

The first plot in **Figure 5** shows the control inputs (57) and (58). The second plot in **Figure 5** shows the control inputs computed using a conventional MRAC framework [6] for a quadrotor tracking the same circular path despite a wind blowing at 6 m/s; numerical simulations show that quadrotors implementing the conventional MRAC framework are unable to fly in the presence of wind gusts faster than 6 m/s. It is clear that our autopilot requires a control effort that is smaller than the effort required by a conventional MRAC-based autopilot to fly in weaker wind.

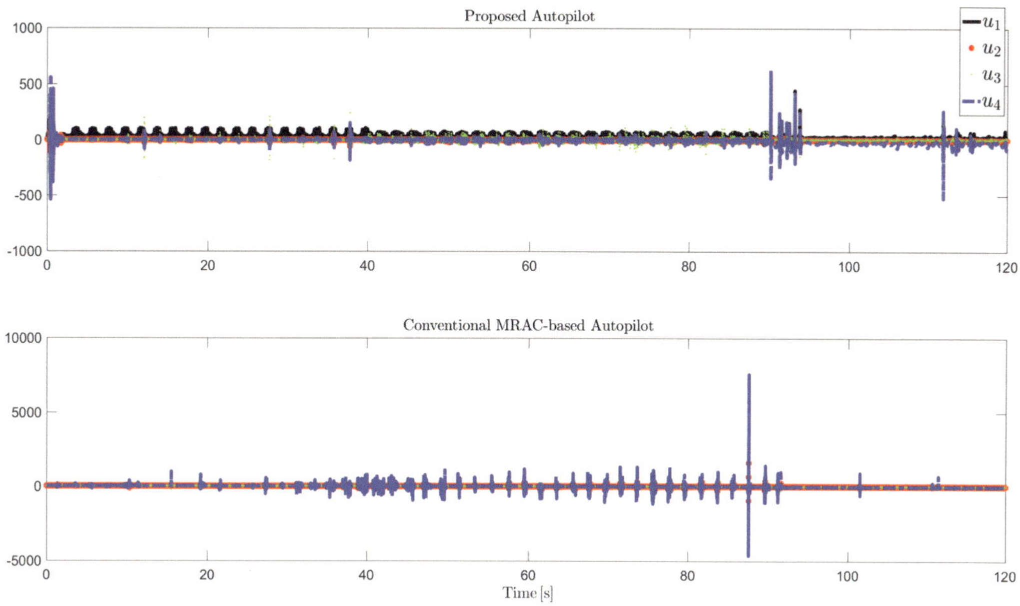

Figure 5. Control input for a quadrotor implementing the proposed control algorithm and control input for an identical quadrotor implementing a conventional MRAC-based autopilot.

9. Conclusion

In this chapter, we presented a robust MRAC architecture, which we employed to design autopilots for quadrotor helicopters. The proposed autopilot is the first to account for the fact that quadrotors are nonlinear time-varying dynamical systems, the exact location of the vehicle's center of mass is usually unknown, and the aircraft reference frame is centered at some point that does not necessarily coincide with the vehicle's barycenter. Moreover, our autopilot does not rely on the assumption that the Euler angles are small at all times and accounts both for the inertial counter-torque and the gyroscopic effect.

The applicability of our theoretical results has been illustrated by a numerical example and it is clearly shown how the proposed autopilot is able to track a given reference trajectory despite the fact that the payload is dropped during the mission, one of the motors is turned off, and the wind blows at the prohibitive velocity of 16 m/s. It is also shown that quadrotors implementing autopilots based on the classical PD framework crash if one of the propellers stops functioning. Lastly, it is shown that our autopilot requires a control effort that is smaller than the effort required by conventional MRAC-based autopilots to fly in less strong wind.

Acknowledgements

This work was supported in part by the NOAA/Office of Oceanic and Atmospheric Research under NOAA-University of Oklahoma Cooperative Agreement #NA16OAR4320115, U.S. Department of Commerce, the National Science Foundation under Grant no. 1700640, and the National Science Foundation REU under Grant Oklahoma EPSCoR-2017-3. Lastly, the author wishes to thank Mr. Keyvan Mohammadi for performing the numerical simulation presented in Section 8.

Author details

Keyvan Mohammadi and Andrea L'Afflitto*

*Address all correspondence to: a.lafflitto@ou.edu

The University of Oklahoma, USA

References

[1] Carrillo LRG, López AED, Lozano R, Pégard C. Quad Rotorcraft Control: Vision-Based Hovering and Navigation. London, UK: Springer; 2012

[2] Mahony R, Kumar V, Corke P. Multirotor aerial vehicles: Modeling, estimation, and control of quadrotor. IEEE Robotics & Automation Magazine. 2012;**19**(3):20-32

[3] Valavanis KP, editor. Introduction. In: Advances in Unmanned Aerial Vehicles: State of the Art and the Road to Autonomy. Dordrecht, the Netherlands: Springer; 2007. p. 3-13

[4] L'Afflitto A, Mohammadi K. Equations of motion of rotary-wing UAS with time-varying inertial properties. AIAA Journal of Guidance, Control, and Dynamics. 2017 In press. DOI:10.2514/1.G003015. https://arc.aiaa.org/doi/abs/10.2514/1.G003015

[5] Bouadi H, Cunha SS, Drouin A, Mora-Camino F. Adaptive sliding mode control for quadrotor attitude stabilization and altitude tracking. In: IEEE International Symposium on Computational Intelligence and Informatics; 2011. pp. 449-455

[6] Dydek ZT, Annaswamy AM, Lavretsky E. Adaptive control of quadrotor UAVs: A design trade study with flight evaluations. IEEE Transactions on Control Systems Technology. 2013;**21**(4):1400-1406

[7] Jafarnejadsani H, Sun D, Lee H, Hovakimyan N. Optimized $\mathscr{L}_1$ adaptive controller for trajectory tracking of an indoor quadrotor. Journal of Guidance, Control, and Dynamics. 2017;**40**(6):1415-1427

[8] Loukianov AG. Robust block decomposition sliding mode control design. Mathematical Problems in Engineering. 2002;**8**(4–5):349-365

[9] Mohammadi M, Shahri AM. Adaptive nonlinear stabilization control for a quadrotor UAV: Theory, simulation and experimentation. Journal of Intelligent & Robotic Systems. 2013;**72**(1):105-122

[10] Zheng E-H, Xiong J-J, Luo J-L. Second order sliding mode control for a quadrotor UAV. ISA Transactions. 2014;**53**(4):1350-1356

[11] Foehn P, Falanga D, Kuppuswamy N, Tedrake R, Scaramuzza D. Fast trajectory optimization for agile quadrotor maneuvers with a cable-suspended payload. In: Robotics: Science and Systems; 2017. pp. 1-10

[12] Faust A, Chiang H-T, Rackley N, Tapia L. Avoiding moving obstacles with stochastic hybrid dynamics using pearl: Preference appraisal reinforcement learning. In: International Conference on Robotics and Automation; 2016

[13] Gao F, Shen S. Quadrotor trajectory generation in dynamic environments using semidefinite relaxation on nonconvex qcqp. In: IEEE International Conference on Robotics and Automation; 2017. pp. 6354-6361

[14] Lin Y, Saripalli S. Path planning using 3D Dubins curve for unmanned aerial vehicles. In: International Conference on Unmanned Aircraft Systems; 2014. pp. 296-304

[15] Bhat SP. Controllability of nonlinear time-varying systems: Applications to spacecraft attitude control using magnetic actuation. IEEE Transactions on Automatic Control. 2005;**50**(11):1725-1735

[16] Narendra K, Annaswamy A. A new adaptive law for robust adaptation without persistent excitation. IEEE Transactions on Automatic Control. 1987;**32**(2):134-145

[17] Lavretsky E, Wise K. Robust and Adaptive Control: With Aerospace Applications. London, UK: Springer; 2012

[18] Haddad WM, Chellaboina V. Nonlinear Dynamical Systems and Control: A Lyapunov-Based Approach. Princeton, NJ: Princeton Univ. Press; 2008

[19] Khalil HK. Nonlinear Systems. Princeton, NJ: Prentice Hall; 2002

[20] Morris AS. Measurement and Instrumentation Principles. Woburn, MA: Elsevier; 2001

[21] Etkin B. Dynamics of Flight: Stability and Control. New York, NY: Wiley; 1982

[22] L'Afflitto A. A Mathematical Perspective on Flight Dynamics and Control. London, UK: Springer; 2017

[23] Hoffmann G, Huang H, Waslander S, Tomlin C. Quadrotor helicopter flight dynamics and control: Theory and experiment. In: AIAA Guidance, Navigation and Control Conference; 2007. pp. 1-20

[24] Gentle J. Matrix Algebra: Theory, Computations, and Applications in Statistics, Springer Texts in Statistics. New York, NY: Springer; 2007

[25] Fantoni I, Lozano R. Non-linear Control for Underactuated Mechanical Systems. Berlin, Germany: Springer; 2002

[26] Zhao B, Xian B, Zhang Y, Zhang X. Nonlinear robust adaptive tracking control of a quadrotor UAV via immersion and invariance methodology. IEEE Transactions on Industrial Electronics. 2015;**62**(5):2891-2902

[27] Islam S, Liu XP, Saddik AE. Adaptive sliding mode control of unmanned four rotor flying vehicle. International Journal of Robotics and Automation. 2015;**30**:140-148

[28] Kotarski D, Beníc Z, Krznar M. Control design for unmanned aerial vehicles with four rotors. Interdisciplinary Description of Complex Systems. 2016;**14**(2):236-245

[29] Liu Z, Hedrick K. Dynamic surface control techniques applied to horizontal position control of a quadrotor. In: International Conference on System Theory, Control and Computing; 2016. pp. 138-144

[30] Hermann R, Krener A. Nonlinear controllability and observability. IEEE Transactions on Automatic Control. 1977;**22**(5):728-740

[31] Isidori A. Nonlinear Control Systems. New York, NY: Springer; 1995

[32] Sussmann HJ, Jurdjevic V. Controllability of nonlinear systems. Journal of Differential Equations. 1972;**12**(1):95-116

[33] Sato K. Algebraic controllability of nonlinear mechanical control systems. SICE Journal of Control, Measurement, and System Integration. 2014;**7**(4):191-198

[34] Haynes GW, Hermes H. Nonlinear controllability via Lie theory. SIAM Journal on Control. 1970;**8**(4):450-460

[35] L'Afflitto A, Haddad WM. Optimal singular control for nonlinear semistabilisation. International Journal of Control. 2016;**89**(6):1222-1239

[36] Bouabdallah S, Noth A, Siegwart R. PID vs LQ control techniques applied to an indoor micro quadrotor. In: International Conference on Intelligent Robots and Systems. Vol. 3; 2004. pp. 2451-2456

Chapter 6

Attitude Control of a Quadcopter Using Adaptive Control Technique

Ramiro Ibarra Pérez, Gerardo Romero Galvan,
Aldo Jonathan Muñoz Vázquez,
Silvia Florida Melo and David Lara Alabazares

Additional information is available at the end of the chapter

http://dx.doi.org/10.5772/intechopen.71382

Abstract

This chapter presents an adaptive control technique to stabilize the attitude dynamics of unmanned aerial vehicle (UAV) type quadrotor in the presence of disturbances and/or uncertainties in the parameters due to changes in the payload, nonlinear actuators, and change in environmental conditions. To address the above problem, MRAC (model reference adaptive control) strategy is used. In this schema, a cost function is defined as a function of the error between the output of the system and a desired response from a reference model. Based on this error, the controller parameters are updated. To guarantee the global asymptotic stability of the system, Lyapunov's theory is applied. Simulation results using MATLAB-Simulink platform are presented to demonstrate the convergence of the controller parameters.

Keywords: Quadrotor UAV, adaptive control, Lyapunov's method, aircraft, attitude control

1. Introduction

Unmanned aerial vehicles (UAVs) are aircrafts that do not require a pilot on board to be controlled. In the beginning, they were solely used for military purposes. One of the first applications of these vehicles was aerial photography. In the 1883, an Englishman named Douglas Archibald provided one of the world's first reconnaissance UAVs. However, it was not until the World War I that UAVs became recognized systems. Since then, they are being widely used in military missions such as surveillance of enemy activity, airfield base security, airfield damage assessment, elimination of unexploded bombs, etc.

In the last decades, advances in technology and costs reduction permitted to adopt this technology in civil applications such as aerial photography, video and mapping, pollution and land monitoring, powerline inspection, fire detection, agriculture, and among other applications [1].

Quadrotors are a kind of mini-UAV's with vertical take-off and landing, controlled completed through four rotors mounted on each end of the crossed axes, which provide lift forces for the quadrotor move, this vehicle is very popular in the research community due to their special features like strong coupling subsystems, unknown physical parameters, and nonparametric uncertainties in inputs and external disturbances. Therefore, a suitable control system for stabilizing the closed loop control system is required; to do this, various control techniques, linear and nonlinear, have been implemented such as control PD [2, 3], control PID [4, 5], control of position and orientation by vision [6], sliding mode control [1, 7], fuzzy logic [8, 9], and adaptive control in [10].

The dynamic behavior of quadrotor has been published in varying complexity [11, 12]; particularly, the model we used is based on [13], where an extended mathematical description by the full consideration of nonlinear coupling between the axes is presented. We assume elastic deformations sufficient stiffness and realistic flight maneuvers were omitted, mass distributions of the quadrotor are symmetrical in the x-y plane, drag factor and thrust factor of the quadrotor are constant, and air density around the quadrotor is constant.

In this work, we develop an adaptive control strategy to stabilize the attitude dynamics of a quadrotor UAV. The adaptive control permit deals with modeling errors and disturbance uncertainty, variations of the mass, inertia, actuators faults, nonlinear aerodynamics, etc.

This chapter is organized as follows: Section 2 presents the mathematical model of the quadrotor obtained using Newton-Euler equations. Section 3 provides brief introduction about the adaptive control theory and the methodology used is based to obtain the adaptive law equations. In Section 4, simulations and analysis are presented to verify the performance of MRAC schema, and finally in Section 5, conclusions of this work are presented.

2. Mathematical model

2.1. ESC + motor + propeller

The rotors of quadrotor are modeled by

$$T_i = K\frac{\omega_M}{s+\omega_M}u_i \tag{1}$$

$$v_i = \frac{\omega_M}{s+\omega_M}u_i \tag{2}$$

For $i=(1,2\ldots,4)$, where u_i represents the PWM input, the gain $K>0$ and ω_M are the rotor bandwidth, v_i is the actuator dynamics. We assume K and ω_M are the same for all rotors [5].

2.2. Description of Quadrotor system

The configurations of quadrotor are described in **Figure 1** and consist of a body-fixed frame denoted by $F_b=\{F_{bx}, F_{by}, F_{bz}\}$, the center mass that coincident with the body-frame origin denoted

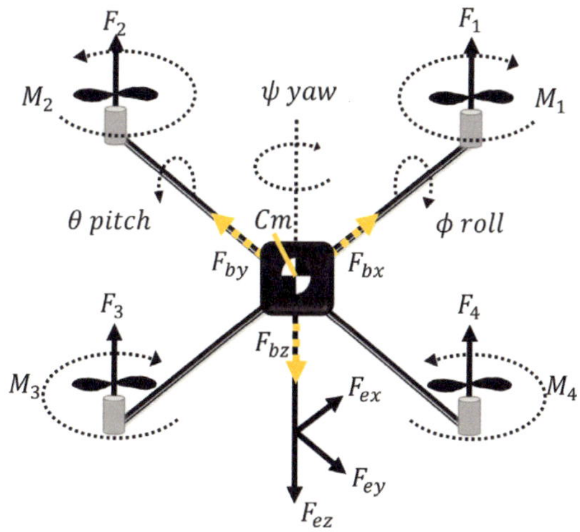

Figure 1. Quadrotor configuration.

by C_m, the earth-fixed inertial frame is denoted by $F_e=\{F_{ex}, F_{ey}, F_{ez}\}$, the position in earth-fixed inertial frame is given by a vector $r_I=(x, y, z)^T$, four rotors denoted each one as M_i with $i=(1, 2, \ldots, 4)$.

To obtain the position of CM in the NED (North-East-Down) coordinate system with respect to the inertial frame F_e, we use the Euler angles (roll ϕ, pitch θ, yaw ψ) combined in a vector $\Omega=(\phi, \theta, \psi)^T$, where roll angle is generated by differential thrust betweet M_2 and M_4, pitch angle is generated by differential thrust between M_1 and M_3, and yaw angle is generated by diferential torque between clock wise and anticlock wise rotors, i.e., $(M_1-M_2+M_3-M_4)$. This kind of device is considered an underactuated system with six degree of freedom (DOF). Rotation matrix R is used to map from F_b to F_e, applying three consecutive rotations denoted by

$$\begin{aligned}
R_z &= \begin{pmatrix} C_\psi & -S_\psi & 0 \\ S_\psi & C_\psi & 0 \\ 0 & 0 & 1 \end{pmatrix} \\
R_y &= \begin{pmatrix} C_\theta & 0 & S_\theta \\ 0 & 1 & 0 \\ -S_\theta & 0 & C_\theta \end{pmatrix} \\
R_x &= \begin{pmatrix} 1 & 0 & 0 \\ 0 & C_\phi & -S_\phi \\ 0 & S_\phi & C_\phi \end{pmatrix} \\
R = \left(R_z, R_y, R_x\right) &= \begin{pmatrix} C_\psi C_\theta & C_\psi S_\theta S_\phi - S_\psi C_\phi & C_\psi S_\theta C_\phi + S_\psi S_\phi \\ S_\psi C_\theta & S_\psi S_\theta S_\phi + C_\psi C_\phi & S_\psi S_\theta C_\phi - C_\psi S_\phi \\ -S_\theta & C_\theta S_\phi & C_\theta C_\phi \end{pmatrix}
\end{aligned} \tag{3}$$

where S and C refer to sin and cos function, respectively.

2.3. Rigid body dynamics

The dynamics of a rigid body under external forces and moments in the f_b can be formulated as follows [14]

$$m\dot{v} = \sum F_i \tag{4}$$

$$J\dot{W} + W \times JW = \sum M_i \tag{5}$$

where $i = (1, 2, \ldots, 4)$, $\sum F_i \in \mathbb{R}^3$ represent the sum of thrust forces from each rotor, $\sum M_i \in \mathbb{R}^3$ represent the sum of moments acting on the vehicle, v is the velocity in the F_b, m is the quadrotor mass, $W = (wx, wy, wz)^T$ determines the angular velocity vector which has its origin in at the center of mass of the rigid body, and J is the inertial tensor of the symmetric rigid body around its center of mass.

To obtain the thrust force contribution of each rotor applied to F_b, we use the following formula

$$F_i = f_{xi} + f_{yi} + f_{zi} = -k\,\omega_i^2 \tag{6}$$

with f_{xi} force from rotor i applied in F_b respect x axes, f_{yi} force from rotor i applied in F_b respect y axes and f_{zi} force from rotor i applied in F_b respect z axes, ω_i as rotor's angle speed, k as proportional gain related with air density, the geometry of the rotor blade and its pitch angles. As f_{xi} and f_{yi} is zero, since they form an angle of 90 degrees respect to the rotor thrust, rewriting Eq. (6), we have

$$F_i = -k\omega_i^2\,[0\,0\,1]^T \tag{7}$$

By transforming F_i to F_e and making use of the principle of linear momentum, the following equations can be introduced for a quadrotor of mass m under gravity g:

$$\ddot{r} = [0\,0\,g]^T + R\sum_1^4 \left(\frac{F_i}{m}\right) \tag{8}$$

The rotatory moment of the body is described by Eq. (4) can be rewrite as

$$J\dot{W} + W \times JW = M_B - M_G \tag{9}$$

where M_B is the vector of external torques and is composed of the thrust differences and drag moments of the individual rotors and under considerations of the rotatory directions and can be calculated by

$$M_B = \begin{pmatrix} Lb\left(\omega_2^2 - \omega_4^2\right) \\ Lb\left(\omega_1^2 - \omega_3^2\right) \\ d\left(\omega_1^2 - \omega_2^2 + \omega_3^2 - \omega_4^2\right) \end{pmatrix} \tag{10}$$

where d is called drag factor and it is related with the air resistance, L is a parameter which represents the length of the lever between center of mass and the four rotors. M_G is the vector of gyroscopic torques generated due to the propellers rotational movements and can be calculated by the follow formula

$$M_G = J_R \left(W \times \begin{pmatrix} 0 \\ 0 \\ 1 \end{pmatrix} \right) (\omega_1 - \omega_2 + \omega_3 - \omega_4) \tag{11}$$

where J_R represents the inertia of rotating rotors. The follow equations permit us to determine the rate of change of the Euler angles in the inertial frame F_e

$$\dot{\Omega} = H(\Phi)W = \begin{bmatrix} 1 & \tan\theta\sin\phi & \tan\theta\cos\phi \\ 0 & \cos\phi & \sin\phi \\ 0 & \dfrac{\sin\phi}{\cos\theta} & \dfrac{\cos\phi}{\cos\theta} \end{bmatrix} W \tag{12}$$

The relationship that maps rotor angular velocities to forces and moments on the vehicle is

$$u = \begin{pmatrix} u_1 \\ u_2 \\ u_3 \\ u_4 \end{pmatrix} = \begin{pmatrix} b & b & b & b \\ 0 & b & 0 & -b \\ b & 0 & -b & 0 \\ d & -d & d & -d \end{pmatrix} \begin{pmatrix} \omega_1^2 \\ \omega_2^2 \\ \omega_3^2 \\ \omega_4^2 \end{pmatrix} \tag{13}$$

where the variable $u_1 = F_1 + F_2 + F_3 + F_4$. The variables u_2 and u_3 correspond the forces from rotors necessary to generate the pitch and roll moments, and u_4 represents the yaw moment.

The translation motion is obtained combining Eqs. (3), (7), (8), and (13), as a result, we have the follow equations

$$\ddot{r} = \begin{bmatrix} \ddot{x} \\ \ddot{y} \\ \ddot{z} \end{bmatrix} = \begin{pmatrix} -(C_\psi S_\theta C_\phi + S_\psi S_\phi)\dfrac{u_1}{m} \\ -(S_\psi S_\theta C_\phi - C_\psi S_\phi)\dfrac{u_1}{m} \\ g - (C_\theta C_\phi)\dfrac{u_1}{m} \end{pmatrix} \tag{14}$$

On the other hand, if we assume small perturbations in hover flight, $\dot{\Omega} \approx W$ and combining Eqs. (9), (11), and (13), we obtain the follow equations

$$\ddot{\phi} = \dot{\theta}\dot{\psi}\left(\frac{J_y - J_z}{J_x}\right) - \left(\frac{J_r}{J_x}\right)\dot{\theta}h(u) + \frac{L}{Jx}u_2$$
$$\ddot{\theta} = \dot{\phi}\dot{\psi}\left(\frac{J_z - J_x}{J_y}\right) + \left(\frac{J_r}{J_y}\right)\dot{\phi}h(u) + \frac{L}{Jy}u_3 \quad (15)$$
$$\ddot{\psi} = \dot{\phi}\dot{\theta}\left(\frac{J_x - J_y}{J_z}\right) - \frac{1}{J_z}u_4$$

2.4. Linearization of quadrotor dynamics

To simplify the earlier equations, we have linearized the rotatory system around hover state assuming small change in Euler angles, $u_1 \approx mg$ in x and y directions, $\cos(\alpha) = 1$, $\sin(\alpha) = \alpha$ and neglected the gyroscopic torques; therefore,

$$\begin{aligned}
\ddot{x} &= -\theta g \\
\ddot{y} &= \phi g \\
\ddot{z} &= u_1 - mg \\
\ddot{\phi} &= \frac{L}{J_x}u_2 \\
\ddot{\theta} &= \frac{L}{J_y}u_3 \\
\ddot{\psi} &= \frac{1}{J_z}u_4
\end{aligned} \quad (16)$$

This work only considers the angular momentum corresponding to the orientation of the vehicle, so that the displacement in (x, y, z) axes will not be used for mathematical analysis.

Therefore, the transfer function for dynamic correspondence of orientation is expressed as

$$\begin{aligned}
\phi(s) &= \frac{L}{s^2 J_x}u_2 \\
\theta(s) &= \frac{L}{s^2 J_y}u_3 \\
\psi(s) &= \frac{1}{s^2 J_z}u_4
\end{aligned} \quad (17)$$

and adding rotor's model, we have

$$G(s, e^{-\tau s}) = \begin{pmatrix} \frac{K_{p1}}{s^2(s + a_\phi)} & 0 & 0 \\ 0 & \frac{K_{p2}}{s^2(s + a_\theta)} & 0 \\ 0 & 0 & \frac{K_{p3}}{s^2(s + a_\psi)} \end{pmatrix} \quad (18)$$

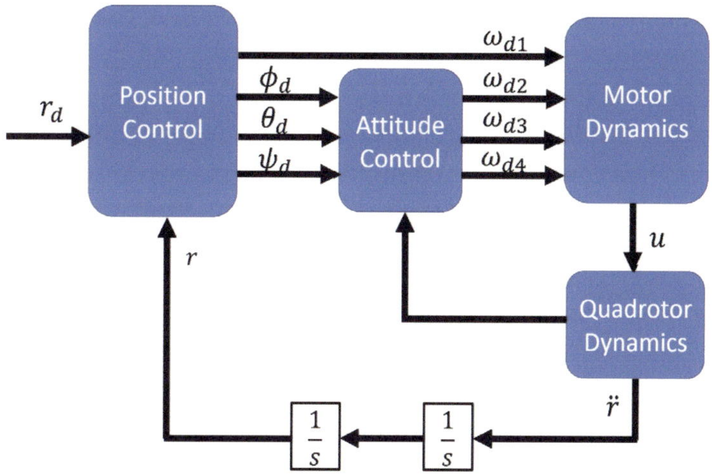

Figure 2. Overall control system.

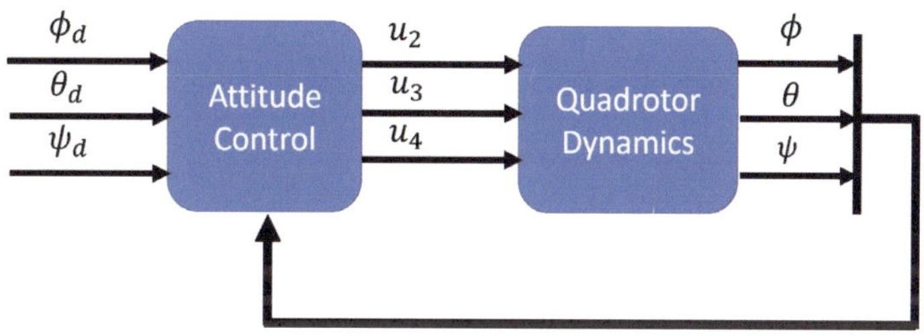

Figure 3. Attitude control system of a quadrotor.

where $K_{p1}=Lk_\phi/J_x$, $K_{p2}=Lk_\theta/J_y$, $K_{p3}=Lk_\psi/J_z$. These parametres were computed for a X4-flyer and have the following values: $K_{p1}=K_{p2}=33.23$, $K_{p3}=16.95$, $a_\phi=a_\theta=a_\psi=4.1$ and $\tau=0.017$. This mathematical model will be used to synthesize the control law to stabilize the closed-loop system [15].

The overall control schema is showed in **Figure 2**, where position and attitude control are presented. This schema consist in two loops, first one is used to perform the quadrotor tracking of desired trajectory r_d, while the second one is used to achieve the desired Euler (ϕ_d, θ_d, ψ_d). Due that we are only interested in attitude control, the above schema is redrawn as is showed in **Figure 3**.

3. Adaptive controller design

The adaptive control is an advance control technique which provides a systematic approach for automatic adjustment of controllers in real time, in order to achieve or to maintain a desired level of control system performance, when the parameters of the plant dynamic model are unknown and/or change in time [16]. Two different approaches can be distinguished: indirect and direct approaches. In the first approach, the plant parameters are estimate online and used to calculate the controller parameters. In the second, the plant model is parameterized in terms of the controller parameters that are estimated directly without intermediate calculations involving plant parameter estimates [17].

The model reference adaptive control or MRAC is a direct adaptive strategy which consists of some adjustable controller parameters and an adjusting mechanism to adjust them. The goal of the MRAC approach is adjusting the controller parameters so that the output of the plant tracks the output of the reference model having the same reference input. The MRAC schema is combine two loops: the inner or primary loop where controller and plant are feedback as in normal loop and outer loop or also called adjustment loop where some adaptive mechanisms and a model reference are used to obtain the some performance [18]. In **Figure 4**, an overall MRAC schema is presented.

This section presents the design of an adaptive controller MRAC using the Lyapunov's stability theory. This allows us to ensure the tracking trajectory of an X4 to our reference model and makes the system insensitive to parameter variation and external disturbances, leading the state error to zero. Based on this, the process model can be represented by state space as follows:

$$\dot{x} = A_p x + B_p u \tag{19}$$

$$y = Cx \tag{20}$$

where A_p and B_p represent the matrix and the vector of unknown constant parameters of the system, u is the output signal of the controller, and x is the state vector.

The reference model is defined as follows:

$$\dot{x}_m = A_m x_m + B_m u_c \tag{21}$$

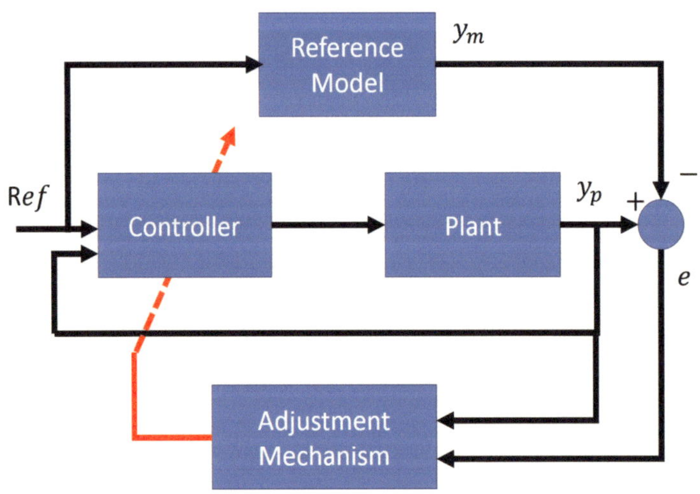

Figure 4. MRAC control schema.

$$y = C_m x_m \tag{22}$$

Then, the control law is selected as:

$$u = L_r u_c - Lx = \theta_4 u_c - \theta_1 x_1 - \theta_2 x_2 - \theta_3 x_3 \tag{23}$$

where L_r and L are the matrix containing the parameters of the controller, which can be freely selected, and u_c is the reference signal.

Substituting (Eq. (22)) in (Eq. (18)), the closed loop system is expressed as follows:

$$\dot{x} = \left(A_p - B_p L\right)x + B_p L_r u_c = A(\theta)\mathrm{x} + B(\theta)u_c \tag{24}$$

Now we introduce the error equation as follows

$$e = x - x_m \tag{25}$$

Differentiating the error with respect to time, we obtain:

$$\frac{de}{dt} = \frac{dx}{dt} - \frac{dx_m}{dt} = Ax + Bu - A_m x_m - B_m u_c \tag{26}$$

and adding $A_m x$ and subtracting to the left side of the equation

$$\frac{de}{dt} = A_m e + (A - A_m)x + (B - B_m)u_c \tag{27}$$

The error goes to zero if A_m is stable and

$$A(\theta) - A_m = 0 \tag{28}$$

$$B(\theta) - B_m = 0 \tag{29}$$

If we assume that the closed-loop system can be described by (Eq. (14)), where the matrices A and B depend on the parameter θ, and it is some combination of L_r and L, then we can define the following Lyapunov function for the parameter adaptation law:

$$V = e^T Pe + tr(A(\theta) - A_m)^T Q_a(A(\theta) - A_m) + tr(B(\theta) - B_m)^T Q_b(B(\theta) - B_m) \tag{30}$$

where $P \in \Re^{3x3}$ is a positive definite matrix and V is a positive definite function. If its first derivative to time V is not positive definite, then V is Lyapunov function. Now we can solve the derivative of V to time t.

$$\frac{dV}{dt} = e^T PA_m e + e^T A_m^T Pe + 2tr(A - A_m)^T \left(Q_a \dot{A} + Pex^T\right) + 2tr(B - B_m)^T \left(Q_b \dot{B} + Peu_c{}^T\right) \tag{31}$$

where Q is a positive definite matrix such that meet the follow equation:

$$A_m^T P + P A_m = -Q \tag{32}$$

Therefore, if we choose a stable A_m matrix, we will get always a P and Q positive definite matrix.

Accordingly, the derivative to time of function V is

$$\frac{dV}{dt} = -e^T Q e \tag{33}$$

Where the function V is a Lyapunov function negative semi-definite ensures the output error between the real system and the reference model will tend to be zero, and the system is asymptotically stable.

Thus, we obtain the following parameter adaptation laws

$$\frac{d\theta_1}{dt} = p_{11}e_1x_1 + p_{12}e_2x_1 + p_{13}e_3x_1 \tag{34}$$

$$\frac{d\theta_2}{dt} = p_{11}e_1x_2 + p_{12}e_2x_2 + p_{13}e_3x_2 \tag{35}$$

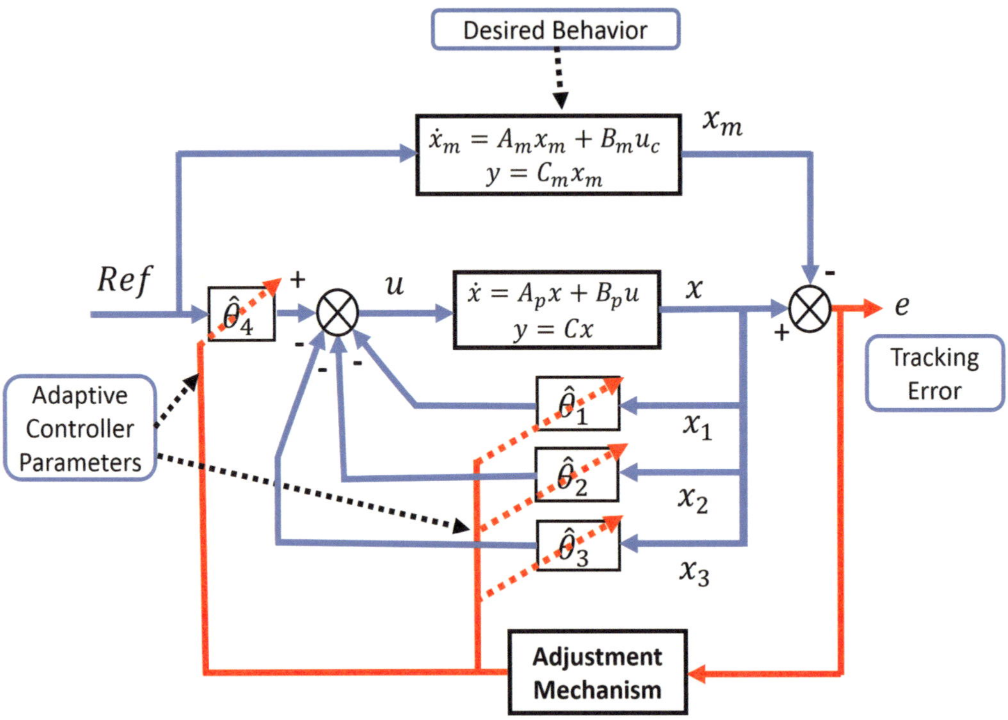

Figure 5. MRAC control of quadrotor: block diagram.

$$\frac{d\theta_3}{dt} = p_{11}e_1x_3 + p_{12}e_2x_3 + p_{13}e_3x_3 \tag{36}$$

$$\frac{d\theta_4}{dt} = -(p_{11}e_1r + p_{12}e_2r + p_{13}e_3r) \tag{37}$$

Then, the resulting control diagram is shown in **Figure 5**.

4. Simulation and analysis

This section presents several simulations test made to prove the performance of MRAC controller to stabilize a mini-UAV quadrotor. As mentioned before, only orientation dynamic (angle position, angular velocity and acceleration) are considered. Analyzing Eq. (18), it is easy to see that roll, pitch, and yaw dynamics are very similar; for this reason, only roll moment is used as example in simulation.

The test begins considering controller parameters are unknown and by using an online adaptive mechanism to determine the values that permit the convergence of plant response to reference model response. It is important to note that the MRAC approach seeks to keep the tracking error $(x - x_m)$ equal to zero by an adjustment of the controller parameters and do not to seek to identify the real parameters of the plant.

Figures 6–8 shows a comparative between the states response of the plant and the state response of the model reference, where can be observer than all states of the plant converge asymptotical to the reference model states. To verify this, the **Figures 9–11** are presented; these figures show the tracking error of states goes to zero. Additionally, in **Figure 12**, the reference input is compared to the plant and model reference output.

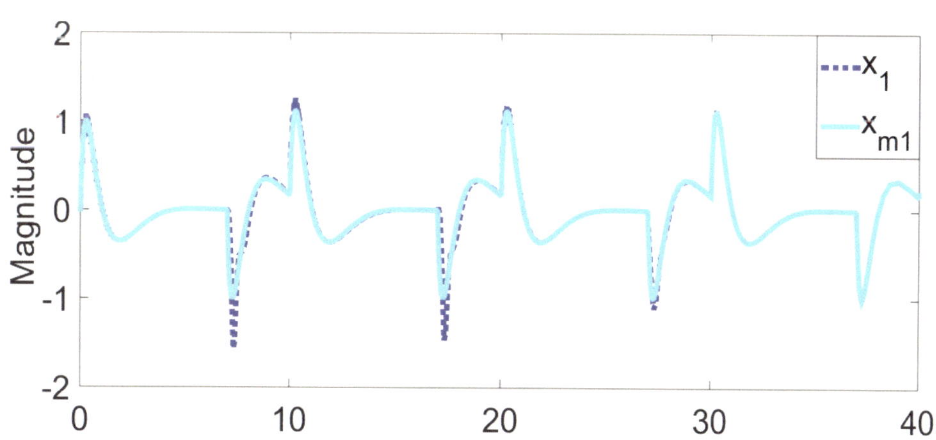

Figure 6. Comparison between x_1 and x_{m1}.

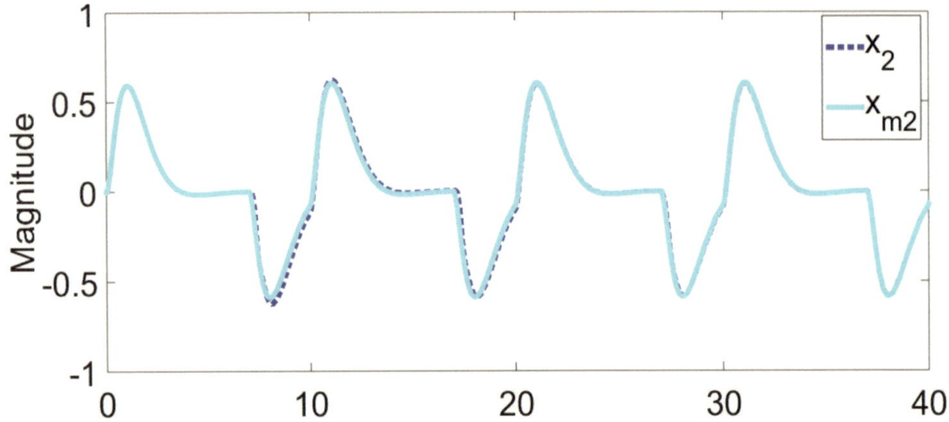

Figure 7. Comparison between x_2 and x_{m2}.

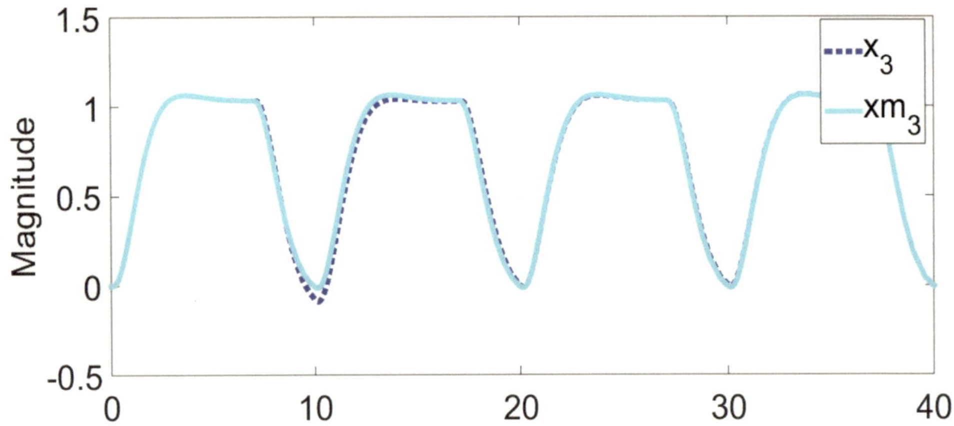

Figure 8. Comparison between x_3 and x_{m3}.

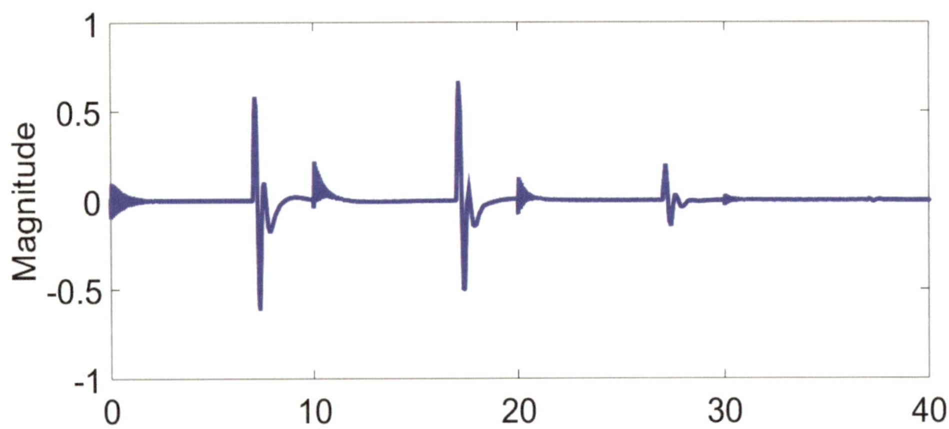

Figure 9. Error between x_1 to x_{m1}.

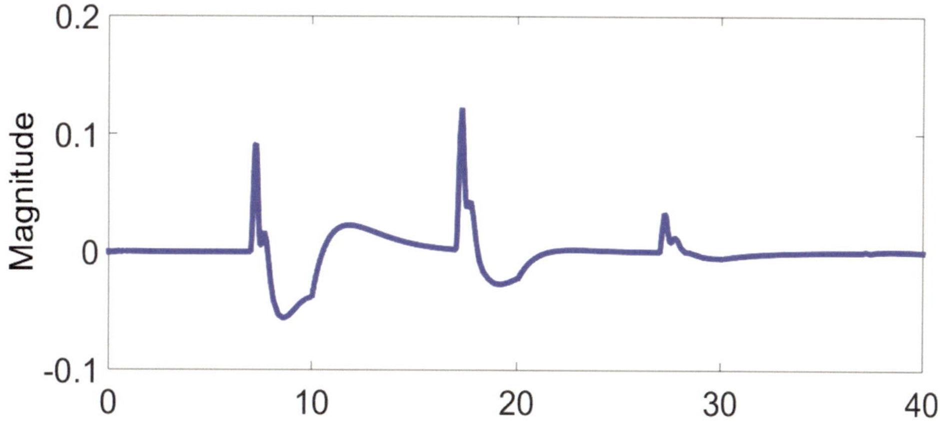

Figure 10. Error between x_2 to x_{m2}.

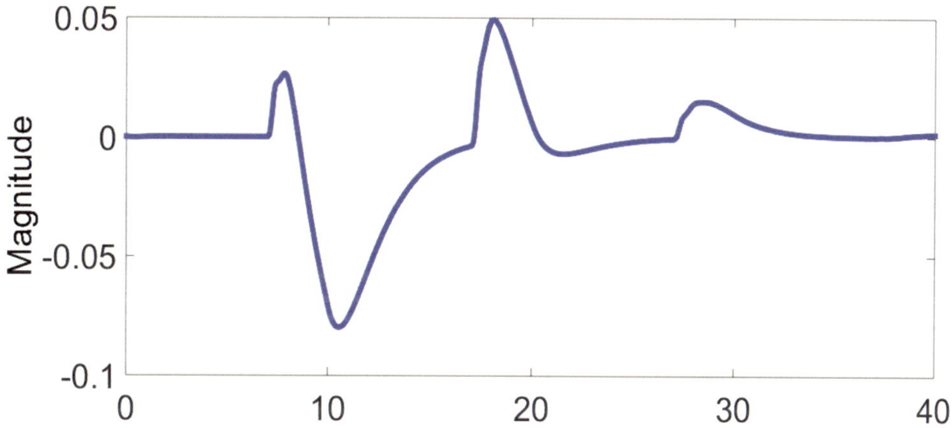

Figure 11. Error between x_3 to x_{m3}.

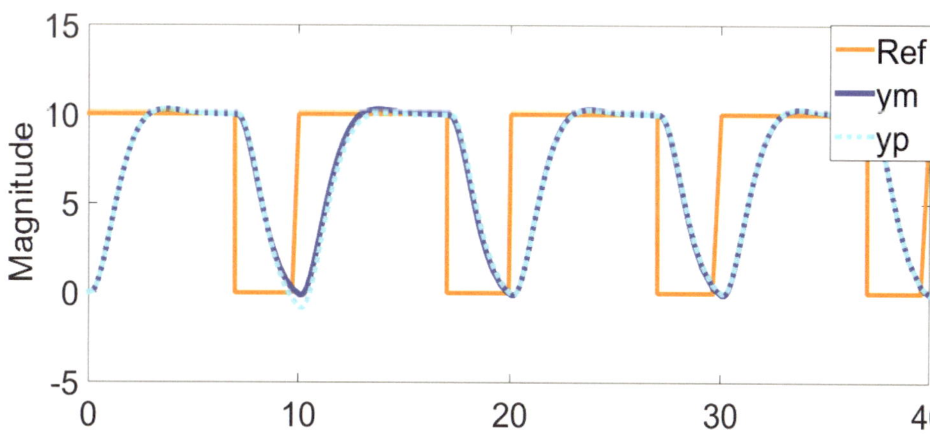

Figure 12. Roll response.

Figure 13 shows the variations of the controller parameters during the adjustment process. This mechanism started many times is necessary to assure the perfect tracking of the plant to the desire response. Finally, in **Figure 14**, the simulation diagram is presented.

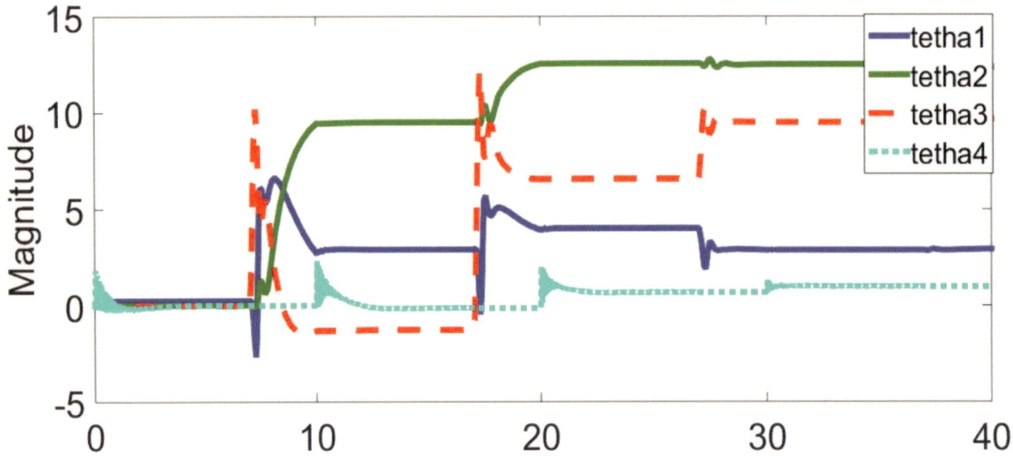

Figure 13. Tuning of controller parameters.

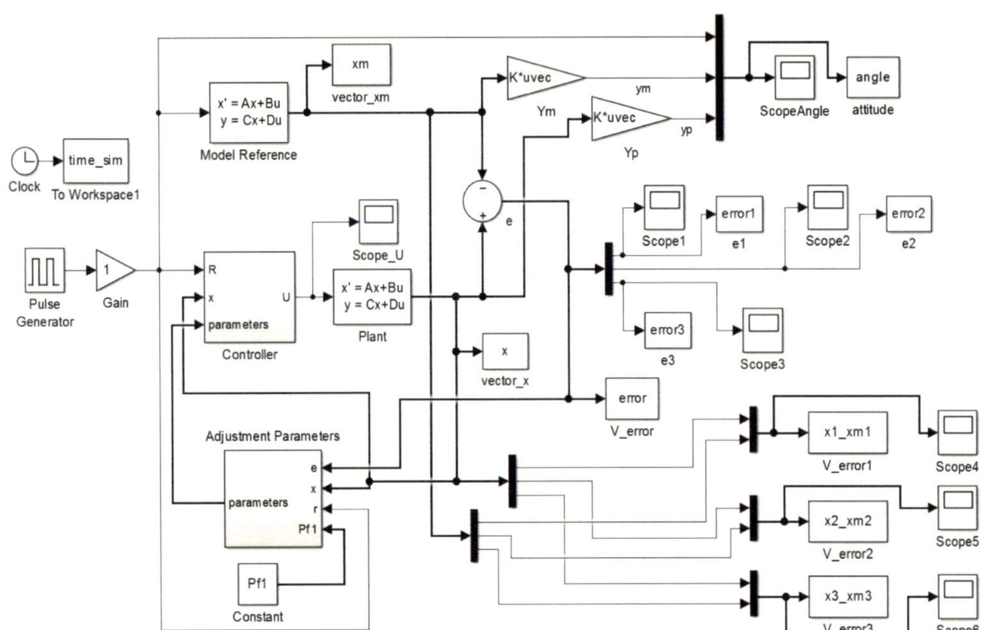

Figure 14. Simulink diagram.

5. Conclusions

This work presents an adaptive control technique to stabilize the attitude of a quadrotor UAV using MRAC schema, which requires no information of the plant model. The asymptotic stability was demonstrated using the well-known Lyapunov's theory, obtaining in this way the adaptation law of the controller parameters. Simulations results demonstrate that the adaptive control approach proposed have a good performance to perform the asymptotic tracking of model reference output. It is important to note that the adaptive mechanism is started only when it is needed.

Author details

Ramiro Ibarra Pérez*, Gerardo Romero Galvan, Aldo Jonathan Muñoz Vázquez, Silvia Florida Melo and David Lara Alabazares

*Address all correspondence to: rm_ibarra@live.com

UAM Reynosa Rodhe, Autonomous University of Tamaulipas, Tamaulipas, México

References

[1] Austin R. Unmanned Aircraft Systems: UAVS Design, Development and Deployment. John Wiley & Sons. United Kingdom; 2010. p. 372 DOI: 10.1002/9780470664797

[2] Jeong S et al. Position Control of a Quad-Rotor System. In: Kim J-H, Matson ET, Myung H, Xu P, editors. Robot Intelligence Technology and Applications. Berlin Heidelberg: Springer-Verlag; 2013; 2012. p. 971-981. DOI: doi.org/10.1007/978-3-642-37374-9_94

[3] Bouabdallah S, editor. Design and control of quadrotor with application to autonomous flying [dissertation]. École Polytechnique Fédérale de Lausanne; 2007

[4] Jithu G, Jayasree PR. Quadrotor modelling and control. International Conference on Electrical, Electronics, and Optimization Techniques (ICEEOT)—2016

[5] Jun L, Yuntang Li. Dynamic analysis and PID control for a quadrotor. In: Proceedings of the 2011 IEEE. 2011

[6] Changlong L, Jian P, Chang Y. PID and LQR trajectory tracking control for an unmanned quadrotor helicopter: Experimental studies. In: Proceedings of the 35th Chinese Control Conference. 2016; DOI: 10.1109/ChiCC.2016.7555074

[7] Salazar S, Romero H, et al. Real-time stereo visual serving control of an UAV having eight rotors, In: 6th International Conference on Electrical engineering, computing science and automatic control, CCE 2009. DOI: 10.1109/ICEEE.2009.5393423

[8] Bouabdallah S, Siegwart R. Backstepping and sliding-mode techniques applied to an indoor micro quadrotor. In: Proceedings of the 2005 IEEE International Conference on Robotics and Automation, ICRA 2005. DOI: 10.1109/ROBOT.2005.1570447

[9] Abeywardena DMW, Amaratunga LAK, Shakoor SAA, Munasinghe SR. A velocity feedback fuzzy logic controller for stable hovering of a quadrotor UAV. In: International Conference on Industrial and Information Systems (ICIIS), 2009. DOI: 10.1109/ICIINFS.2009.5429800

[10] Theerasak S, Pined L, Wonlop C, et al. Path tracking of UAV using self-tuning PID controller based on fuzzy logic, In: Proceedings of SICE Annual Conference 2010

[11] M Imran R, Akhtar S. Adaptive control of a quadrotor with unknown model parameters In: 9th International Bhurban Conference on Applied Sciences and Technology (IBCAST) 2012. DOI: 10.1109/IBCAST.2012.6177518

[12] Bouabdallah S, Murrieri P, Siegwart R. Design and control of an indoor micro quadrotor. Proceedings of IEEE International Conference on Robotics and Automation, ICRA '04. 2004. DOI: 10.1109/ROBOT.2004.1302409

[13] Sanca AS, Alsina PJ, Cerqueira JJ. Dynamic modelling of a quadrotor aerial vehicle with nonlinear inputs. In: IEEE Latin American Robotic Symposium, 2008. LARS '08. 2009. DOI: 10.1109/LARS.2008.17

[14] Schreier M. Modeling and adaptive control of a quadrotor. International Conference on Mechatronics and Automation (ICMA), 2012. 2012; DOI: 10.1109/ICMA.2012.6282874

[15] Tao G, Selfridge Justin M. A multivariable adaptive controller for a Quadrotor with guaranteed matching conditions. In: American Control Conference (ACC) Portland, Oregon, USA. 2014. DOI: 10.1109/ACC.2014.6859355

[16] Lara D, Romero G. Onboard system for flight control of a small UAV. In: World Automation Congress (WAC), 2012. 04 October 2012. Puerto Vallarta, Mexico. Mexico: IEEE. 2012.

[17] Landau ID, Lozano R, M'Saad M, Karimi A. Introduction to adaptive control. In: Adaptive Control: Algorithms, Analysis and Applications. London: Springer; 2011. p. 1-33. DOI: 10.1007/978-0-85729-664-1

[18] Ioannou P, Sun J. Robust Adaptive Control. Mineola, New York: Dover Publications; 2013. p. 848

Chapter 7

Matlab-Simulink-Based Compound Model Reference Adaptive Control for DC Motor

Marian Găiceanu

Additional information is available at the end of the chapter

http://dx.doi.org/10.5772/intechopen.71758

Abstract

The objective of this chapter is to develop a compound Model Reference Adaptive Control (MRAC) of the dc motor by using the Matlab/Simulink software. The purpose of the chapter is to serve as a tutorial for the students or researchers in the field correlating step by step the presented theory with the Matlab/Simulink programming environment. The supraunitary relative degree model reference adaptive control is proposed as a solution to the parameters variation of the electric drives. The numerical simulation results confirm the robustness of the proposed solution at unmodelled dynamics or parameter variation of the process. The conventional control of the dc drive based on the cascaded loops is also treated in this chapter.

Keywords: dc motor, PI control, adaptive control, Matlab-Simulink, supraunitary relative degree

1. Introduction

The conventional control of the dc motor supposes the use of the modulus and symmetric criteria, adapted to the fast processes (Kessler variant), by the optimum choosing of the tuning parameters of the corresponding controllers. Therefore, the cascaded control loops are the most appropriate choice for the conventional dc drive system. The advantages of using the conventional drive are: the complex process is divided into simple subprocesses having only one significant time constant compensated by an independent controller; it assures the minimum time response; thanks to the feedback path of the control loop, the stability of the dc drive system is maintained; the null steady state error; the fast compensation of the dc drive perturbations.

The main disadvantage of the conventional control is its inadaptability to the parameters variations of the processes. Taking into account that the normal operation of the processes is assured through the electric motors, the values of the electric parameters are not constant. It is well-known that the parameters of the controllers depend on the parameters of the process. Therefore, a change of the process parameters conducts to the deterioration of the drive system performances. Moreover, the uncertainness of the parameters and the structural mismatch of the processes are the main barriers of avoiding the use of the conventional control. In these particular cases the conventional control, even the robust control is inefficient due to the invariance of the controller parameters. An adaptive control is desired in order to maintain the imposed performances on the electric drive. Consequently, an additional adaptive loop is inserted to the conventional feedback system. The purpose of inserting the adaptive loop is to update the controller parameters related to the parameter variations of the process or to the modified structure of the physical process. According to the exogenous variation (reference and perturbation), the adaptive mechanism of the regulation part is performed by the minimization of the specific performance criterion. The result of the designed regulation algorithm consists of a new set of the controller parameters. The closed loop adaptive systems (widely spread) assure the appropriate control correlation with the process uncertainness (modified parameters or structure). The adaptive loop delivers the real time control based on the designed on-line controller.

The signals in **Figure 1** are defined as follows: r- the reference signal; y_m- the output signal of the reference model; u- the adaptive control; y-the output of the process (the measurable signals); e_0 – the tracking error vector; θ_p – the parameters of the process.

In this manner, the structural mismatch and the parametric uncertainness are compensated. In the mechatronic domain or the rolling mill the adaptive control is mandatory.

The name of the adaptive control structure comes from the use of the on-line parameters estimator. Based on it, the process parameters are provided by measuring the control, the process output and the tracking error. Taking into account the reference signal and the process parameters, the adaptive algorithm delivers the appropriate estimation of the control.

There are two types of the adaptive control structures [2]. The adaptive control can be in explicit (indirect) form or in implicit (direct) form. The indirect adaptive control delivers the estimation of the process parameters. The direct adaptive control delivers the estimation of the controller parameters.

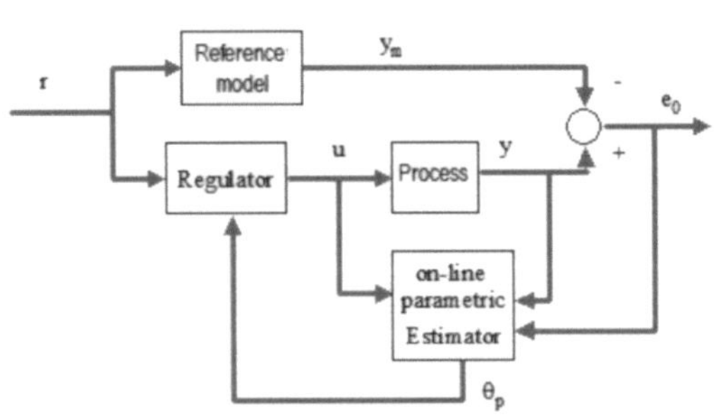

Figure 1. The principle of the adaptive control [1].

In this chapter the adaptive control with supraunitary relative degree is introduced through the dc drive system.

The design procedure of the adaptive control is based on the strictly positive real (SPR) concept.

In order to attain the SPR condition, the equivalent transfer function should satisfy the following tasks:

1. strictly stable (the poles are situated in the left half plane of the complex space);
2. the relative degree must be 0 or 1;
3. $\mathrm{Re}[H_m(j\omega)] \geq 0$, for any $\omega > 0$, i.e., the transfer function of the reference model, $H_m(s)$, should be minimum of phase (without dead time and zero situated in the right half plane).

The model reference adaptive control (MRAC) contains three aspects:

1. MRAC in simple form;
2. variable structure form; and
3. compound MRAC.

In order to obtain a perfect tracking, $\lim_{t \to 0} e_0(t) \to 0$, the asymptotic limit of the tracking error vanishes, and the relative degree of the reference model should be greater or at least of the same order with the relative degree of the process:

$$n^*_{\mathrm{m}} \geq n^*_{\mathrm{p}} \tag{1}$$

2. Conventional dc drive system

In this Section the control methodology of the conventional dc drive system is presented.

The drive system consists of the full controlled dc motor connected to the load. At the high power, the controlled six pulses full bridge ac-dc power converter is involved; at the low – medium power the full bridge dc-dc power converter connected in series with the uncontrolled ac-dc power rectifier is used. From the point of view of commutation, there is a substantial difference between the above presented solutions: while at low and medium power, the force commutation is used (based on the high frequency switching power transistors), at high power, the natural commutation is used (based on thyristors).

In order to design the conventional control for fast processes, the entire system should be modeled under certain assumptions:

2.1. Assumptions for mathematical modeling of the dc motor

The magnetizing flux is maintained at the rated value; the permeability of the ferromagnetic core is infinite; the unsaturated magnetic circuit is considered, a compensated dc motor is taken into consideration; there are auxiliaries poles; the brushes are situated on the neutral axis; and the brush droop voltage is neglected.

2.2. Assumptions for the ac-dc/dc-dc power converters

The uninterrupted conduction is taken into account (a high value for the additional armature inductance is designed), the conduction droop voltage on the power semiconductors and the switching time are negligible.

2.3. Assumptions for the load

The load torque is considered as mathematical model of the process, the load is reduced appropriately to the dc motor shaft (by using both equivalent inertia moment and reduced equivalent speed by taking into account the specific transmission ratio).

2.4. Assumption for the conventional control

Taking into account that the mathematical model of the dc motor is of the second order, the two state variables are deducted (armature current, i_A, and the speed, ω_m). The mathematical model is characterized by two different time constants: T_m>>T_A (the electromechanical time constant, T_m, is greater with one order than the electromagnetic ones, T_A). This conclusion leads to the cascaded control with two loops: the armature current loop (the inner loop), and the angular velocity (the outer loop). By considering that the angular velocity should be controlled and the armature current should be limited at the maximum value, the angular velocity is the dc motor output. Taking into account that the dc motor is supplied by the six pulses ac-dc full bridge power converter, the modulus criterion is applied for tuning inner loop, and the symmetrical optimum for the outer loop (Kessler variant). The adequate operational block diagram is deducted according to **Figure 2**.

The dc drive is constituted by two parts: the power and the control. A unified [0, 10]V voltage system is taken into account with respect to the control part. Therefore, for the maximum value of 10 V the maximum allowable speed at the rated flux is obtained (i.e., $n^* = 1{,}2\ n_r$, by taken into consideration the 20% speed overshoot introduced by the Kessler criterion).

It is well-known that the symmetrical criterion supposes the ramp reference for the speed loop. In order to transform the step reference analogue signal from the potentiometer into ramp signal, an adequate filter has been designed with the following transfer function:

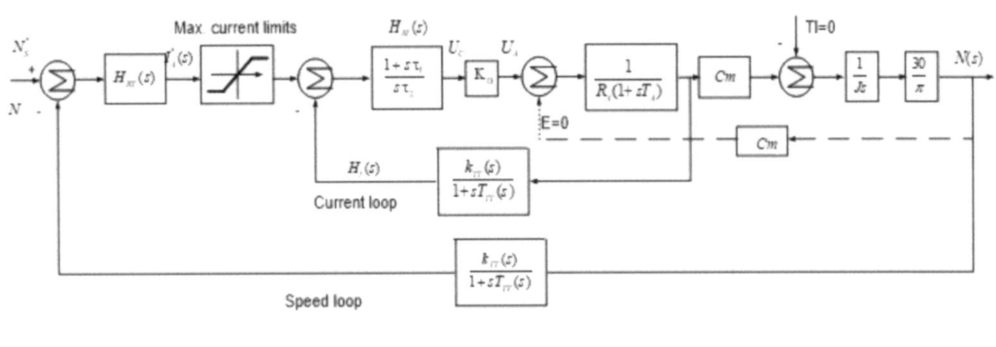

Figure 2. The conventional control of the dc drive.

http://dx.doi.org/10.5772/intechopen.71758

$$H_{fn}(s) = \frac{1}{1 + sT_{Fn}}, \tag{2}$$

where T_{Fn} is the time constant of the speed transducer.

The corresponding Simulink block diagram is shown below in **Figure 3**.

The control design procedure starts with the armature current loop design (inner loop). The imposed performances for the inner closed loop are:

I. *the steady state regime:*

  a. null steady state error $\varepsilon_{st} = 0$;

  b. the rejection of the perturbation influence (load torque is the main perturbation).

II. *the dynamic regime:*

  a. current overshoot, $\sigma = 4.3\%$;

  b. response time depends on the cutting frequency of the loop: $t_r = 2.35/\omega_{cI}$;

  c. phase margin of the current loop: $\gamma = 63°26'$.

In order to satisfy the above mentioned performances the modulus criterion is used. The Kessler criterion guarantees of the above mentioned performances on condition that the open loop transfer function of the inner loop has the form:

$$H_{cutI}(s) = \frac{1}{2sT_{\Sigma I}(1 + sT_{\Sigma I})} \tag{3}$$

where, $T_{\Sigma I}$ – the parasitic time constant of the current loop is considerably lower than the armature time constant:

$$T_{\Sigma I} = T_{FI} + T_{TI} << T_A \tag{4}$$

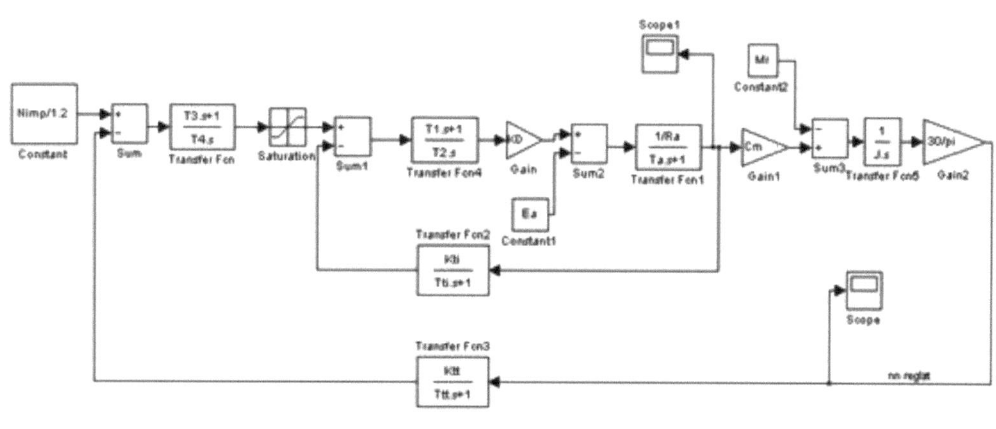

Figure 3. Simulink block diagram of the dc drive system.

in which T_{FI} is the current filter time constant, and T_{TI} is the current transducer time constant.

According to modulus criterion, the Proportional Integral (PI) controller is suitable to control the armature current:

$$H_{RI}(s) = \frac{1 + s\tau_1}{s\tau_2}, \tag{5}$$

with the appropriately controller parameters:

$$\tau_2 = T_A \tag{6}$$

$$\tau_2 = 2T_{\Sigma I} K_d K_{TI} \frac{1}{R_A}. \tag{7}$$

Eq. (8) contains the mathematical model of the dc-dc power converter:

$$K_d = \frac{U_{Ar}}{10}, \tag{8}$$

and the attenuation factor of the current transducer (obtained by imposing the time response of the current loop) has the form:

$$K_{TI} = \frac{10}{I_{A\max}} \left[\frac{V}{A}\right]. \tag{9}$$

The closed loop armature current transfer function is as follows:

$$H_{oI}(s) = \frac{1}{k_{TI}(1 + 2sT_{\Sigma I})} \tag{10}$$

The equivalent block diagram of the dc drive system is shown in **Figure 4**.

Taking into account the reduced block diagram of the dc drive system (**Figure 4**), the symmetrical optimum criterion could be applied.

According to symmetrical optimum criterion, by using the following open loop transfer function (Kessler) the required performances of the closed loop system are attained:

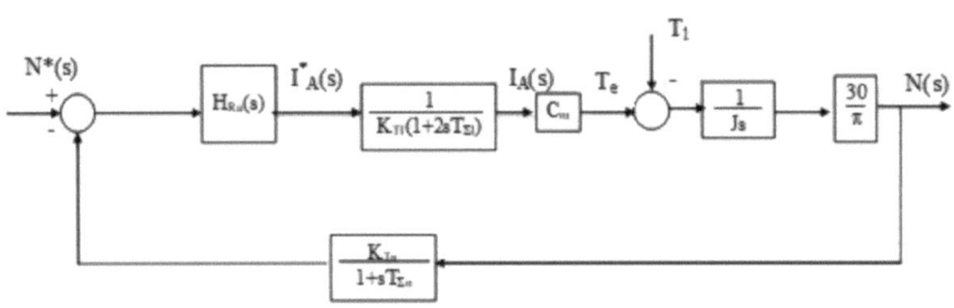

Figure 4. The operational model of the dc drive system.

http://dx.doi.org/10.5772/intechopen.71758

$$H_{cut,n}(s) = \frac{1 + 4sT_{\Sigma n}}{8s^2 T_{\Sigma n}^2 (1 + sT_{\Sigma n})}. \tag{11}$$

By comparing the equivalent open loop transfer function (**Figure 4**) with the imposed open loop transfer function (11), the PI speed controller is obtained:

$$H_{Rn}(s) = \frac{1 + s\tau_3}{s\tau_4}. \tag{12}$$

Moreover, the following speed controller parameters are obtained:

$$\tau_3 = 4T_{\Sigma n}, \tag{13}$$

$$\tau_4 = \frac{8T_{\Sigma n}{}^2 C_m \cdot 30}{K_{TI} J \pi K_{Tn}}. \tag{14}$$

In order to well understand the tuning procedure of the speed controller, the additional parameters are provided:

$$T_{\Sigma n} = 2T_A + T_{TN}, \tag{15}$$

$$C_m = \frac{T_{lr}}{I_{Ar}}, \tag{16}$$

$$K_{Tn} = \frac{10}{1.2 \cdot n_r} \left[\frac{rot/\min}{V} \right], \tag{17}$$

in which $T_{\Sigma n}$ is the parasitic time constant of the speed loop, T_{lr} is the rated load torque, C_m- the mechanical constant of the dc motor, and T_{Tn} is the time constant of the speed transducer (it is obtained by imposing the time response of the speed loop).

The numerical implementation of the dc drive system is based on the Simulink block diagram shown in **Figure 5**.

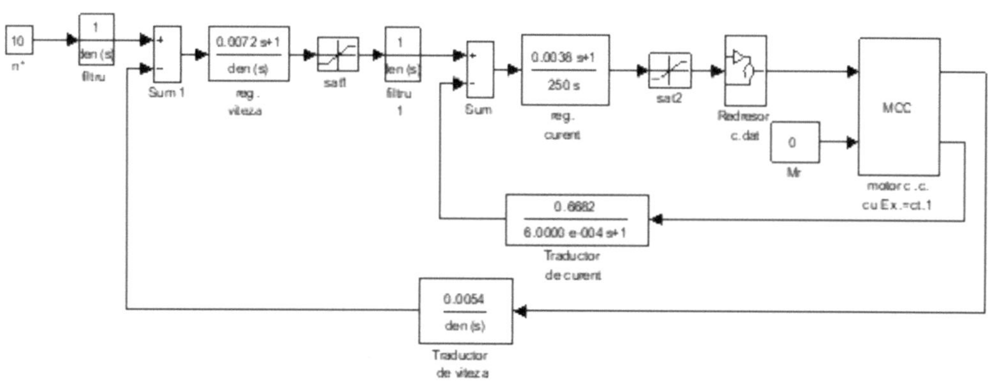

Figure 5. The numerical implementation of the dc drive system.

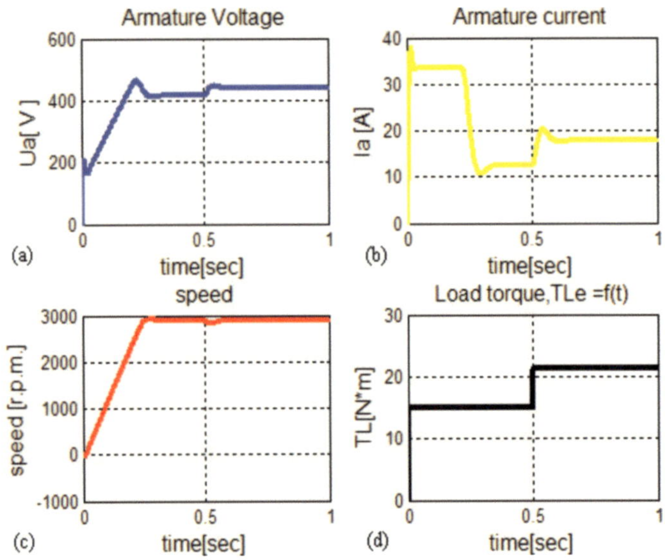

Figure 6. The dc drive numerical simulations results under load step variation.

By considering a dc motor with the following nameplate data: rated power $P_r = 5.1$ kW, the maximum armature voltage $U_r = 440$ V, the nominal armature current $I_{Ar} = 17.8$ A, the reduced moment of inertia $J = 0.02$ kgm^2, the viscous force $F_v = 0.0006$ Nms/rad, rated speed $n_r = 2700$ rpm, the simulation results are obtained (**Figure 6a–d**). The motor data can be obtained based on the nameplate values by using the detailed Matlab software provided in [3].

Figure 6a–d shows the $0.7T_r$ load starting simulation results of the dc conventional control based on the dc-dc full bridge power converter. **Figure 6a** contains the obtained armature voltage of the dc motor. The armature current varies according to **Figure 6b**, the speed varies as in **Figure 6c** under rated value of the load torque $T_l = 22.8$ Nm (**Figure 6d**). The load torque is applied at $t = 0.5$ s.

3. Adaptive control

There are three assumptions available [4, 5]: the mathematical model of the process is linear, strictly proper and of minimum phase, having the supraunitary relative degree $n^*_p = 2n^*_p = 2$; the reference model has the relative degree greater than one ($n^*_m = 2$), is stable and of minimum phase; the reference signal should be bounded limit, being a continuous function. The second order mathematical model of the dc motor is used in this chapter. This supposes the transfer function of the process has the form:

$$H_p(s) = k_p \frac{N_p(s)}{D_p(s)}. \tag{18}$$

At the same time, the pole excess is known:

$$\partial\left[D_p(s)\right] - \partial\left[N_p(s)\right] = n_p^* > 1. \tag{19}$$

Taking into account that the relative degree of the process is supraunitary, the second order of the reference model is chosen, $n^*_{\mathrm{m}} = n^*_{\mathrm{p}}$.

$$H_m(s) = k_m \frac{N_m(s)}{D_m(s)}. \tag{20}$$

Due to the supraunitary relative degree, the strictly real positive condition for the reference model cannot be accomplished. This condition that the tracking error differs from the identification error [2–6] implies, in this case, the use of the augmented error.

The augmented error depends on the gain factor knowing.

3.1. The case of knowing only the sign of k_p factor

In order to obtain a stable system the following signals vector is inserted:

$$\mathbf{v}_1 = \begin{bmatrix} \mathbf{v_u} \\ \mathbf{v_y} \\ \mathbf{y_p} \end{bmatrix}, \tag{21}$$

The dynamic filters ($\mathbf{\Lambda}$, $\mathbf{h}$) are placed on the command $\mathbf{v}_u$ and on the output of the process $\mathbf{v}_{yp}$:

$$\begin{cases} \overset{\circ}{\mathbf{v}}_{\mathbf{u}}(t) = \mathbf{\Lambda}\mathbf{v_u}(t) + \mathbf{hu}(t) \\ \overset{\circ}{\mathbf{v}}_{\mathbf{y}}(t) = \mathbf{\Lambda}\mathbf{v_y}(t) + \mathbf{h}\mathbf{y_p}(t) \end{cases} \tag{22}$$

The solution of the dynamical filter is implemented in Matlab as in **Figure 7** (applied only for the first equation).

The ($\mathbf{\Lambda}$, $\mathbf{h}$) pair is chosen in controllable canonical form, $\mathbf{\Lambda} \in \mathfrak{R}^{(n_p-1)\times(n_p-1)}$, $\mathbf{h} \in \mathfrak{R}^{n_p}$, such that:

$$\det(\varepsilon\mathbf{I} - \mathbf{\Lambda}) = N_m(\varepsilon)\cdot\lambda_1(\varepsilon), \tag{23}$$

in which: $\lambda_1(s)$ is an arbitrary Hurwitz polynomial having the degree [7]:

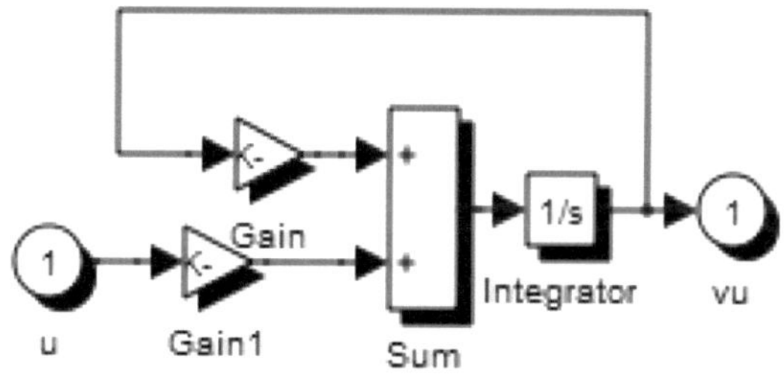

Figure 7. **L-h** ($\mathbf{\Lambda}$, **h**) vu filter block.

$$\partial[N_m(s)\lambda_1(s)] = n_p - 1. \quad (24)$$

Therefore, the $\lambda_1(s)$ polynomial is a design component.

The Hurwitz polynomial L(s) is chosen such that the transfer function $H_m(s)L(s)$ becomes SPR. The degree of the L(s) is $\partial[L(s)] = n^*_m - 1$. If the L(s) is Hurwitz polynomial, then L^{-1} is stable.

The parameters vectors $\mathbf{v}, \xi \in \Re^{2n_p}$ consist of

$$\mathbf{v} = \begin{bmatrix} v_u \\ v_{yp} \\ y_p \\ r \end{bmatrix}, \xi = L^{-1}(s)[\mathbf{v}(t)], \quad (25)$$

The $\mathbf{v}_1$, $\xi_1 \in \Re^{2n_p - 1}$ are defined as follows:

$$\mathbf{v}_1 = \begin{bmatrix} v_u \\ v_{yp} \\ y_p \end{bmatrix}, \quad \xi_1 = \begin{bmatrix} \xi_n \\ \xi_y \\ \xi_p \end{bmatrix}. \quad (26)$$

The auxiliary error is computed on-line:

$$e_a = \theta^T \xi - L^{-1}(s)\left[\theta^T(t)\mathbf{v}(t)\right], \quad (27)$$

where:

$$\xi(t) = L^{-1}(s)[\mathbf{v}(t)]. \quad (28)$$

The augmented error is defined as:

$$e_c = e_0 + H_m(s)L(s)\left[K_1 e_a - \xi_1^T \xi_1 e_c\right], \quad (29)$$

and the on-line gradient adjustable parameter K_1 depends only by the augmented error:

$$\overset{o}{K}_1 = -e_c e_a \quad (30)$$

3.2. The parametric adjustment laws for the compound adaptive control

3.2.1. Gradient

The gradient law [2–4] is expressed as:

$$\overset{o}{\theta}_g = -\gamma_g sign(K_p) e_c \xi(t) \frac{1}{1 + \xi^T \xi} \quad (31)$$

$$e_a = \left(\theta_g + \theta_v\right)^T \xi - L^{-1}(s)\left[\left(\theta_g + \theta_v\right)^T \mathbf{v}(t)\right] \quad (32)$$

$$\xi(t) = L^{-1}(s)[\mathbf{v}(t)]; \text{where } \xi = \begin{bmatrix} \xi_u \\ \xi_{yp} \\ \xi_p \end{bmatrix}. \quad (33)$$

3.2.2. *Deduction of the variable structure parameter* θ_v

The parameter θ_v can be deducted by using the following [2]:

$$\theta_v = \overline{\theta}_v\left[\left(e^{Ke_c\xi} - 1\right)/\left(e^{Ke_c\xi} + 1\right)\right]sign\left(K_p\right) \tag{34}$$

$$\overset{\circ}{\overline{\theta}}_v = -\lambda\overline{\theta}_v - \gamma_v|\xi e_c|. \tag{35}$$

The block diagram of obtaining the augmented error is depicted in **Figure 8**, in which Φ is the vector of the parameter estimation errors

$$\Phi = \theta - \theta^0 \tag{36}$$

The vector of the parameter θ is obtained by using the compound structure: $\theta = \theta_g + \theta_v$.

In the adaptive control, there is a commutation function; usually the *signum* function conducts toward a sliding mode regime such that the evolution to the equilibrium point is very fast. Therefore, the compound adaptive law is used:

$$u(t) = \left(\theta_g + \theta_v\right)^T\mathbf{v} + \left[-\gamma_p sign\left(K_p\right)\xi e_0 + \overset{\circ}{\overline{\theta}}_v\frac{e^{K\xi e_0} - 1}{e^{K\xi e_0} + 1}\right]^T\xi. \tag{37}$$

The adaptive control provides robust characteristics to external disturbances and to unmodelled dynamics.

3.3. The stability of the solution

The perturbation of the dc drive system can leads to the instability of the system. The signals in the variable structure law are bounded. Therefore, the adaptive control assures a global stability [4].

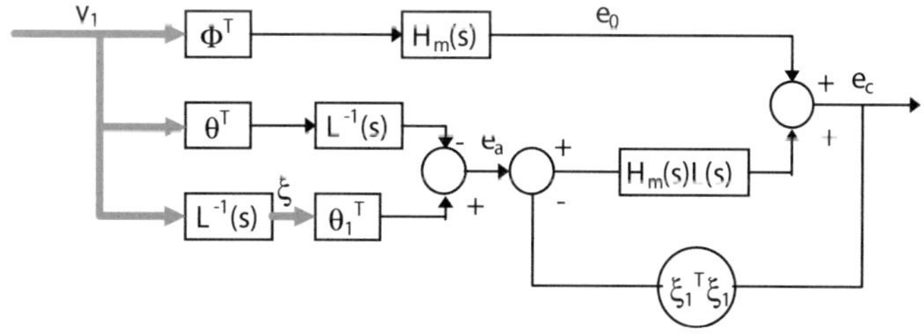

Figure 8. The block diagram of the augmented error determination e_c.

4. Numerical simulation results

Taken into account the dc machine from the conventional control, operating at the constant flux, a speed cycle is applied in order to test the compound adaptive control. The speed cycle contains the dynamic regimes (starting, braking, reversing) and the steady state regime (**Figures 9–22**).

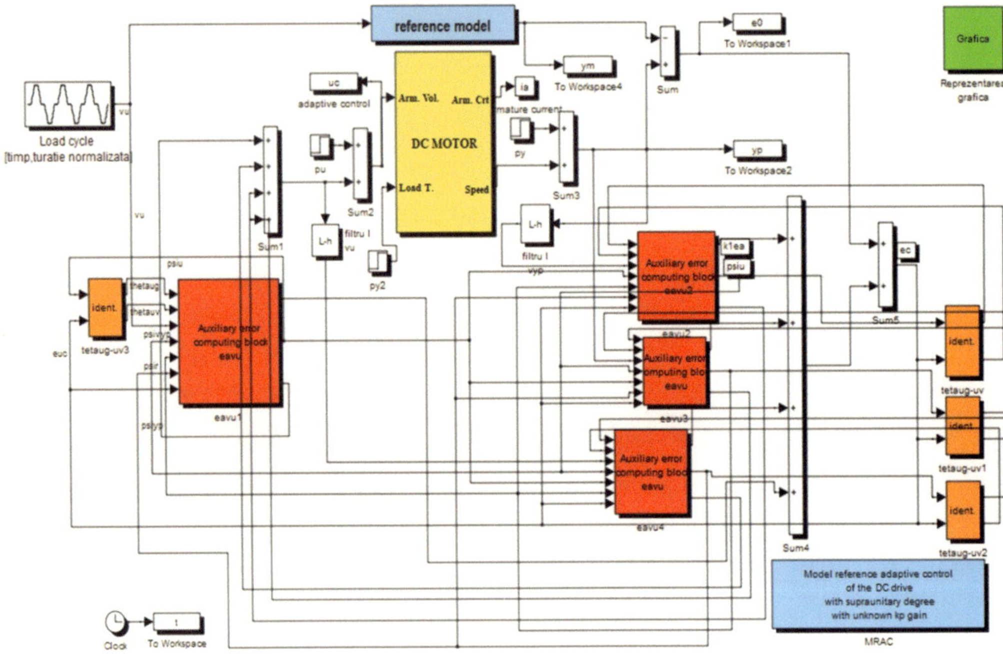

Figure 9. The Simulink implementation of the compound adaptive dc drive with supraunitary degree and with unknown gain.

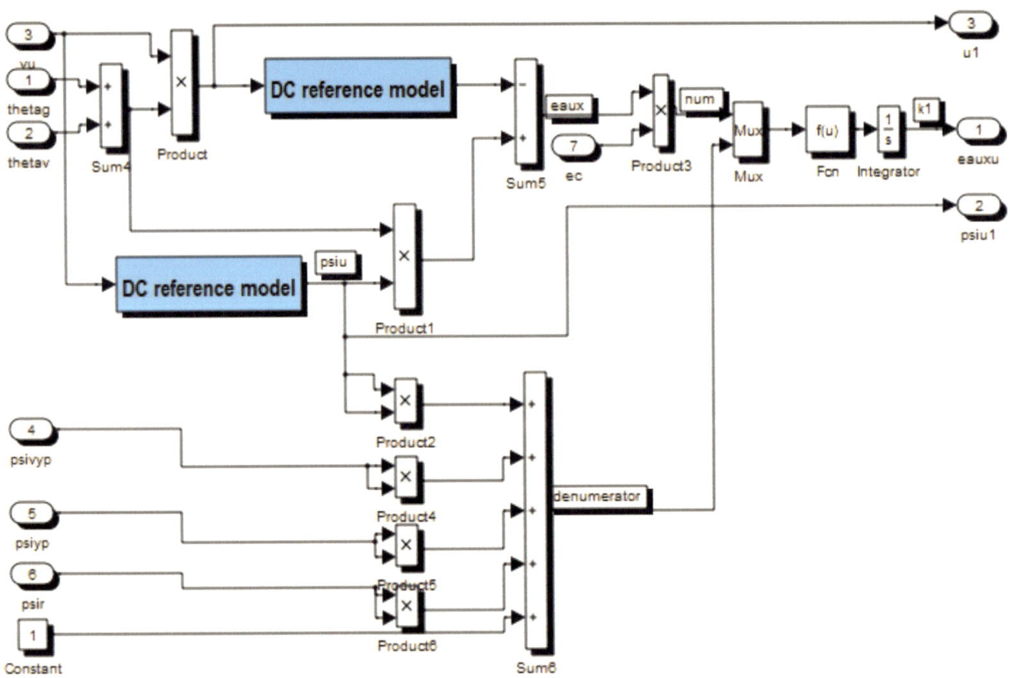

Figure 10. The Simulink implementation of the augmented error deduction *ea_vu*.

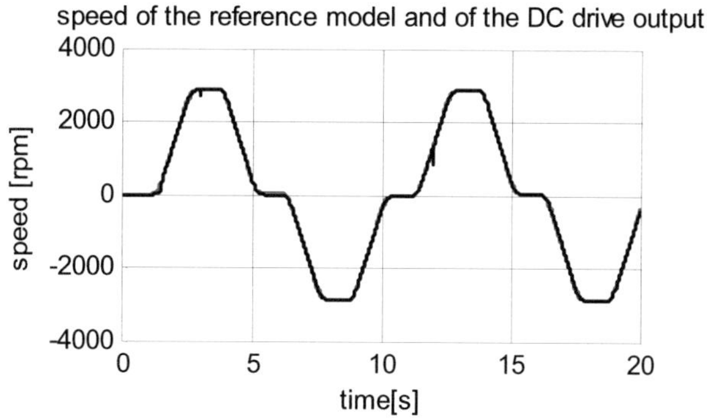

Figure 11. The speed cycle. At $t = 12$ s, the rated load torque is applied.

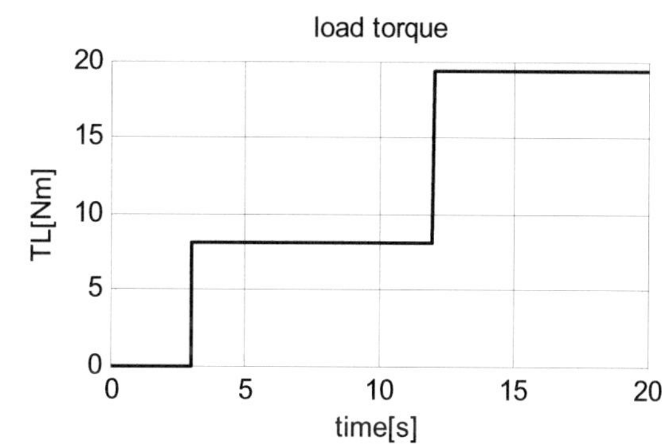

Figure 12. The load torque. At $t = 12$ s, the rated load torque is applied.

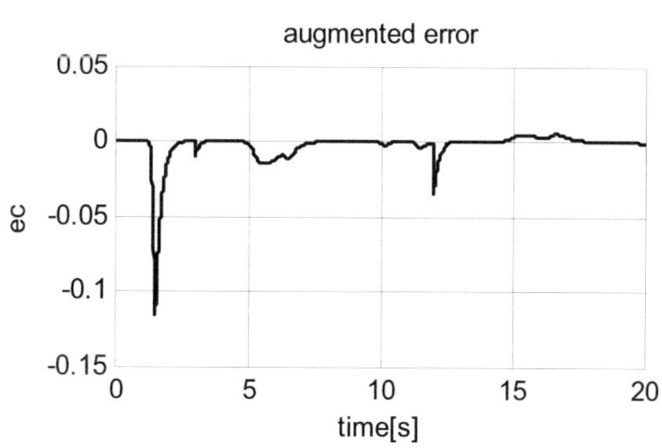

Figure 13. The augmented error. At $t = 12$ s, the rated load torque is applied.

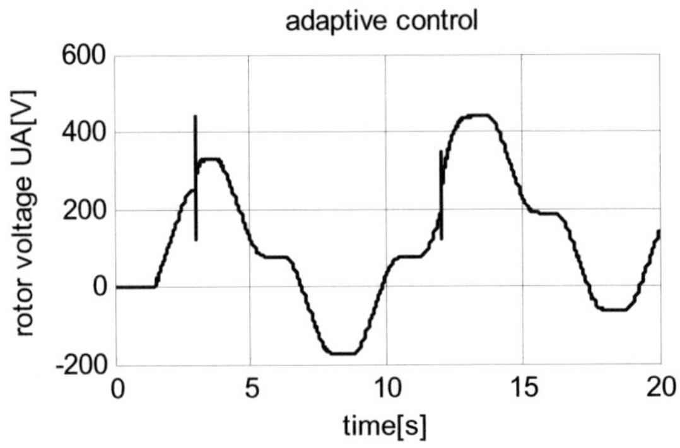

Figure 14. The adaptive control. At $t = 12$ s, the rated load torque is applied.

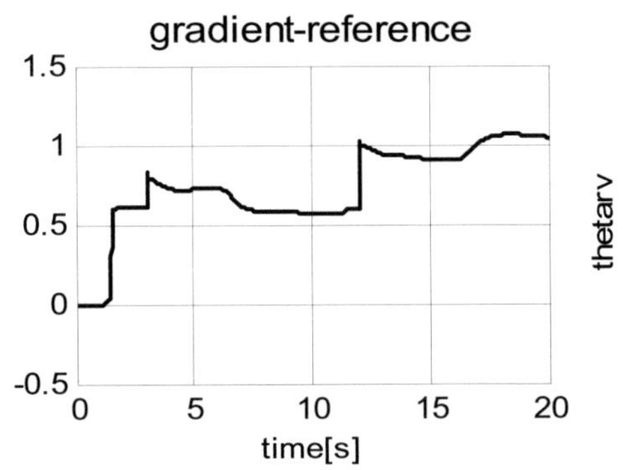

Figure 15. The adaptive mechanism of the parameter vector θ obtained by using the compound structure: θ_{rg} gradient reference.

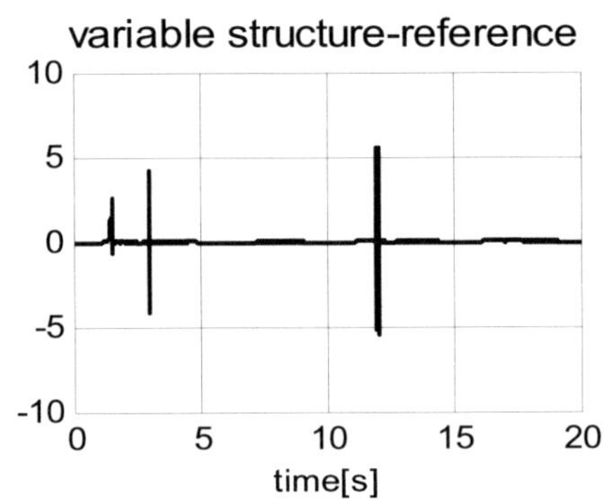

Figure 16. The adaptive mechanism of the parameter vector θ obtained by using the compound structure: θ_{rv} variable structure.

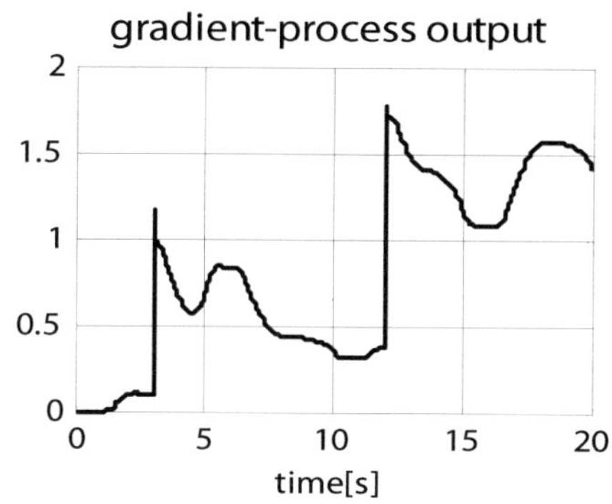

Figure 17. The adaptive mechanism of the parameter vector θ obtained by using the compound structure: θ_{yg} gradient-process output.

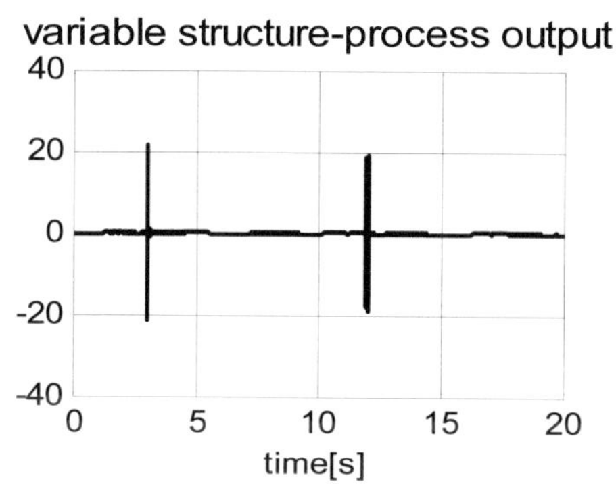

Figure 18. The adaptive mechanism of the parameter vector θ obtained by using the compound structure: θ_{yv} variable structure process output.

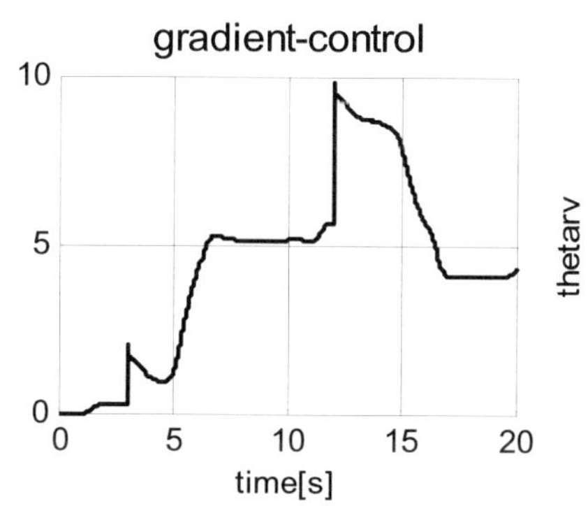

Figure 19. The adaptive mechanism of the parameter vector θ obtained by using the compound structure: θ_{ug} gradient control.

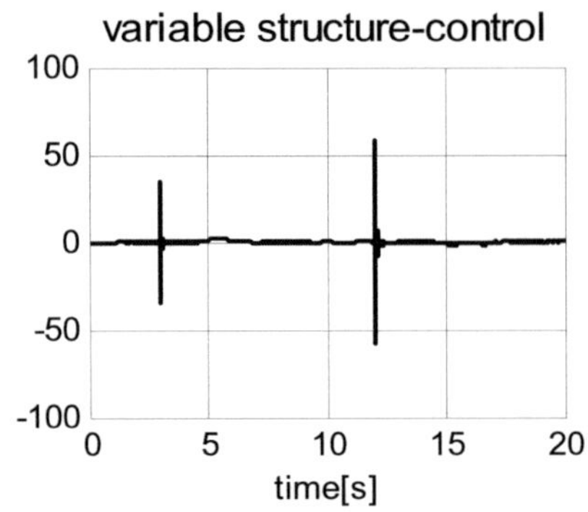

Figure 20. The adaptive mechanism of the parameter vector θ obtained by using the compound structure: θ_{uv} variable structure control.

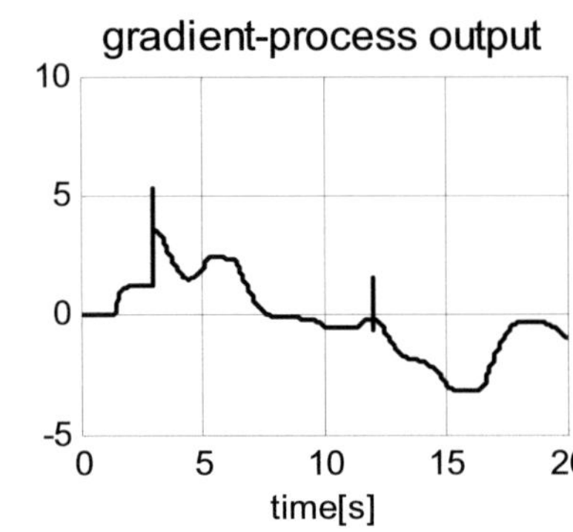

Figure 21. The adaptive mechanism of the parameter vector θ obtained by using the compound structure: θ_{pg} gradient-process output.

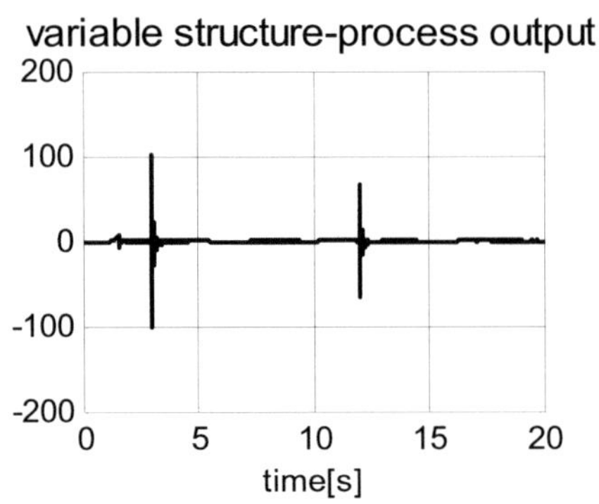

Figure 22. The adaptive mechanism of the parameter vector θ obtained by using the compound structure: θ_{pv} variable structure process output.

5. Conclusions

The conventional control and the compound adaptive control have been investigated by using a dc drive system. The complete methodology of tuning the controller parameters for the conventional control is provided. Under the assumptions mentioned in this chapter, the dc drive system has been implemented in Matlab-Simulink software. The adequate numerical simulation results have been obtained (**Figure 6**) for. Therefore, the regulation capability of the PI controller is tested under a variation of the load torque for a starting. Both, the dynamic and steady state regimes are investigated. In **Figure 6b**, the maximum load torque of the dc motor is used for a starting under the 70% load conditions. The maximum limit is maintained during the dynamic regime, the torque decreases in steady state due to the armature current decreasing. The constant parameter values have been considered. In case of the gradient and variable structure laws, the adaptive system is more robust to parameter uncertain or to unmodelled dynamics of the dc drive, and the increased regulation performances are obtained. The supraunitary relative degree model reference adaptive control has been considered.

Acknowledgements

This work was supported by a grant of the Romanian National Authority for Scientific Research, CNDI–UEFISCDI, project number PN-II-PT-PCCA-2011-3.2-1680.

Author details

Marian Găiceanu

Address all correspondence to: marian.gaiceanu@ieee.org

Integrated Energy Conversion Systems and Advanced Control of Complex Processes Research Center, Dunarea de Jos University of Galati, Romania

References

[1] Didactical Pedagogical House, editor. Automatizări electronice. 1993rd ed. Bucuresti: Dumitrache, I.; 1993

[2] Filipescu A. Robustness in compound adaptive controlling with sigmoid function for supraunitary relative degree of the plant model. In: The 9'th Symposium on Modelling and Identification systems, SIMSIS9'96; 1996; Galati. Romania; 1996. pp. 45-52

[3] Gaiceanu M. Embedded control of the DC drive system for education. In: Yildirim S, editor. Design, Control and Applications of Mechatronic Systems in Engineering. Croatia: InTechOpen; 2017. DOI: 10.5772/67461. Available from: https://www.intechopen.com/

books/design-control-and-applications-of-mechatronic-systems-in-engineering/embedded-control-of-the-dc-drive-system-for-education

[4] Narendra KS, Annaswamy AM, editors. Stable Adaptive Systems. 1989th ed. Englewood Cliffs, NJ: Prentice Hall; 1989

[5] Gaiceanu M, Solea R, Codres B, Eni C. Efficient DC drive system by using adaptive control. In: Book Group Author(s): IEEE Conference: International Conference on Optimization of Electrical and Electronic Equipment (OPTIM) Location: ROMANIA Date: MAY 22-24, 2014 Sponsor(s): IEEE Ind Elect Soc; IEEE Ind Applicat Soc; IEEE Power Elect Soc; Transilvania Univ Brasov, 2014 INTERNATIONAL CONFERENCE ON OPTIMIZATION OF ELECTRICAL AND ELECTRONIC EQUIPMENT (OPTIM); MAY 22-24, 2014; Brasov. Romania: IEEE; 2014. pp. 381-388

[6] Gaiceanu M, Eni C, Coman M. The model reference adaptive control of the DC electric drive system. Advanced Materials Research. 2014;**875-877**:2030-2035

[7] Filipescu A. Two new adjustment laws for variable structure and compound robust adaptive control. In: SIMSIS9'96; 24-25 Oct 1996; Galati. pp. 53-60

Chapter 8

Model Reference Adaptive Control of Quadrotor UAVs: A Neural Network Perspective

Nikhil Angad Bakshi

Additional information is available at the end of the chapter

http://dx.doi.org/10.5772/intechopen.71487

Abstract

Adaptive models and online learning are two equivalent topics under different umbrellas of research – control systems and machine learning. This chapter will tackle one such application of a neural network-based model reference adaptive controller on a quadrotor unmanned aerial vehicle while stating the general principles behind each design decision so the knowledge can be generalized to other practical applications. The application-oriented presentation of this chapter will run parallel to most research and development processes in the field, where the physical system or a simulator is usually available and a simple control system (such as PID) has already been implemented as a baseline. The black-box nature of a neural network can truly be leveraged to improve performance after reading this chapter. Several practical considerations when approaching such a problem have been discussed together with their general and implemented solutions. The simulation results for the problem have been presented to demonstrate the success of this control strategy.

Keywords: model reference adaptive control, neural networks, UAV, quadrotor, machine learning, robotics, online learning, MLP networks

1. Introduction

Artificial intelligence is a term that, very paradoxically, holds a prerequisite for the absence of intelligence, which the designer must then overcome. Intelligence can be viewed as the ability of a system to *perceive* its environment, *reason* upon the acquired *knowledge* and perform an *action* or task based on this information to meet its *objective*.

When the possible states of the environment are predictable the designer can create an intelligent system that *performs* well for all possible situations. However the world is a messy place and more often than not the environment is unpredictable and knowing or encoding such information into the system a priori is impractical.

The set of *actions* of a system and its *objective*, on the other hand, is usually known a priori, so it logically follows that one should design a system that should be able to *learn* how to deal with new situations to meet its *objective* given the limited set of *actions*.

Formally, the system or agent should improve its performance as measured by a metric (P) on a task (T) with increasing experience (E). This brings us to *machine learning* (ML).

At this point, let us note that an *adaptable control system* is one that modifies the *control law* so that the system remains stable and the *control objective* is met.

Whether one looks at it from the perspective of ML, in that the system is initially poor at meeting the objective and hence it changes system parameters to improve or from the control theory perspective that the environment or system has changed and the control objective is not met demanding a change in the control law, we are describing a similar situation.

This chapter is written using the *attitude* and *altitude* controller of a quadrotor unmanned aerial vehicle (UAV) as a running example however every idea will be presented generally at first and then tied back to the practical example in consideration.

2. Quadrotor system

A quadrotor UAV (refer **Figure 1**) has four motors that independently provide thrust (as indicated by F_1, F_2, F_3 *and* F_4) and based on these thrusts the UAV can change its *attitude* (roll ϕ, pitch θ and yaw ψ) and *altitude* (z). This chapter will focus on the inner loop control of the *attitude* and *altitude* variables and is based on the work of Bakshi and Ramachandran [1].

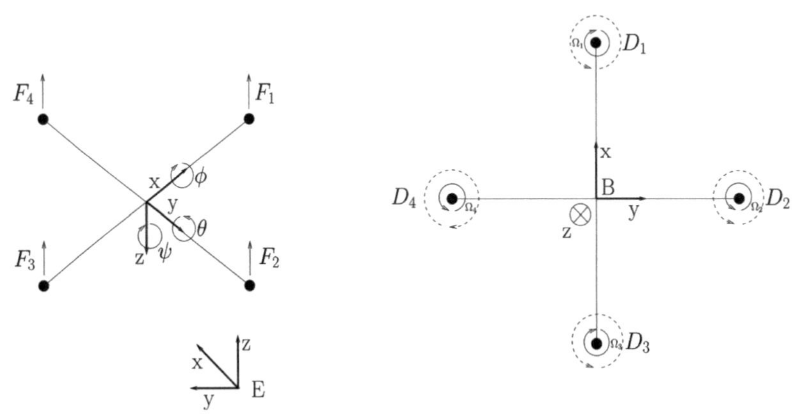

Figure 1. Quadrotor coordinate system [2].

The horizontal coordinates of the UAV (x, y) can be controlled using an outer loop on this inner loop but that is not covered here, we shall assume that the environment of the quadrotor is boundless in all directions except the datum in z, i.e. the ground.

The quadrotor model that has been used is based on work done by Bouabdallah et al. [2]. Following Bouabdallah, the earth-fixed frame E and the body-fixed frame B are as seen in **Figure 1**. The aerodynamic forces and moments considered in the model are based on the work of Gary Fay as in Ref. [3].

A dynamic model for the quadrotor is used for the purpose of simulation in this chapter so that the control strategy that has been presented here can be thoroughly evaluated.

2.1. Quadrotor parameters

The quadrotor parameters are based on the work of Bouabdallah [4] (**Table 1**).

Design Variable	Value	Units	Description
m	0.53	kg	Mass
L	0.23	m	Craft diameter
J_r	6×10^{-5}	$kg.m^2$	Rotor inertia
I_{xx}	6.23×10^{-3}	$kg.m^2$	Moment of inertia along x
I_{yy}	6.23×10^{-3}	$kg.m^2$	Moment of inertia along y
I_{zz}	1.12×10^{-2}	$kg.m^2$	Moment of inertia along z
b	3.13×10^{-5}	$N.s^2$	Thrust factor in hover
d	7.50×10^{-5}	$N.s^2$	Drag factor in hover
N	2	—	Number of blades
R	0.15	m	Propeller radius
c	0.04	m	Chord
θ_0	0.26	rad	Pitch of incidence
θ_{tw}	0.05	rad	Twist pitch
a	5.70	—	Lift slope
C_d	0.05	—	Airfoil drag coefficient
A_c	0.01	m^2	Helicopter center hub area
ρ	1.29	kg/m^3	Air density
ϑ	1.80×10^{-5}	$Pa.s$	Air viscosity
V	3.04×10^{-4}	m^3	Volume

Table 1. Quadrotor parameters [4].

2.2. Moments and forces on the quadrotor system

See **Table 2**.

Rolling moments	Body gyro effect	$\dot{\theta}\dot{\psi}(I_{yy} - I_{zz})$
	Propeller gyro effect	$J_r\dot{\theta}\Omega_r$
	Roll actuators action	$l(-T_2 + T_4)$
	Hub moment due to sideward flight	$h\left(\sum_{i=1}^{4} H_{yi}\right)$
	Rolling moment due to forward flight	$(-1)^{i+1}\sum_{i=1}^{4} R_{mxi}$
Pitching moments	Body gyro effect	$\dot{\varphi}\dot{\psi}(I_{zz} - I_{xx})$
	Propeller gyro effect	$J_r\dot{\varphi}\Omega_r$
	Pitch actuators action	$l(T_1 - T_3)$
	Hub moment due to forward flight	$h\left(\sum_{i=1}^{4} H_{xi}\right)$
	Rolling moment due to sideward flight	$(-1)^{i+1}\sum_{i=1}^{4} R_{myi}$
Yawing moments	Body gyro effect	$\dot{\varphi}\dot{\theta}(I_{xx} - I_{yy})$
	Inertial counter-torque	$J_r\dot{\Omega}_r$
	Counter-torque unbalance	$(-1)^{i}\sum_{i=1}^{4} Q_i$
	Hub force unbalance in forward flight	$l(H_{x2} - H_{x4})$
	Hub force unbalance in sideward flight	$l(-H_{y1} + H_{y3})$
Forces along z axis	Actuators action	$c\psi c\varphi\sum_{i=1}^{4} T_i$
	Weight	mg
Forces along x axis	Actuators action	$c\psi c\varphi\sum_{i=1}^{4} T_i$
	Hub force in x axis	$-\sum_{i=1}^{4} H_{xi}$
	Friction	$\frac{1}{2}C_x A_c \rho\dot{x}\lvert\dot{x}\rvert$
Forces along y axis	Actuators action	$(-c\psi s\varphi + s\psi s\theta c\varphi)\left(\sum_{i=1}^{4} T_i\right)$
	Hub force in y axis	$-\sum_{i=1}^{4} H_{yi}$
	Friction	$\frac{1}{2}C_y A_c \rho\dot{y}\lvert\dot{y}\rvert$

Table 2. Quadrotor moments and forces summary [4].

2.3. Equations of motion

$$I_{xx}\ddot{\varphi} - \dot{\theta}\dot{\psi}\left(I_{yy} - I_{zz}\right) + J_r\dot{\theta}\Omega_r + l(-T_2 + T_4) - h\left(\sum_{i=1}^{4} H_{yi}\right) + (-1)^{i+1}\sum_{i=1}^{4} R_{mxi} \tag{1}$$

$$I_{yy}\ddot{\theta} - \dot{\varphi}\dot{\psi}\,(I_{zz} - I_{xx}) - J_r\dot{\varphi}\Omega_r + l(T_1 - T_3) - h\left(\sum_{i=1}^{4} H_{xi}\right) + (-1)^{i+1}\sum_{i=1}^{4} R_{myi} \tag{2}$$

$$I_{zz}\ddot{\psi} - \dot{\theta}\dot{\varphi}\left(I_{xx} - I_{yy}\right) + J_r\dot{\Omega}_r + (-1)^i\left(\sum_{i=1}^{4} Q_i\right) + l(H_{x2} - H_{x4}) + \ l\left(-H_{y1} + H_{y3}\right) \tag{3}$$

$$m\ddot{z} - mg - (c\psi c\varphi)\sum_{i=1}^{4} T_i \tag{4}$$

$$m\ddot{x} - (s\psi s\varphi + c\psi s\theta c\varphi)\sum_{i=1}^{4} T_i - \sum_{i=1}^{4} H_{xi} - \frac{1}{2}C_x A_c \rho \dot{x}|\dot{x}| \tag{5}$$

$$m\ddot{y} - (-c\psi s\varphi + s\psi s\theta c\varphi)\sum_{i=1}^{4} T_i - \sum_{i=1}^{4} H_{yi} - \frac{1}{2}C_y A_c \rho \dot{y}|\dot{y}| \tag{6}$$

2.4. Rotor dynamics

A first order closed loop transfer function is used to reproduce the dynamics between the propeller's speed set point and its true speed as in Ref. [2].

$$G(s) = \frac{0.936}{0.178s + 1} \tag{7}$$

2.5. Summary of model

The following is a breakdown of the basic understanding of a quadrotor UAV system required for the purposes of this chapter:

a. The sum of the four thrusts (as indicated by F_1, F_2, F_3 *and* F_4 in **Figure 1**) along the z axis is responsible for countering the weight of the craft. Any surplus or shortage of the total thrust along z will result in motion in the z-direction.

b. An imbalance in the forces F_2 *and* F_4 will result in a rolling motion along the x-axis, similarly imbalance in F_1 *and* F_3 will result in a pitching motion along the y-axis. Note that the very act of rolling or pitching tilts the craft such that the motor thrusts are no longer effected purely in the z-direction, causing the craft to descend, therefore the total thrust must be summarily increased so that the component along z is maintained.

c. The rotations of motors 1 and 3 are in the same direction and the opposite of motors 2 and 4. To achieve a yawing motion increased thrust must be applied to a diametrically opposite pair (say 1 and 3) and reduced thrust on the other pair (2 and 4).

d. Various second order effects come into play, which have been modeled but understanding them is not crucial to this chapter.

e. This model dynamically calculates the thrust and drag coefficients which results in increased accuracy with the real world scenario.

3. Objective

The basic inner loop controller of any helicopter deals with maintaining a specified height above ground, i.e. *altitude*, and maintaining a particular pose, or *attitude*. The *attitude* in turn allows the helicopter to translate in the x-y plane assuming *altitude* is held constant.

The standard approach is decentralized or cascaded PID controllers for the various control variables (in this case: roll ϕ, pitch θ, yaw ψ, altitude z), these controllers will have to be tuned for each particular quadrotor UAV. In general, any non-adaptive controller will need to be tuned to a particular quadrotor.

In this chapter we employ neural networks to design an adaptive controller that is system unspecific, i.e. it should work for any quadrotor system. It learns the system parameters online, i.e. in-flight. The challenge is to keep the system stable during online learning.

4. Indirect model reference adaptive control

Indirect adaptive control is when the controller estimates the plant to predict its state variables and these are used to modify the controller parameters.

Direct adaptive control is when there is no plant identification, instead the controller parameters are modified on the basis of the error the system has with the reference.

MRAC is a direct adaptive control method (refer **Figure 2**), however in this chapter we shall be taking a mixed approach to the problem.

In MRAC we define a reference model that responds to the input signal (r) as we would like our plant to respond. The controller generates a control signal (u) based on its control law which it expects will make the plant output (y) follow the reference output (y_{ref}). Depending on the deviation (or error), the adjustment mechanism will update the control law (by modifying parameters) of the controller. This process is repeated iteratively so that the plant follows the reference.

The beauty of the approach taken in this chapter is that we needn't formalize the control logic. We will delve deeper into neural networks before returning to the problem at hand. For the

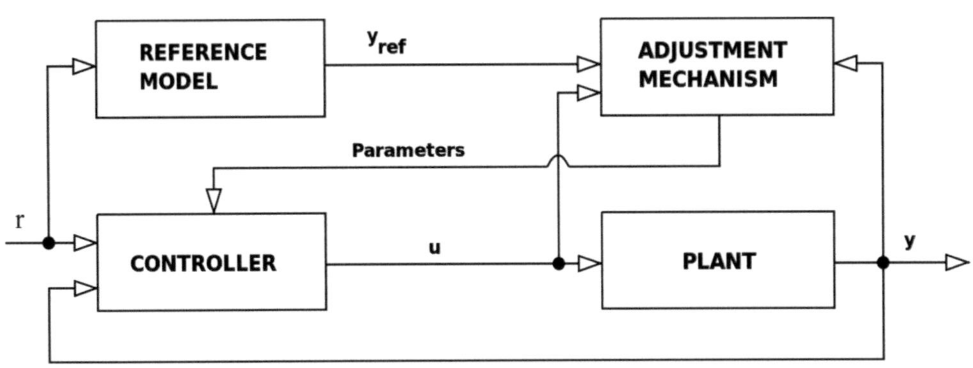

Figure 2. MRAC block diagram. Image courtesy of Wikipedia under CC license.

time being we will leave the 'reference model, 'adjustment mechanism' and the 'controller' as black boxes, they will be revisited in Section 6.

5. Neural networks

5.1. Introduction to machine learning

A formal definition of ML is:

> *A computer program is said to learn from experience E with respect to some task T and some performance measure P, if its performance on T, as measured by P, improves with experience E.*
> *-Tom Mitchell*[1] *(1998)*

Neural networks are one such machine learning algorithm. This sub-section will briefly cover the two broad categories of machine learning algorithms. Bear in mind that this chapter will elaborate on neural networks used in a *supervised learning* setting.

5.1.1. Supervised learning

In supervised learning, the data is tagged with the correct values for prediction/classification associated to it. The algorithm learns by minimizing the error between its results and the correct results. This is the most common form of machine learning and the easiest, however labeled data is not easy to come by as its curation and tagging is usually expensive.

Neural networks, support vector machines (SVMs), linear/logistic regression and decision trees are a few examples of supervised learning algorithms and the applications could be classification of images, weather prediction, sentiment detection, face recognition, etc.

5.1.2. Unsupervised learning

In unsupervised learning, the task is to find patterns or meaning in untagged data such as classifying similar data together without actually knowing what those classes may represent (clustering) or we take data in some low level/uncompressed representation and learn a high level/compressed representation with minimum information loss or we have a lot of data which mostly subscribes to a particular pattern and we would like to detect the outliers (anomaly detection).

K-means clustering, autoencoders (NN based) and principal component analysis are a few algorithms used for unsupervised learning tasks.

5.2. History and intuition of neural networks

In the twentieth century scientists were able to definitively identify the primary unit of the brain – the neuron. One theory of the time was that information is not pre-loaded in the brain of a newly born child, only the basic structure and connections between the neurons in its brain exist, the brain learns to function by strengthening/weakening various neural pathways.

[1]Machine Learning, Tom Mitchell, McGraw Hill, 1997.

Therefore it was theorized that the neuron, which either fires or does not, can be modeled as a function that has multiple inputs, a single output (which may be connected to several other neurons) and only 'fires' when the sum of inputs exceeds a certain threshold. The connections between neurons have weights, which strengthen or weaken a connection between two neurons.

> *"Neurons that wire together, fire together"*
>
> *-Donald Hebb (1949)*

The above quotation may be familiar. It is based on Donald Hebb's theory to explain neural adaptation and learning and this forms the basis of learning in modern-day artificial neural networks.

5.3. Formalization

As shown in **Figure 3**, a basic neural network consists of layers of nodes (or neurons) where each node has a connection to all the nodes of the next layer, and takes input from each of the nodes in the previous layer. Each of the connections has a real number weight associated with it. Every neuron does some simple computation (we will restrict ourselves to the sigmoid activation and linear activation) on the sum of its inputs to yield an output value.

The above definition is that of multilayer perceptron (MLP) network which is the most basic form of a feed forward neural network.

Let us assume that $\boldsymbol{x}$ is the input (vector) for this neural network, the dimension of $\boldsymbol{x}$ is 6×1. Let the weight of the connections between the input layer and the first hidden layer be represented by the matrix $\theta^{(1)}$, each value in the matrix will be referenced as $\theta_{ij}^{(1)}$. The dimension of $\theta^{(1)}$ is 6×4.

$$\sigma(z) = \frac{1}{1 + e^{-z}} \tag{8}$$

Eq. (8) shows us the sigmoid activation function that will be applied on every node in the hidden layers. If z is a vector then applying the sigmoid activation function will result in a

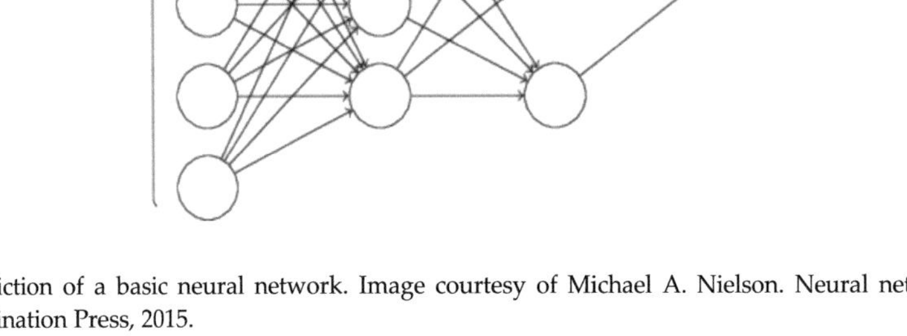

Figure 3. A depiction of a basic neural network. Image courtesy of Michael A. Nielson. Neural networks and deep learning. Determination Press, 2015.

http://dx.doi.org/10.5772/intechopen.71487

vector of the same dimension. Sigmoid function is a continuous differentiable function, which is bounded between 0 and 1.

$$\mathbf{a}^{(1)} = \sigma\left(\left(\theta^{(1)}\right)^T \boldsymbol{x} + \boldsymbol{b}^{(1)}\right) \tag{9}$$

Eq. (9) shows us the first step in *forward propagation.* $\mathbf{a}^{(1)}$ is known as the activation of the first hidden layer. As a sanity-check we can see the dimension of $\mathbf{a}^{(1)}$ is 4×1, which is congruent with what we expect.

Neural networks have a bias unit (not shown in **Figure 3**), which is a neuron that is always firing and is connected to every node in the next layer but does not take input from the previous layer. Mathematically it can be represented as the additive term $\boldsymbol{b}^{(1)}$ of dimension 4×1 shown in Eq. (9).

$$\mathbf{a}^{(2)} = \sigma\left(\left(\theta^{(2)}\right)^T \mathbf{a}^{(1)} + \boldsymbol{b}^{(2)}\right) \tag{10}$$

$$y_{predicted} = \boldsymbol{a}^{(3)} = \mathbf{z}^{(3)} = \left(\theta^{(3)}\right)^T \mathbf{a}^{(2)} + \boldsymbol{b}^{(3)} \tag{11}$$

Eqs. (10) and (11) complete the forward propagation. Eq. (11) can include a sigmoid activation too if the desired output is between 0 and 1 like in a classification problem. No activation function is also known as linear activation.

The neural network learns by optimizing an objective function (or cost function) such as the squared-error cost function for dealing with regression problems as in this text.

$$J(\theta) = \frac{1}{2m} \sum_{i=1}^{m} \left({y_{predicted}}^{(i)} - \boldsymbol{y}^{(i)} \right)^2 \tag{12}$$

where $\boldsymbol{y}^{(i)}$ is vector of the target values of output layer. In our example it is 1×1 but in general the neural network can have multiple outputs. In an offline setting we have all our data beforehand, indicated here by m examples and we compute cost iterated over all $\boldsymbol{y}^{(i)}$.

As seen in **Figure 4** the sigmoid function is approximately linear between −2 and 2 and is almost a flat line beyond −4 and 4. This has 3 implications:

1. It gives us a convenient differentiable function that is nearly binary
2. If the input of a neuron is too extreme the neuron becomes saturated and therefore information gets attenuated going through the network if it is too deep.
3. In the case of saturated nodes, backpropagating, which involves computing the gradient, becomes ineffective, as the gradient at the extremes is nearly zero.

Due to the latter two points it is beneficial to ensure that the weights of the network are small. If the weights have large values then the network is sure those connections are very important which makes it hard for it to learn otherwise. Therefore, it makes sense to keep weights small so the network is responsive to new information. To accomplish this we incorporate the

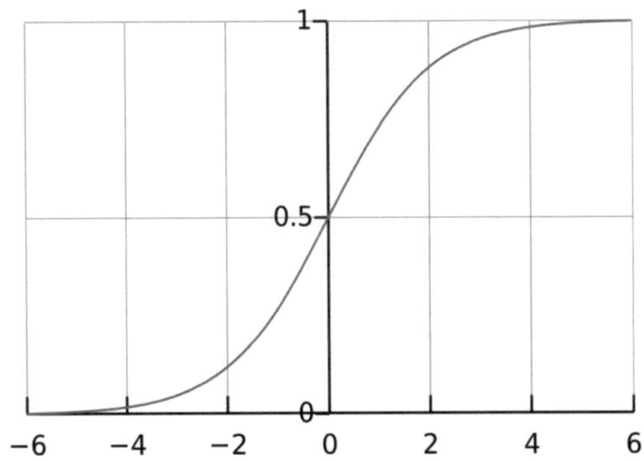

Figure 4. Graph of sigmoid function.

weights into the cost function. In L1 *regularization* we add the modulus of the weights to the cost function in Eq. (12). In L2 regularization we add the squares of the weights to the cost function resulting in Eq. (13).

$$J(\theta) = \frac{1}{2m}\sum_{i=1}^{m}\left(y_{predicted}{}^{(i)} - y^{(i)}\right)^2 + \frac{\lambda}{2}\cdot\sum_{l=1}^{L-1}\sum_{i=1}^{s_l}\sum_{j=1}^{s_{l+1}}\left(\theta_{ij}^{(l)}\right)^2 \tag{13}$$

where L is the total number of layers in the neural network, s_l is the number of nodes in the lth layer. Note that regularization is not done on the weights from the bias node. λ is the regularization parameter that helps us control the extent to which we want to penalize the neural network for the size of its weights.

Gradient descent is used to train the neural network, which means running several iterations making small adjustments to the parameters θ in the direction that minimizes the cost function. The weight update equation is shown in Eq. (14).

$$\theta^{(l)} := \theta^{(l)} - \alpha.\frac{\delta J}{\delta\theta^{(l)}}, l = 1, \ldots, L - 1 \tag{14}$$

where α is the learning rate that controls the size of the adjustment, it must not be too small, else the learning will take very long, and it must not be too large, else the network will not converge. $\theta^{(l)}$ is a matrix and therefore the derivative term is also a matrix.

The *backpropagation* algorithm is used to calculate the derivative term. The intuition is to calculate the error term at every node in the network from the output layer backwards and use this to compute the derivative. We shall denote the error vector as $\boldsymbol{\delta}^{(l)}$ where l denotes the layer number. Eq. (15) denotes the error in the output layer of our example network.

$$\boldsymbol{\delta}^{(4)} = \boldsymbol{y} - \boldsymbol{a}^{(3)} \tag{15}$$

Except for the output layer the error term is defined as:

$$\boldsymbol{\delta}^{(l)} = \left(\theta^{(l)}\right)^{T} \boldsymbol{\delta}^{(l+1)} \diamond \sigma'\left(\boldsymbol{a}^{(3)}\right), \quad l \in \{2, \ldots, L-1\} \tag{16}$$

where $\sigma'(.)$ denotes the derivative of the sigmoid function, $\diamond$ signifies an element-wise multiplication and L is the total number of layers in the network, here $L = 4$. Note that $\boldsymbol{\delta}^{(1)}$ can be calculated but error on our inputs does not have any significance.

$$\sigma'\left(\boldsymbol{a}^{(3)}\right) = \sigma\left(\boldsymbol{a}^{(3)}\right) \diamond \left(\mathbf{1} - \sigma\left(\boldsymbol{a}^{(3)}\right)\right) \tag{17}$$

Eq. (18) is the derivative of the cost function (without the regularization term) computed using the errors previously found. The mathematical proof[2] of backpropagation is beyond the scope of this chapter. The final gradient averaged over all training examples with the regularization term is Eq. (19)

$$\frac{\delta J}{\delta \theta^{(l)}} = \boldsymbol{\delta}^{(l+1)} \left(\boldsymbol{a}^{(l)}\right)^{T} \tag{18}$$

$$\frac{\delta J}{\delta \theta^{(l)}} = \frac{1}{m} \cdot \sum\nolimits_{i=1}^{m} \left(\boldsymbol{\delta}^{(l+1)} \left(\boldsymbol{a}^{(l)}\right)^{T}\right) + \lambda \cdot \theta^{(l)} \tag{19}$$

The weights of the network are *randomly initialized* to small positive or negative real values. If one were to initialize all the weights to the same value (say zero) then the gradient calculated at every node in a layer would be the same, and we'd end up with a neural network with lots of redundancies. Note that if there were no hidden layers this would not be the case but the power of the algorithm significantly goes down without them.

5.4. Limitations

Neural networks have large time and space requirements. Assume an n hidden layer fully connected neural network with m neurons in each hidden layer. We have $(n-1) \times m \times m$ parameters just in the hidden layers. This number is for the basic MLP network and more sophisticated implementations (*deep learning*) will have even more parameters.

For example, ILSVRC[3] evaluates algorithms for object detection and image classification at large scale to measure the progress of computer vision for large scale image indexing for retrieval and annotation. Over the years, deep neural networks have been used to solve the problem statement with better accuracy every year. AlexNet [5] had 60 million parameters and took two weeks to train on 2 GPUs in 2012 with 16.4% classification error using *convolutional neural networks*. GoogLeNet [6] had 4 million parameters achieving classification error of 6.66% in 2014 with the advent of their inception module in the *convolutional neural network*. So, even as the situation continues to improve, neural networks are still time and memory intensive.

---

[2] The concise proof can be found in chapter 2 of the book by Michael A. Nielson, "Neural Networks and Deep Learning", Determination Press, 2015

[3] http://www.image-net.org/challenges/LSVRC/

The second problem is predictability. Just as the human brain is an enigma that man has been trying to understand and find patterns in, the fact remains, humans are still unpredictable to quite an extent. Similarly, as far as machine learning algorithms go, neural networks are a black box and no one fully understands the functions that have been mapped in them once trained. Therefore no one can predict when they might fail and it is not always possible to justify the results produced as opposed to a rule-based classification method such as *decision trees*. Yet, neural networks have proved to be a great tool and are widely used by organizations today.

6. System design

Figure 5 depicts the block diagram of our system. The feedback is implied through the conglomerate controller/plant emulator artificial neural network (ANN) using the backpropagation algorithm.

The plant (quadrotor) block is simulated based on the model described in Section 2.

Two errors are generated:

$$\text{Model error: } y_{plant} - y_{model} \tag{20}$$

$$\text{Reference error: } y_{reference} - y_{plant} \approx y_{reference} - y_{model} \tag{21}$$

When approaching convergence, both errors tend to zero and the approximation in the reference error becomes increasingly accurate. The extended ANN has two functions – making an adaptive estimate of the next state of the plant given the current state and computing the control signals required by the plant to minimize reference error.

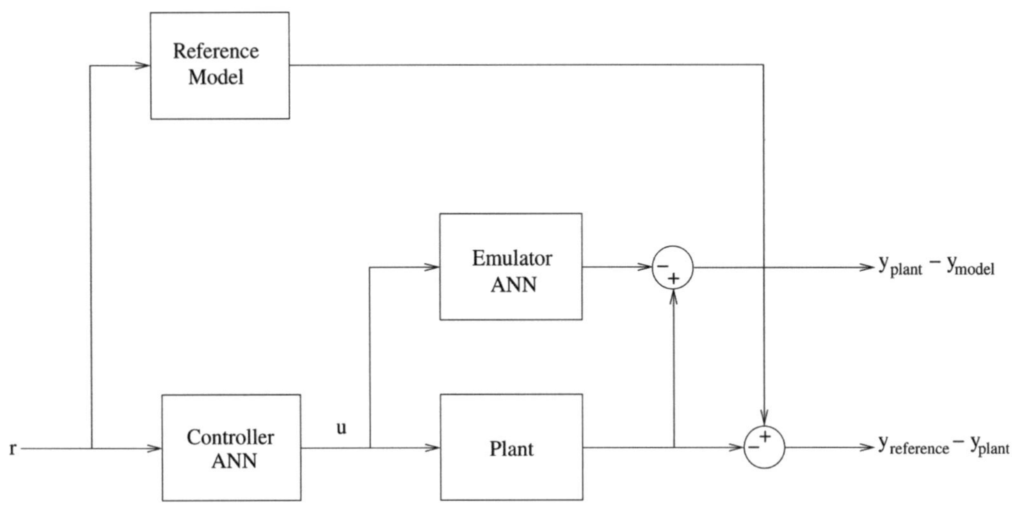

Figure 5. Block diagram of model reference adaptive control using artificial neural networks.

6.1. Selecting the reference model

The reference model consists of transfer functions for roll, pitch, yaw and z of the quadrotor. The transfer functions are as follows:

$$\text{roll and pitch: } G(s) = \frac{0.5}{s + 0.5} \tag{22}$$

$$\text{yaw: } G(s) = \frac{0.5}{s + 0.5} \tag{23}$$

$$\text{z: } G(s) = \frac{1}{s + 1} \tag{24}$$

The transfer functions are chosen to have a quick response to changing set points. Quadrotors are non-linear systems, yet we model the reference using first order transfer functions because we hold machine learning algorithms to the standard of an expert human operator. An expert helicopter pilot would not oscillate to attain a target *altitude* or *attitude* and neither should our controller.

6.2. Designing the plant emulator block

The plant emulator ANN predicts the next state of the plant given the control signal of the controller and the current state of the plant, thereby providing a channel for backpropagation to compute the errors on the control signals.

The plant emulator ANN needs to be pre-trained to a certain extent to ensure stability of the quadrotor while the controller ANN is learning in the air. Additionally the design of the plant emulator should be optimal for the application – accurate enough to model complexities and agile enough to respond quickly. Refer to **Figure 6** for the final plant emulator ANN.

To verify a good design for the plant emulator ANN, data was collected over several simulations run with varying set points for roll, pitch, yaw and z, separately and simultaneously. The control signals and plant states were mapped gathering a dataset of 8000 entries. This data was used to train the plant emulator ANN; hence set points were not mapped. In these simulations, de-centralized PID controllers were used for roll, pitch and yaw channels while a cascaded PID controller was used for the z channel.

The standard procedure in an offline setting is to divide the available tagged data into three parts (randomizing them if each entry is independent, which is not the case here) - the training set (~60%), the cross validation set (~20%) and the test set (~20%). The error on the cross validation set is monitored to select the hyperparameters (like λ or α) of the network and the test error (generalization error) is used to gauge its performance on unseen data. For the purpose of offline pre-training in this chapter, the true test is the actual system; hence we divide the data into training and cross validation sets only to make those design choices that are fixed in-flight.

6.2.1. Selecting the inputs and outputs of the network

The naive approach to this problem would be to demand all states of the plant – x, y, z, ϕ, θ *and* ψ, their derivates and double derivatives as output and to give as input, the previous

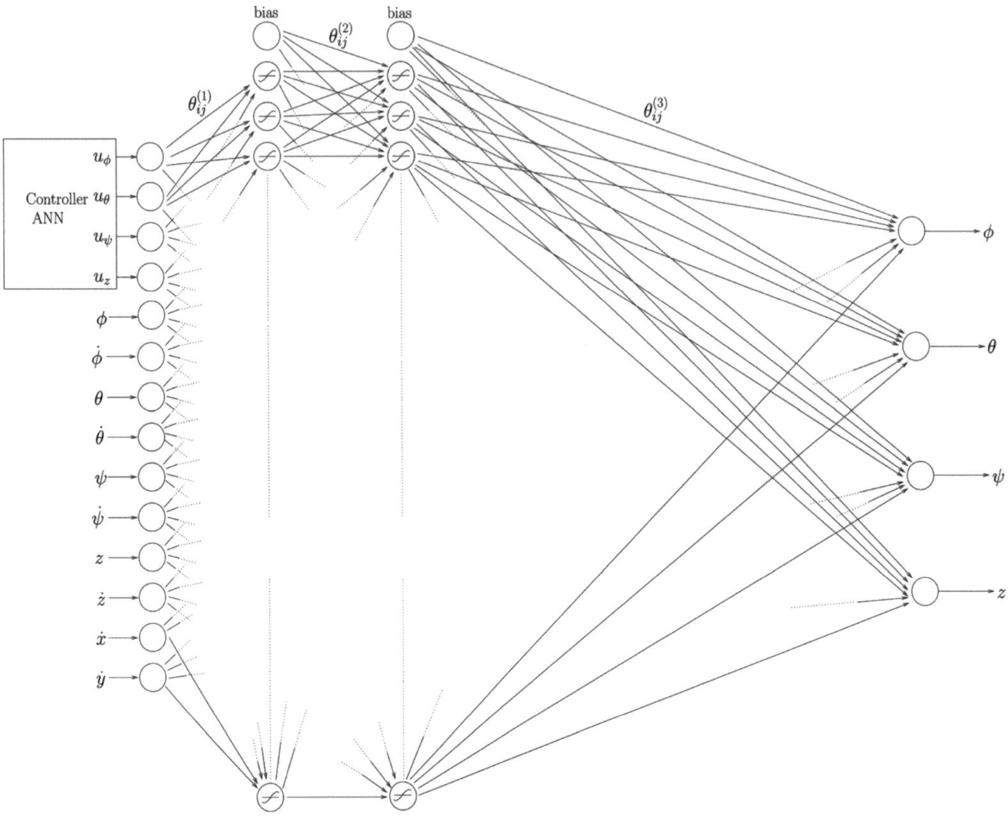

Figure 6. The double hidden layer plant emulator artificial neural network: First layer is the input, second and third layers are for computation, they comprise nodes with sigmoid activation applied and fourth layer is the output.

states of these 18 variables and the control signals. This playing-it-safe approach is costly as we place an undue burden on our algorithm.

Firstly – what is the output required from our network? The plant emulator should accurately predict the next state of the control variables of the system. Therefore the output should be the four control variables - z, ϕ, θ *and* ψ.

Secondly – what information would a human expert require to calculate the output? The previous states of z, ϕ, θ *and* ψ, their derivatives and the control signals would be required. Note that we do not need to give system information like the mass/inertia as input. The neural network will deduce such information based on the data. Interestingly, these 12 inputs are insufficient, $\dot{x}$ and $\dot{y}$ are required to model the dynamic thrust and drag coefficients by the model and therefore they must be given as inputs here.

If the initial choice of input/output variables is suboptimal leading to poor performance of the network, this step must be revisited.

6.2.2. *Selecting depth of the network – accuracy vs. agility*

A two hidden layer network was selected due to its ability to model most non-linear practical systems [7]. Increasing depth increases complexity of the network, which increases number of parameters and reduces agility of the network.

The field of *deep learning* is dedicated to using neural networks with deeper architectures, which are very powerful as we saw in Section 5.4. However, in deeper networks some of the design principles change, for example, in Section 5.3 the possibility of attenuation of information was pointed out in sigmoid-based neural networks, therefore in deeper implementations the *rectified linear unit* (ReLU) activation function is preferred. Architectural changes are also prevalent in deep learning such as *convolutional/recursive/recurrent* neural networks. Refer to Ref. [8] for more detailed reading on deep learning.

6.2.3. *Selecting the width of the network*

Performance with rectangular configurations in neural networks has been found to be equal to or better than performance with pyramid or inverted pyramid configurations [7] and therefore we have same number of hidden units in both hidden layers of our network.

As part of pre-training, the neural network performance was mapped against number of hidden units as seen in **Figure 7** and the elbow of the curve was found at 44 nodes (without

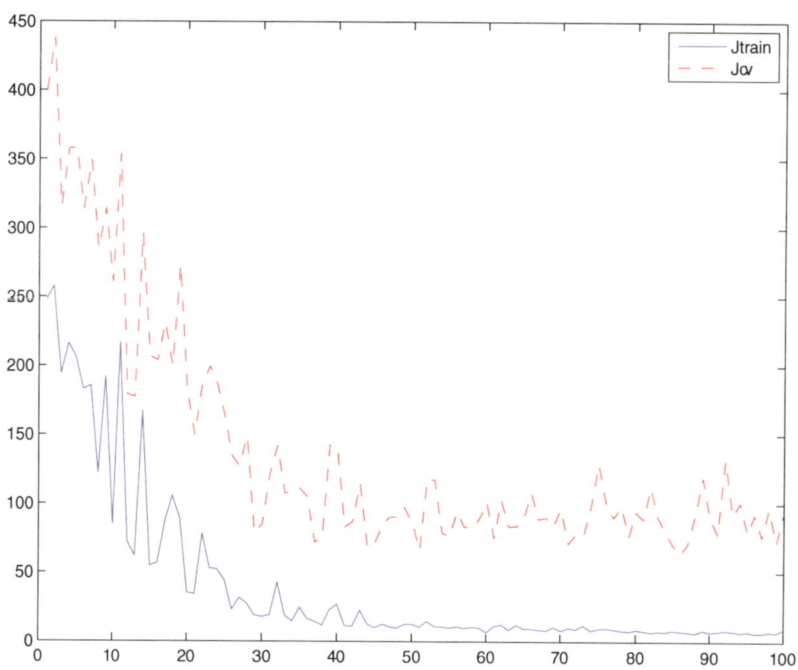

Figure 7. Costs on training and cross validation sets against hidden layer size.

the bias node) in each hidden layer. The elbow is the point beyond which the performance is unlikely to improve however the speed is sure to slow down with every node increased.

For this pre-training we stipulate to the cost function in Eq. (13) as we are training offline.

6.2.4. Cost function for online learning

To allow for real valued outputs, the ANN output layer has linear activation (read: no activation) applied while the hidden layers have sigmoid activation applied. The squared error cost function was used with regularization as shown in Eq. (25)

$$J = \frac{\left(y_{plant} - y_{model}\right)^2}{2} + \frac{\lambda}{2} \cdot \sum_{l=1}^{L-1} \sum_{i=1}^{s_l} \sum_{j=1}^{s_{l+1}} \left(\theta_{ij}^{(l)}\right)^2 \tag{25}$$

Notice the difference with Eq. (13), there is no term in m as the learning happens with one data point at a time, i.e. online. The $\theta_{ij}^{(l)}$ term, which is a subset of the set of weights, must not be confused with θ the pitch angle of the quadrotor.

6.2.5. Backpropagation and learning

The backpropagation equations are as follows:

$$\text{error in final layer: } \boldsymbol{\delta}^{(L)} = y_{model} - y_{plant} \tag{26}$$

$$\text{error in hidden layers: } \boldsymbol{\delta}^{(l)} = \left(\theta^{(l)}\right)^T \boldsymbol{\delta}^{(l+1)} \diamond \boldsymbol{a}^{(l)} \diamond \left(1 - \boldsymbol{a}^{(l)}\right), l \neq L \tag{27}$$

$$\text{derivative: } \frac{\delta J}{\delta \theta^{(l)}} = \boldsymbol{\delta}^{(l+1)} \left(\boldsymbol{a}^{(l)}\right)^T + \lambda \cdot \theta^{(l)} \tag{28}$$

where $\diamond$ signifies element-wise multiplication. Notice the derivative term is Eq. (19) with $m = 1$. This distinction differentiates *stochastic* gradient descent from *batch* gradient descent (when we have all our data beforehand.) The parameter update equation is the same as Eq. (14).

The weights learned in offline pre-training were used to initialize the weights when actual testing was done. While the error was large to start with, the weights were biased in the correct direction. This is essential as the plant emulator ANN is the conduit for backpropagation to generate the error in the control signals and random corrections to the control signal based on a purely untrained plant emulator ANN right after take-off will destabilize and ground the craft before it has a chance to complete learning.

6.3. Designing the controller block

Figure 8 depicts the controller ANN. This segmented neural network does not resemble standard MLP networks as it is highly modified. This section will be structured differently from the previous one as it deals mainly with practical aspects and online learning.

We have a much broader understanding of a subject than we may be able to comfortably express in math. The easy approach to ML is to expect the neural network to learn

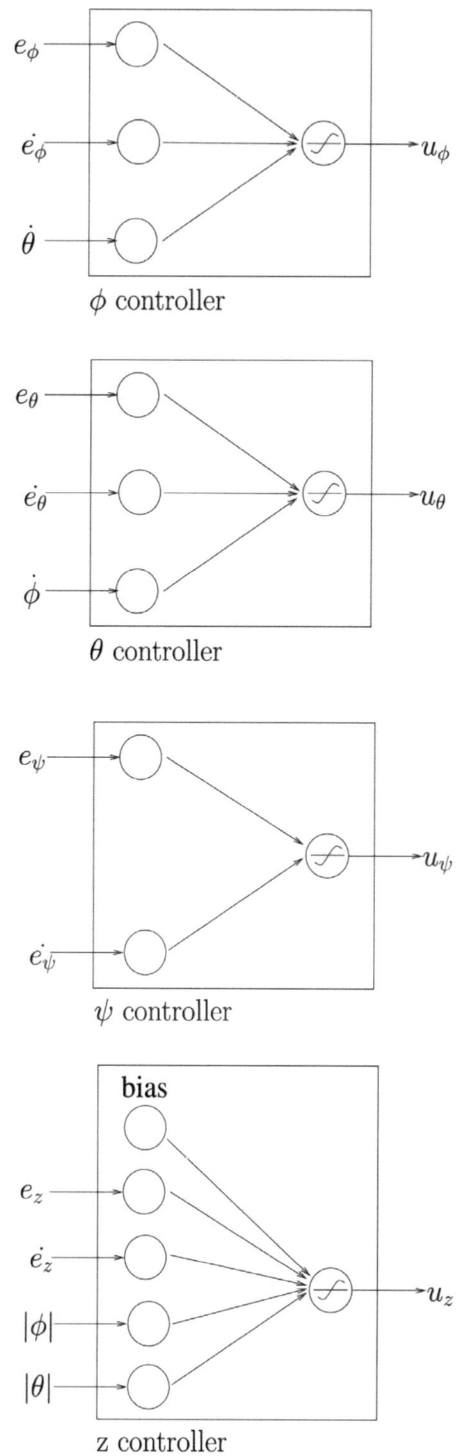

Figure 8. Segmented controller ANN with no hidden layers and constant multiplied constant offset sigmoid activation on output layer.

everything and we only provide the framework. However, this approach is unlikely to get the best results. For example, we know that a change in the z set point should not cause a change in the roll control signal, we can either make a fully connected neural network with hidden layers and expect the neural network to *learn* this by reducing the cross weights to zero or we can simply eliminate those connections ourselves. Taking the former approach revealed the learning was too slow and the quadrotor would take-off, somersault and crash as the changing z set point would cause the roll and pitch control signals to change too.

Such intuitions are application specific. It would greatly benefit the algorithm by reconciling our human understanding with the control system being designed.

The initial design of the controller ANN was very similar to the plant emulator ANN, it had two hidden layers, the inputs were the state of the plant and the state of the reference model, the outputs were the control signals which were unbounded and real valued. To summarize: 8 inputs, 4 outputs and 2 hidden layers.

At the very outset, it is clear that we should input the reference error directly rather than expecting the ANN to waste precious time learning that it has to minimize the difference between the state of the plant and the state of the reference model. Additionally an optimal number of nodes in the hidden layers cannot be experimented with as the testing is directly on the system (read: simulation) and the quadrotor would crash almost immediately after take-off. The adaptability of the controller is insufficient to keep up with the dynamically unstable nature of the quadrotor and the unbounded control signal gives noisy and, often, spurious commands to the motors thereby destabilizing the system.

A systematic approach was used to solve all the above issues:

1. The errors in the control variables were directly fed to the ANN, the second order effects were represented by appropriately adding inputs as per Eqs. (1–6).
2. The hidden layers were removed to make the ANN responsive and learn fast.
3. The output nodes were fed only those inputs that affected the state variable they controlled. Thus independence was achieved.
4. To make the control signal bounded, a constant-multiplied, constant-offset sigmoid function was used in the output layer.
5. A training regime was formulated so that each segment of the controller learned sequentially instead of simultaneously allowing for stability during online learning.
6. Feature scaling, gradient scaling and directed initialization were implemented specifically to aid in learning in the online setting.
7. The derivative term of the reference error was added to the cost function.

6.3.1. Controller summary

The controller ANN comprises four segments. The output node activations are:

$$y_n = A.\frac{1}{1+e^{-(\theta_n)^T X_n}} + B = A.\sigma\left((\theta_n)^T X_n\right) + B \tag{29}$$

where A and B are real constants and $n \in \{1, 2, 3, 4\}$ signifies each de-centralized controller.

1. **Z Controller**: This has a bias unit and four inputs - reference error in z, derivative of reference error in z, roll and pitch. Roll and pitch changes during a stable hover cause the quadrotor to deviate, to avoid this the absolute value of the corresponding angles are fed into the z controller. Its output is the z control signal. $A = 8$, $B = 4$.
2. **Roll Controller**: This has three inputs - reference error in roll, derivative of reference error in roll and derivative of pitch. The derivative of the pitch is to counter the coupling effect of roll and pitch. Its output is the roll control signal. A = 0.01, B = −0.005.
3. **Pitch Controller**: This has three inputs - reference error in pitch, derivative of reference error in pitch and derivative of roll. The derivative of the roll is to counter the coupling effect of roll and pitch. Its output is the pitch control signal. A = 0.01, B = −0.005.
4. **Yaw Controller**: This has two inputs - reference error in yaw and derivative of reference error in yaw. Yaw variation due to roll and pitch changes are minute enough to be compensated for by the controller. Its output is the yaw control signal. A = 0.6, B = −0.3.

6.3.2. Cost function and backpropagation

$$J = \frac{\left(y_{reference} - y_{plant}\right)^2}{2.\sigma_r^2} + \frac{\left(\dot{y}_{reference} - \dot{y}_{plant}\right)^2}{2.\sigma_d^2} + \frac{\lambda}{2}\cdot\sum_{c}^{\varphi\theta\psi z}\sum_{i=1}^{s_l}\sum_{j=1}^{s_{l+1}}\left(\theta_{ij}^{(c)}\right)^2 \tag{30}$$

σ_x denotes the standard deviation of the respective component and should not be confused with the sigmoid activation function. With this cost function, the rules have been bent. The derivative terms are not outputs of the neural network, yet we are attempting to minimize them. This modification proved to be a difference-maker in achieving online stability.

$$\text{error in final layer: } \boldsymbol{\delta}^{(L)} = \frac{y_{plant} - y_{ref}}{\sigma_r^2} + \frac{\dot{y}_{plant} - \dot{y}_{ref}}{\sigma_d^2} \tag{31}$$

$$\text{error in hidden layers: } \boldsymbol{\delta}^{(l)} = \left(\theta^{(l)}\right)^T \boldsymbol{\delta}^{(l+1)} \circ \boldsymbol{a}^{(l)} \circ \left(\mathbf{1} - \boldsymbol{a}^{(l)}\right), l \neq L \tag{32}$$

$$\text{error in control signals: } \boldsymbol{\delta}^{(c)} = A\cdot\left(\theta^{(c)}\right)^T \boldsymbol{\delta}^{(1)} \circ \boldsymbol{a}^{(c)} \circ \left(\mathbf{1} - \boldsymbol{a}^{(c)}\right) \tag{33}$$

$$\text{derivative: } \frac{\delta J}{\delta \theta^{(c)}} = \boldsymbol{\delta}^{(c)}\left(\boldsymbol{X}^{(c)}\right)^T + \lambda\cdot\theta^{(c)}, \quad c \in \{\varphi, \theta, \psi, z\} \tag{34}$$

λ is optional depending on the λ value used for the plant emulator ANN, i.e. if λ_{plant} is high (0.1) then $\lambda_{controller}$ can be set to zero as early stopping (the network stops learning as gradient becomes negligible) occurs within 6000 iterations. Of the 14 inputs to the plant emulator ANN, only error in control signals are calculated. Here we see that the backpropagation calculation remains the same with two approximations:

1. y_{model} is the output of the network, not y_{plant} yet we backpropagate using y_{plant}
2. $\dot{y}_{plant}$ is being backpropagated through the network meant for y_{plant}

The reasons for the first approximation are twofold: firstly, we want the controller ANN to learn as if the plant emulator ANN got its prediction correct for better learning, and secondly, as the system converges y_{model} will very nearly equal y_{plant} making the approximation asymptotically accurate.

The reason for the second approximation is that it was experimentally found that without information of velocities the controller was not able to distinguish whether the plant was converging to or diverging from the reference when error was the same. This information was encoded into the controller by this modification and supplying the derivative of the reference error to each controller segment. Overhauling the network to incorporate the derivative terms as part of the output of the plant emulator ANN would make the system slower due to the increased complexity as it would have to learn the relation between a state and its derivative. This way the two error terms are additive and embody our intuition well - let us assume the error in z is moderate but the error in velocity of z is large, a larger correction would be effected in that iteration of learning, speeding it along. Conversely, if the reference error is large and the craft is approaching the reference too fast, the additive nature of the derivative term would negate any further positive adjustment due to the error in z thereby covering up the gap in intelligence.

6.4. Speed-ups for online learning

1. **Feature Scaling**: The inputs to the controller are divided by their respective standard deviations, since the values expected vary in magnitude and the weights of the network are regularized regardless of magnitude of input. In order to calculate the standard deviation in an online setting without losing significant information to truncation in data storage the following equations are used (as presented in Ref. [9]):

$$\text{Initialize } M_1 = x_1 \text{ and } S_1 = 0 \tag{35}$$

$$\text{Compute } M_k = M_{k-1} + \frac{(x_k - M_{k-1})}{k}, \quad S_k = S_{k-1} + (x_k - M_{k-1}) \cdot (x_k - M_k) \tag{36}$$

$$\text{Current value } \sigma = \frac{S_k}{k-1}, k \,\epsilon\, [2, n] \tag{37}$$

The scaling is done selectively on the reference error and derivative of the reference error in each controller segment. This is done to scale up and give equal weight to the two errors while not magnifying higher order effects in the control signal.

2. **Directed Initialization**: Since there are no hidden layers and a single output for each segment there is no risk of redundancy in uniform initialization. Initializing all weights to

zero does not affect speed of convergence in comparison to random initialization in case of z and yaw. Therefore either can be used. For roll and pitch, learning is prone to instability and therefore we set the initial values of the weights to 1 or -1 depending on the sign of the weight required for stability. We have termed this *directed initialization*. This simplification of establishing parity in weights is effective as the inputs have been scaled.

3. **Gradient Scaling**: With the above two modifications, chances that the nodes, post sigmoid activation, will be saturated are high. Therefore, the gradient is divided by its σ (standard deviation) (calculated as shown in Point 1.) Gradient scaling is not compulsory for z.

4. **Sequence of Training**: The z and yaw controller weights are zero or random initialized. The pitch controller weights are directed initialized. The controllers are then trained sequentially. The order followed is:

   a. The z set point is varied from zero to a non-zero hover value.

   b. The pitch set point is varied from $-\frac{\pi}{3}$ to $\frac{\pi}{3}$.

   c. The roll controller weights are set to the same values as the pitch controller.

   d. The yaw set point is varied from $-\pi$ to π.

7. Performance evaluation

The testing was carried out as outlined above and the results were plotted. **Figure 9** depicts the first episode of training of the z controller, which converges within 10 seconds (at 100 Hz) and accurately follows the reference thereafter. $\alpha = 0.01$.

Figure 10 depicts the first episode of training of the pitch controller. Deviation from the reference is only due to the limitation on the maximum thrust the controller can demand (5 mN) due to bounding sigmoid function on control signal. $\alpha = 10$.

Figure 11 depicts the first episode of training of the yaw controller. Oscillations steadily decrease as training proceeds and the entire range of $-\pi$ to π is covered. $\alpha = 3$.

Figure 12 depicts simultaneous control of all four states, with controller weights continued forward from the first episode of training and plant ANN weights reset to initial estimate derived from offline training (accounting for the initial overshoot in z.) This result shows that the controller tracks well under simultaneous control signals from all four channels.

Figure 13 demonstrates the controller's robustness to mass and moment of inertia variations. At $t = 25$ sec the mass of all four motors are increased by 50 g leading to a 40% increase in total mass. The plant emulator learns these changes mid-flight enabling the controller to maintain tracking. The deviations in yaw are due to the inability of motors to meet increased thrust demands with increased moment of inertia. At $t = 75$ sec the mass and moments of inertia are reset to original values.

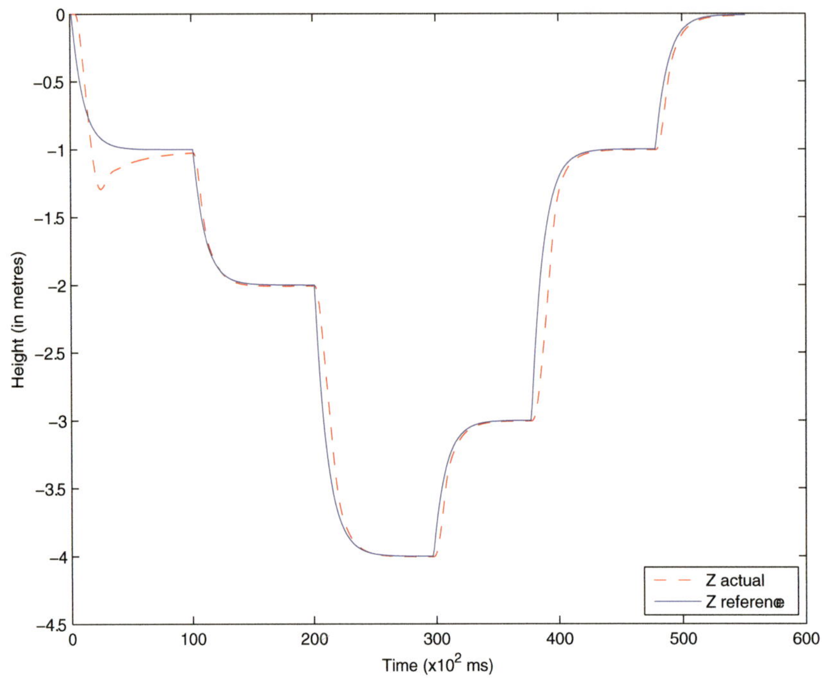

Figure 9. Starting with random weights the controller stabilizes the quadrotor to follow the reference.

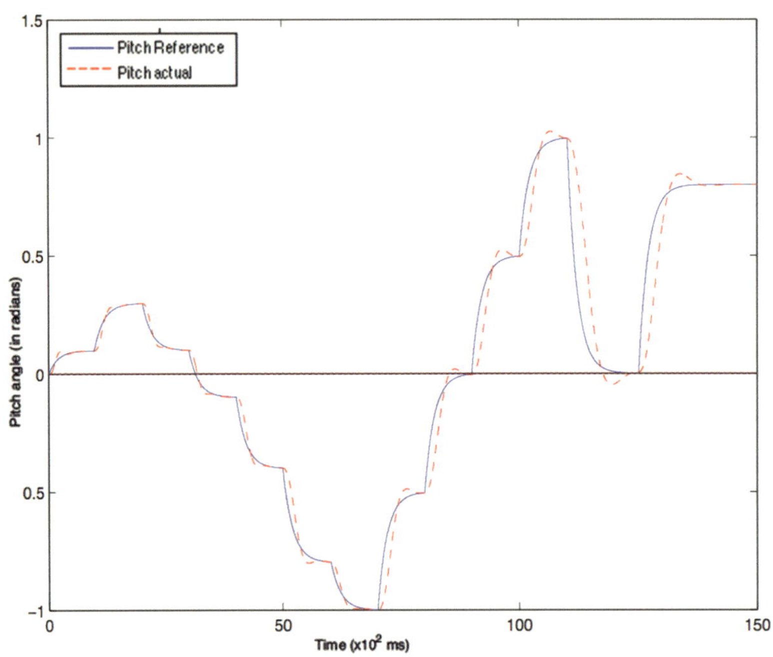

Figure 10. Directed initialized weights of pitch controller follows reference fairly well; any deviations occur due to thrust saturation.

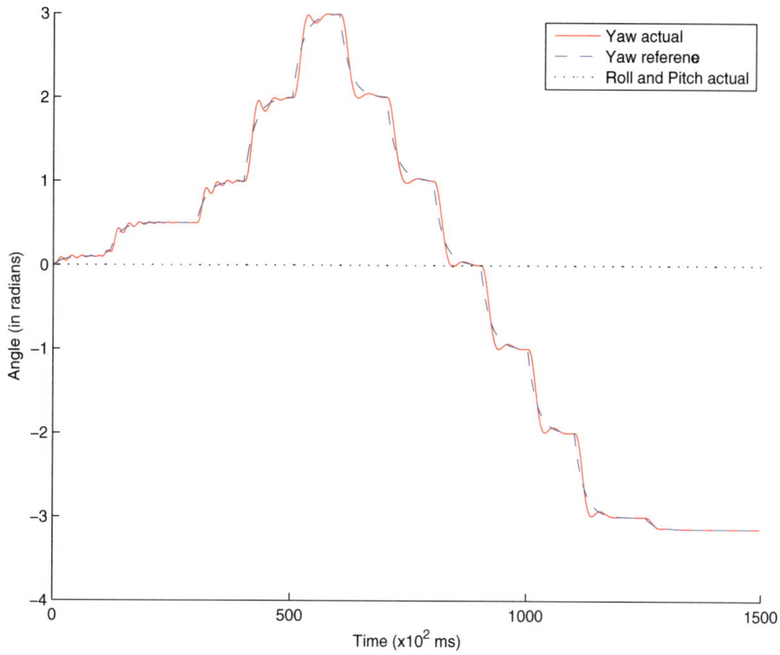

Figure 11. Zero initialized yaw controller reduces oscillations over time while tracking throughout.

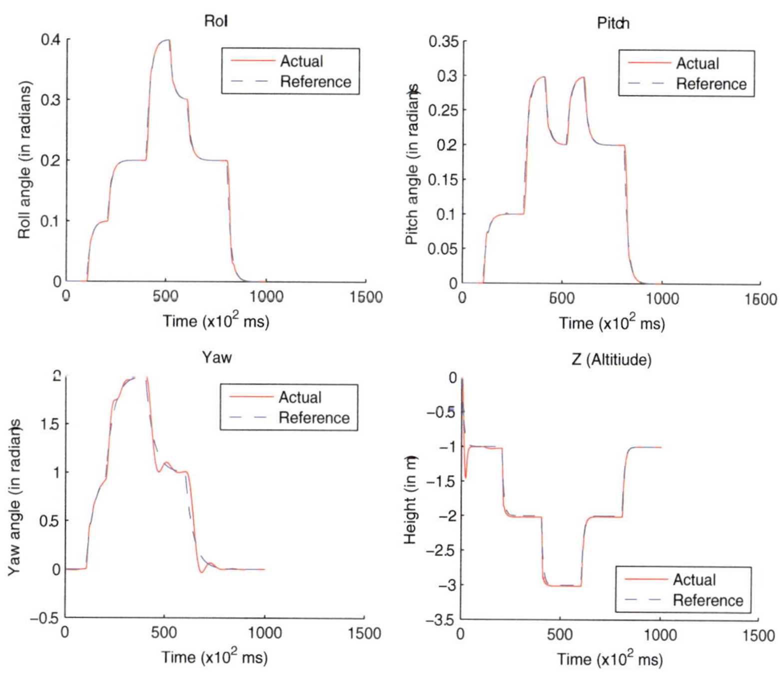

Figure 12. Simultaneous control of roll, pitch, yaw and z shows tracking with minimal error; yaw deviates slightly due to thrust saturation with large roll and pitch, however overshoot is acceptable.

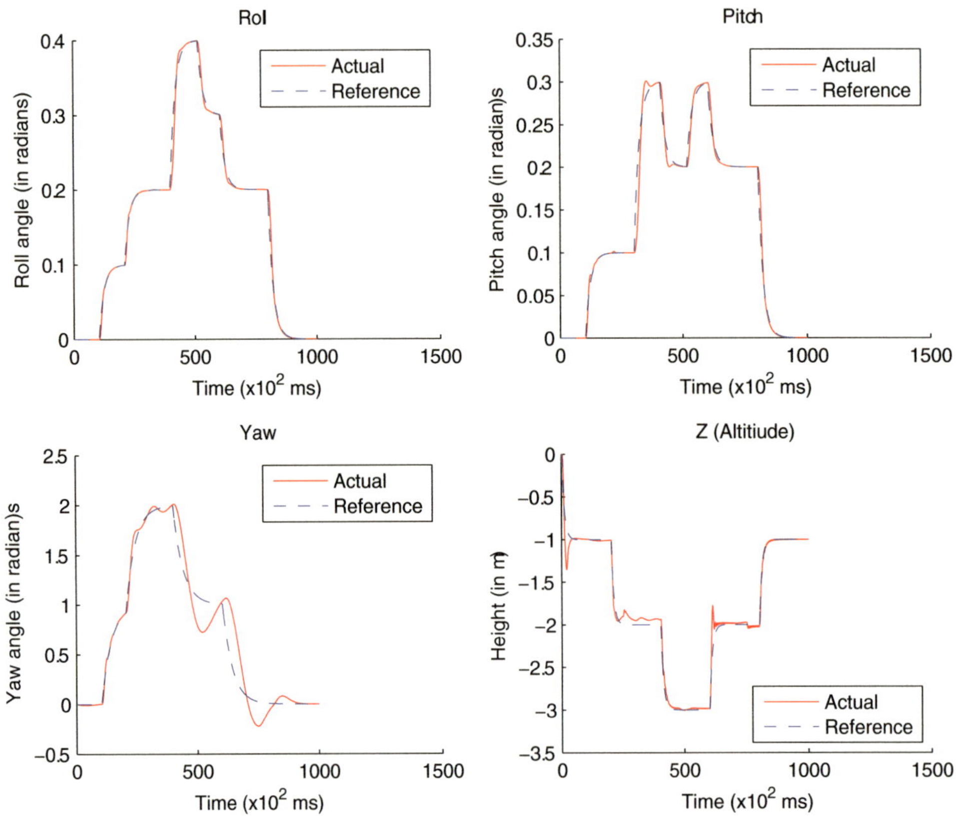

Figure 13. At $t = 25$ sec, the mass of all four motors is increased by 50 g; at $t = 75$ sec, mass and moments of inertia are reset to original values; results indicate the adaptive nature of the controller.

8. In closing

The tracking results of the quadrotor UAV system show the versatility of the control strategy. Once a framework is designed it can be applied to different systems with the same dynamics, i.e. can be applied to any quadrotor UAV, within certain limits, and will converge to stability, indicating *robustness*. It also withstands varying system parameters during operation due to changing environmental conditions or payloads, indicating *adaptability*.

A well-established and highly cited SIMULINK model was used for the simulations to prove the feasibility and good performance of this control strategy on a quadrotor UAV system. The system dynamics incorporated in this model include dynamically modeled thrust and drag coefficients, more reflective of a real-world scenario. Going forward, a disturbance rejection study can be done and the controller can be run on hardware to test it under real-world conditions.

Design choices for the neural network in terms of depth, width and choice of inputs were made based on real (read: simulated) data. The methodical process for this was outlined and applied to the quadrotor UAV system to justify the decisions made.

In real world systems, such as in UAVs where on-board processing is limited, processing time is a major factor and several methods for speed-ups were discussed which are computationally light. Even dynamically unstable systems such as UAVs could be stabilized using an untrained controller in-flight by devising a training regime.

Acknowledgements

Thanks to Mr. Raghu Ramachandran for his tireless efforts day in and day out, without whom the project upon which this chapter is based would not have been successful.

Thanks to all the people at Navstik Labs for the healthy work culture they nurture, setting the foundation for greatness. Thanks to Mr. Nitin Gupta (CEO and Founder of Navstik Labs) for giving me the opportunity to start something new and for his support along the way.

Thanks to Prof. Andrew Ng, whose ML lectures[4] have been referred to in introducing machine learning and neural networks.

Thanks to Dr. Bina Bakshi, my mother, whose love, care and support kept me going through the challenging times.

Author details

Nikhil Angad Bakshi[1,2]*

*Address all correspondence to: nabakshi@iitkgp.ac.in; nikhilangadbakshi@gmail.com

1 Indian Institute of Technology, Kharagpur, India

2 Navstik Labs, Pune, India

References

[1] Bakshi NA, Ramachandran R. Indirect model reference adaptive control of quadrotor UAVs using neural networks. In: Intelligent Systems and Control (ISCO), 2016 10th International Conference; 7–8 Jan. 2016; Coimbatore, India: IEEE; DOI: 10.1109/ISCO.2016.7727123

[2] Bouabdallah S, Siegwart R. Full control of a quadrotor. In: Intelligent Robots and Systems, 2007. IROS 2007. IEEE/RSJ International Conference; Oct. 2007; pp. 153-158. DOI: 10.1109/IROS.2007.4399042

---

[4] http://cs229.stanford.edu/materials.html, https://www.coursera.org/learn/machine-learning

[3] Fay G. Derivation of the Aerodynamic Forces for the Mesicopter Simulation. California, USA: Stanford University; 2001

[4] Bouabdallah S. Design and control of quadrotors with application to autonomous flying [Thesis]. EPFL; 2007. DOI: 10.5075/epfl-thesis-3727

[5] Krizhevsky A, Sutskever I, Hinton GE. ImageNet Classification with Deep Convolutional Neural Networks. Advances in Neural Information Processing Systems 25; 2012. pp. 1097–1105. URL: http://papers.nips.cc/paper/4824-imagenet-classification-with-deep-convolutional-neural-networks.pdf

[6] Szegedy C, Liu W, Jia Y, Sermanet P, Reed S, Anguelov D, Erhan D, Vanhoucke V, Rabinovich A. Going Deeper with Convolutions. Computer Vision and Pattern Recognition (CVPR); 2015. URL: http://arxiv.org/abs/1409.4842

[7] Bengio Y. Practical Recommendations for Gradient-Based Training of Deep Architectures. CoRR.2012;abs/1206.5533. DOI: arXiv:1206.5533

[8] Goodfellow I, Bengio Y, Courville A. Deep Learning. MIT Press; 2016. URL: http://www.deeplearningbook.org

[9] Welford BP. Note on a method for calculating corrected sums of squares and products. Technometrics. 1962:419-420. DOI: 10.2307/1266577

Chapter 9

Adaptive Nonlinear Regulation Control of Thermoacoustic Oscillations in Rijke-Type Systems

William MacKunis, Mahmut Reyhanoglu and Krishna Bhavithavya Kidambi

Additional information is available at the end of the chapter

http://dx.doi.org/10.5772/intechopen.70683

Abstract

Adaptive nonlinear control of self-excited oscillations in Rijke-type thermoacoustic systems is considered. To demonstrate the methodology, a well-accepted thermoacoustic dynamic model is introduced, which includes arrays of sensors and monopole-like actuators. To facilitate the derivation of the adaptive control law, the dynamic model is recast as a set of nonlinear ordinary differential equations, which are amenable to control design. The control-oriented nonlinear model includes unknown, unmeasurable, nonvanishing disturbances in addition to parametric uncertainty in both the thermoacoustic dynamic model and the actuator dynamic model. To compensate for the unmodeled disturbances in the dynamic model, a robust nonlinear feedback term is included in the control law. One of the primary challenges in the control design is the presence of input-multiplicative parametric uncertainty in the dynamic model for the control actuator. This challenge is mitigated through innovative algebraic manipulation in the regulation error system derivation along with a Lyapunov-based adaptive control law. To address practical implementation considerations, where sensor measurements of the complete state are not available for feedback, a detailed analysis is provided to demonstrate that system observability can be ensured through judicious placement of pressure (and/or velocity) sensors. Based on this observability condition, a sliding-mode observer design is presented, which is shown to estimate the unmeasurable states using only the available sensor measurements. A detailed Lyapunov-based stability analysis is provided to prove that the proposed closed-loop active thermoacoustic control system achieves asymptotic (zero steady-state error) regulation of multiple thermoacoustic modes in the presence of the aforementioned model uncertainty. Numerical Monte Carlo-type simulation results are also provided, which demonstrate the performance of the proposed closed-loop control system under various sets of operating conditions.

Keywords: thermoacoustics, robust, adaptive, nonlinear control

1. Introduction

Rijke-type instability is a widely investigated example of a thermoacoustic phenomenon, which describes the generation of potentially unstable pressure oscillations that results from the dynamic coupling between unsteady heat transfer and acoustics [1, 2]. The resulting oscillations in Rijke-type systems can degrade performance and even cause structural damage in combustion systems. Based on this fact, thermoacoustic instability is a primary challenge that must be addressed in the design and manufacture of land-based gas turbines and aircraft engines [3–11]. Other applications for which thermoacoustic oscillations are a concern include boilers, furnaces, ramjet engines, and rocket motors. The myriad practical engineering applications impacted by Rijke-type instability necessitate the design of reliable control systems to regulate the potentially catastrophic effects of thermoacoustic oscillations.

Control design methods for thermoacoustic oscillation suppression systems can be separated into two main categories: passive control and active control approaches. Passive control methods [12–18] can employ acoustic dampers, such as Helmholtz resonators [13] or acoustic liners [12], or they can be achieved by physically redesigning the system by changing the location of the heat source, for example. Passive approaches offer the virtues of simplicity and inexpensive maintenance; however, the performance of passive control methods can only be ensured over a relatively narrow range of operating conditions [4]. To expand the usable range of operating conditions, active control methods offer the capability to automatically adjust the level of control actuation in response to sensor stimuli.

Active control methods are usually implemented in closed-loop configurations, where sensor measurements are utilized in a feedback loop to automatically drive the input signal to the actuators. **Figure 1** provides an example of functional schematic of a closed-loop thermoacoustic oscillation suppression system. The two primary strategies for achieving closed-loop active control of thermoacoustic oscillations include (1) using a monopole-like acoustic source such as a loudspeaker to control the acoustic field [19] or (2) using a secondary fuel injector to control the unsteady heat release rate [20, 21]. Several active control approaches to suppress thermoacoustic oscillations have been presented in recent research literature.

Standard linear control systems for thermoacoustic oscillation suppression are based on stabilizing the closed-loop system through causing the dominant eigenmodes to exponentially decay. However, for realistic thermoacoustic systems where the eigenmodes are nonorthogonal, controlling only the dominant eigenmode can result in the excitation of other modes as a result of the coupling between the acoustic modes. To address this challenge, a transient growth controller is presented in [22–24], which achieves strict dissipativity. Experimental or numerical empirical methods for thermoacoustic oscillation control have been widely considered, but more systematic approaches such as robust and adaptive control have gained popularity in more recent research. Active linear control methods have been widely investigated for applications considering simplified thermoacoustic dynamic models [1, 25]. However, by leveraging the tools of nonlinear control, effective suppression control of thermoacoustic oscillations can be achieved over a wide range of operating conditions and dynamic model uncertainty.

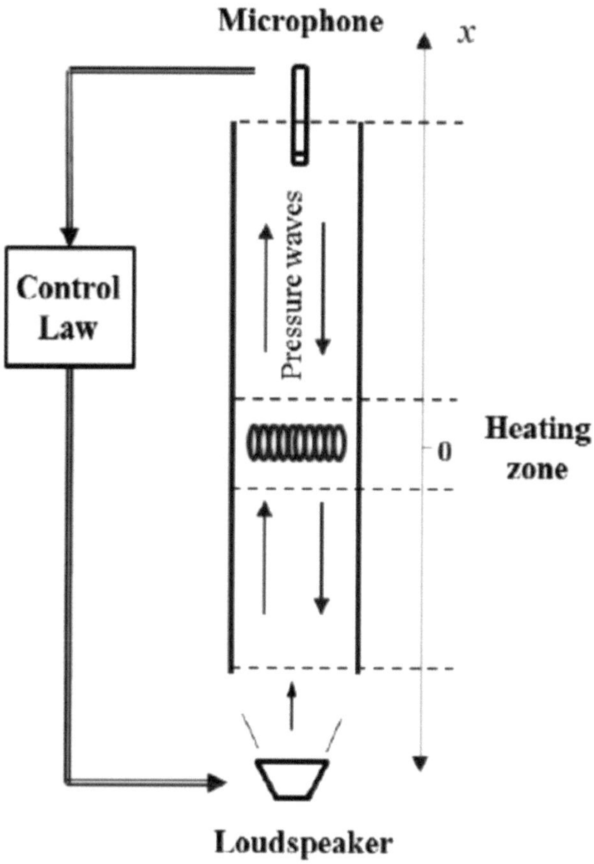

Figure 1. A functional implementation diagram of a thermoacoustic oscillation control system, including a microphone for sensing and a loudspeaker for actuation.

Physically speaking, thermoacoustic oscillation suppression can be achieved by disrupting the inherent dynamic coupling between the unsteady heat release and the acoustic waves. By designing active control systems to alter the interaction between the acoustic waves and the unsteady heat release, the amplitude of the thermoacoustic oscillations can be forced to decrease, instead of increase. Additional challenges in designing control systems for thermoacoustic oscillations can be incurred as a result of parametric model uncertainty and unmodeled operating conditions. The recent result in [26] presents a nonlinear active control method, which is proven to asymptotically regulate thermoacoustic oscillations in Rijke-type systems that do not include parametric model uncertainty and unmodeled nonlinearities. The design of active closed-loop control systems for thermoacoustic oscillation suppression that achieve reliable performance over a wide range of operating conditions and model uncertainty remains very much an open problem.

In this chapter, an observer-based nonlinear active closed-loop control method is presented, which achieves asymptotic suppression of self-excited thermoacoustic oscillations in a Rijke-type system, where the system dynamic model includes unmodeled nonlinearities and parametric uncertainty in the system dynamics and actuator dynamics. To achieve the result, a well-accepted thermoacoustic model is utilized, which employs arrays of sensors and monopole-like actuators. To facilitate the control design, the original dynamic equations are recast in a control-amenable form, which explicitly includes the effects of unmodeled, nonvanishing external disturbances and linear time delay. A sliding-mode observer-based nonlinear control law is then derived to regulate oscillations in the thermoacoustic system. A primary challenge in the control design is the presence of input-multiplicative parametric uncertainty in the control-oriented model. This challenge is handled through innovative algebraic manipulation in the regulation error system derivation along with a Lyapunov-based adaptive law. A rigorous Lyapunov-based stability analysis is used to prove that the closed-loop system achieves asymptotic regulation of a thermoacoustic system consisting of multiple modes. Numerical Monte Carlo-type simulation results are also provided, which demonstrate the performance of the proposed closed-loop active thermoacoustic oscillation suppression system.

2. Thermoacoustic system model

The thermoacoustic system model that will be utilized in this chapter consists of a horizontal Rijke tube with multiple actuators. The model is identical to that studied in our previous work in [22–24, 27].

Consider the system shown in **Figure 2**, where the actuators are modeled as multiple monopole-like moving pistons. It will be assumed that $K \geq 1$ actuators are available for control purposes. To facilitate the following observer and control design and analysis, a block diagram is also provided in **Figure 3**.

To facilitate the subsequent model development, nondimensional system variables are defined as

$$u = \frac{\tilde{u}}{u_0}, \quad p = \frac{\tilde{p}}{\gamma M_0 p_0}, \quad \dot{Q}_s = \frac{\dot{\tilde{Q}}_s}{\gamma p_0 u_0}, \tag{1}$$

$$x = \frac{\tilde{x}}{L_0}, \quad t = \frac{\tilde{t}c_0}{L_0}, \quad \frac{\delta(x - x_f)}{L_0} = \tilde{\delta}(\tilde{x} - \tilde{x}_f), \tag{2}$$

where the above tilde notation denotes the dimensional quantities and the subscript 0 denotes the mean values. In Eqs. (1) and (2), $x \in \mathbb{R}$ denotes the location along the duct (the actuators are located at $x_{ak} \in \mathbb{R}$, for $k = 1, \ldots, K$, and the heat source is located at x_f), $t \in \mathbb{R}_{\geq 0}$ denotes nondimensional time, $p(x, t) \in \mathbb{R}$ is the acoustic pressure, $u(x, t) \in \mathbb{R}$ denotes the velocity, $M \in \mathbb{R}$ is the Mach number, $c \in \mathbb{R}$ is the speed of sound, $L_0 \in \mathbb{R}$ is the length of the duct, and $\gamma \in \mathbb{R}$ is the ratio of specific heats.

By using the nondimensionalized variables defined in Eqs. (1) and (2), the thermoacoustic system with K actuators can be expressed as

$$\frac{\partial u}{\partial t} + \frac{\partial p}{\partial x} = 0, \tag{3}$$

$$\frac{\partial p}{\partial t} + \zeta p + \frac{\partial u}{\partial x} = (\gamma - 1)\dot{Q}_s \delta(x - x_f) + \gamma \sum_{k=1}^{K} \alpha_{ak} v_{ak} \delta(x - x_{ak}), \tag{4}$$

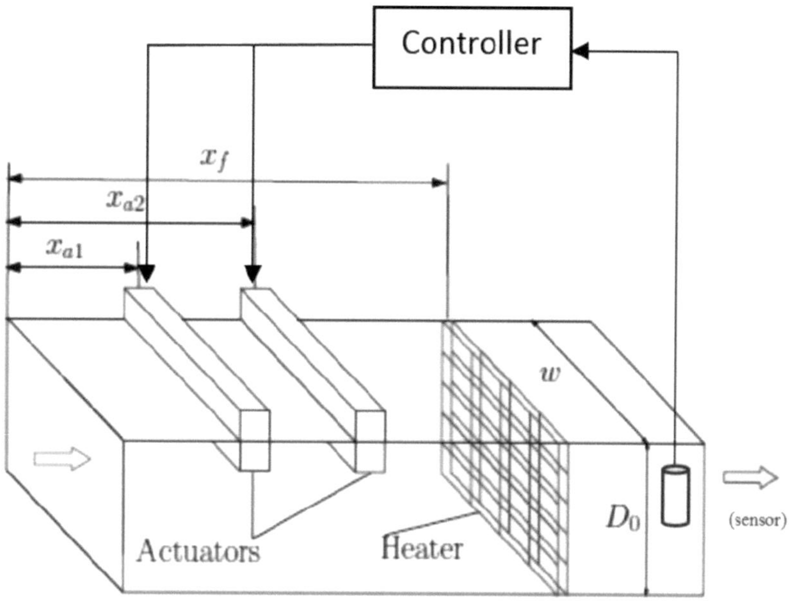

Figure 2. A control-oriented schematic of a combustion system with actuators modeled as monopole-like moving pistons.

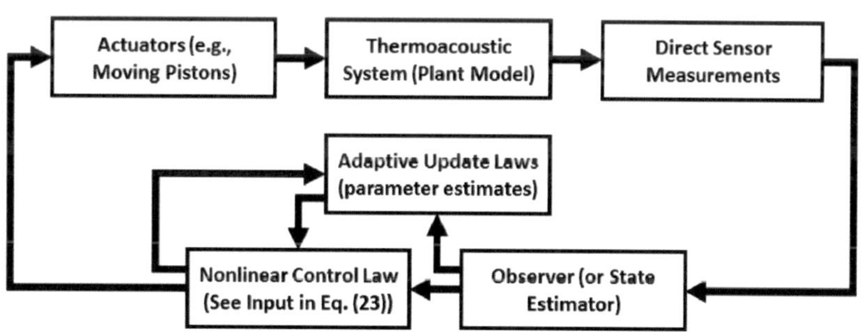

Figure 3. A block diagram illustrating the main components of the proposed robust and adaptive thermoacoustic oscillation control system.

In the expressions in Eqs. (3) and (4), $\zeta \in \mathbb{R}$ denotes a damping coefficient; thus, the term ζp physically expresses losses resulting from the effects of friction and thermo-viscous damping, and $\alpha_{ak} \in \mathbb{R}$ represents a dimensionless area ratio that can be explicitly defined as $\alpha_{ak} = S_{ak}/S$ for $k = 1, \ldots, K$. The nondimensional heat release rate $\dot{Q}_s \in \mathbb{R}$ is explicitly defined as [19].

$$\dot{Q}_s = \mathcal{K}\left[\sqrt{\left|\frac{1}{3} + u_f(t-\tau)\right|} - \sqrt{\frac{1}{3}}\right], \tag{5}$$

where

$$\mathcal{K} = \frac{2L_w(T_w - \overline{T}_0)}{\sqrt{3u_0}S\gamma p_0}\sqrt{\pi\lambda c_v \rho_0 \frac{d_w}{2}}. \tag{6}$$

In Eq. (6), d_w, L_w, and $T_w \in \mathbb{R}$ denote the diameter, length, and temperature of the heated wire, respectively; $\rho \in \mathbb{R}$ denotes air density; $T \in \mathbb{R}$ is temperature; λ is thermal conductivity; $c_v \in \mathbb{R}$ denotes the specific heat capacity at constant volume; and $\tau \in \mathbb{R}$ represents the time delay between the initial velocity field perturbation (i.e., the actuation) and the resulting effect on the heat release. Readers are referred to [22] for details on the numerical values of the physical parameters used for the thermoacoustic model being considered in this chapter. The gas is assumed to be inviscid, perfect, and nonconductive.

The acoustic pressure p and velocity u inside the duct can be expressed as a superposition of the duct natural modes as

$$p(x,t) = -\sum_{j=1}^{N} \frac{\sin(j\pi x)}{j\pi}\eta_j(t), \tag{7}$$

$$u(x,t) = \sum_{j=1}^{N} \cos(j\pi x)\eta_j(t), \tag{8}$$

where $N \in \mathbb{N}$ denotes the number of modes considered in the numerical discretization.

The actuation signal $v_{ak} \in \mathbb{R}$ of the kth monopole-like source (e.g., a loudspeaker) [28] can now be expressed as

$$v_{ak} = \mathcal{R}_k u(x_{ak}) + \mathcal{S}_k p(x_{ak}), \tag{9}$$

where $\mathcal{R}_k$ and $\mathcal{S}_k \in \mathbb{R}$ are dimensionless control parameters of the actuators. After using the expressions in Eqs. (4)–(8), the discretized governing equations are obtained as

$$\frac{\ddot{\eta}_j}{j\pi} + j\pi\eta_j + \zeta_j\frac{\dot{\eta}_j}{j\pi} = -2(\gamma-1)\dot{Q}_s(x_f, t-\tau)\sin(j\pi x_f) - 2\gamma\sum_{k=1}^{K}\alpha_{ak}v_a(x_{ak}, t)\sin(j\pi x_{ak}). \tag{10}$$

In Eq. (10), ζ represents the overall damping in the system [29, 30]. For the model under consideration here, the terms p and $\frac{\partial u}{\partial x}$ are taken to be zero at the ends of the duct. Moreover, it is assumed that no acoustic energy is dissipated in the thermal and viscous boundary layers at the duct walls [29–31].

3. Control-oriented model derivation

To facilitate the presentation of the main ideas, we consider a thermoacoustic system with two modes (i.e., $N=2$). However, the theoretical development presented here can be directly extended to address N modes.

In the following discussion, the vector of modes (i.e., the state vector) will be annotated as $\eta(t) \triangleq [\eta_1(t), \eta_2(t)]^T$, and define

$$\mathbf{\Psi}(x) = [\cos(\pi x), \cos(2\pi x)]^T, \tag{11}$$

$$\mathbf{\Phi}(x) = [\sin(\pi x), \sin(2\pi x)]^T. \tag{12}$$

Assuming that $|u_f(t-\tau)| < 1/3$, the heat release rate can be approximated as

$$\dot{Q}_s(x_f, t-\tau) \approx \frac{\sqrt{3}\mathcal{K}}{2}\mathbf{\Psi}_f^T \eta(t-\tau), \tag{13}$$

where $\mathbf{\Psi}_f \triangleq \mathbf{\Psi}(x_f)$.

By following a derivation procedure similar to that presented in [24], the dynamics of the duct natural modes can be expressed as

$$\mathbf{M}\ddot{\eta} + \left(\mathbf{D} - (\gamma-1)\mathcal{K}\sqrt{3}\tau\mathbf{\Phi}_f\mathbf{\Psi}_f^T\right)\dot{\eta} + \left(\mathbf{M}^{-1} + (\gamma-1)\mathcal{K}\sqrt{3}\mathbf{\Phi}_f\mathbf{\Psi}_f^T\right)\eta - \mathbf{h}(\eta, \dot{\eta}) - \tau_d = \mathcal{B}\mathbf{v}, \tag{14}$$

where

$$\mathbf{M} = \text{diag}\left\{\frac{1}{\pi}, \frac{1}{2\pi}\right\}, \quad \mathbf{D} = \text{diag}\left\{\frac{\zeta_1}{\pi}, \frac{\zeta_2}{2\pi}\right\}. \tag{15}$$

In Eq. (14), $\eta(t) = [\eta_1(t), \eta_2(t)]^T \in \mathbb{R}^2$ is a vector containing the natural modes, $\mathbf{h}(\eta, \dot{\eta}) \in \mathbb{R}^2$ is an unknown nonlinear function, and $\tau_d(t) \in \mathbb{R}^2$ is a general unknown bounded disturbance. To facilitate the control development in the following analysis, the dynamic equation in Eq. (14) is rewritten in the control-oriented form:

$$\mathbf{M}\ddot{\eta} + \mathbf{C_d}\dot{\eta} + \mathbf{K}\eta - \mathbf{h}(\eta, \dot{\eta}) - \tau_d = \mathcal{B}\mathbf{v}, \tag{16}$$

where the uncertain constant terms $\mathbf{C_d}$ and $\mathbf{K} \in \mathbb{R}^{2\times 2}$ are defined as

$$\mathbf{C_d} = \mathbf{D} - (\gamma-1)\mathcal{K}\sqrt{3}\tau\mathbf{\Phi}_f\mathbf{\Psi}_f^T, \tag{17}$$

$$\mathbf{K} = \mathbf{M}^{-1} + (\gamma-1)\mathcal{K}\sqrt{3}\mathbf{\Phi}_f\mathbf{\Psi}_f^T. \tag{18}$$

Also, in Eq. (16), the uncertain constant control input gain matrix $\mathcal{B} \in R^{2\times 2}$ is defined via the relationship:

$$\mathcal{B}\mathbf{v} = -2\gamma\left[\sum_{k=1}^{K} \alpha_{ak}\mathcal{R}_k\boldsymbol{\Phi}(x_{ak})\boldsymbol{\Psi}^T(x_{ak})\right]\eta + 2\gamma\left[\sum_{k=1}^{K} \alpha_{ak}\mathcal{S}_k\boldsymbol{\Phi}(x_{ak})\boldsymbol{\Phi}^T(x_{ak})\mathbf{M}\right]\dot{\eta} \tag{19}$$

where $\mathbf{v}(t) \in \mathbb{R}^2$ is a subsequently defined auxiliary control signal.

Remark 1 *Note that* Eq. (19) *highlights one of the primary challenges in the control design presented in this chapter. Specifically, the input-multiplicative parametric uncertainty in $\mathcal{B}$ presents a nontrivial control design challenge, which will be mitigated in the proposed control method through the use of a novel Lyapunov-based adaptive law.*

Assumption 1 The unknown nonlinear disturbance $\tau_d(t)$ satisfies

$$\|\tau_d(t)\| \leq \zeta, \qquad \forall t \geq 0, \tag{20}$$

where $\zeta \in \mathbb{R}$ denotes a positive bounding parameter.

4. Control development

In this section, a rigorous regulation error system development will be utilized to develop a nonlinear control system, which will be proven to effectively compensate for the inherent parametric uncertainty in the dynamic model of the thermoacoustic system in addition to the uncertain actuator model. Moreover, the proposed controller compensates for unmodeled, norm-bounded disturbances present in the dynamic model (e.g., the disturbances could represent unmodeled nonlinearities resulting from time delays due to the finite heat release rate).

4.1. Open-loop error system

The robust and adaptive nonlinear control design presented here is motivated by the desire to eliminate the transient growth of acoustical energy in a thermoacoustic dynamic system. To present the control design methodology, we consider a simplified $N = 2$ mode scenario, which will be shown to regulate the modes $\eta_1(t)$ and $\eta_2(t)$ to zero in the sense that

$$\|\eta(t)\|, \|\dot{\eta}(t)\| \to 0. \tag{21}$$

To mathematically describe the regulation control objective, an auxiliary regulation error signal $\mathbf{r}(t) \in \mathbb{R}^2$ is defined as

$$\mathbf{r} = \dot{\eta} + \alpha\eta, \tag{22}$$

where $\alpha \in \mathbb{R}$ denotes a positive, constant control gain. After taking the time derivative of Eq. (22), multiplying the result by $\mathbf{M}$, and using Eq. (16), the regulation error dynamics can be expressed as

$$\mathbf{M}\dot{\mathbf{r}} = -\mathbf{D_d}\dot{\eta} - \mathbf{K}\eta + \mathbf{h}(\eta, \dot{\eta}) + \mathcal{B}\mathbf{v} + \tau_d. \tag{23}$$

To address the case where the constant matrices $\mathbf{D_d} = \mathbf{C_d} + \mathbf{M}\alpha$, $\mathbf{K}$, and $\mathcal{B}$ are uncertain, the dynamics can be linearly parameterized as

$$\mathbf{Y}_1\theta_1 = -\mathbf{D_d}\dot{\eta} - \mathbf{K}\eta, \tag{24}$$

$$\mathbf{Y}_2\theta_2 = \mathcal{B}\mathbf{v}. \tag{25}$$

In Eqs. (24) and (25), $\mathbf{Y}_1(\eta, \dot{\eta}) \in \mathbb{R}^{2\times p_1}$ and $\mathbf{Y}_2(\mathbf{v}) \in \mathbb{R}^{2\times p_2}$ are measurable regression matrices, and $\theta_1 \in \mathbb{R}^{p_1}$ and $\theta_2 \in \mathbb{R}^{p_2}$ are vectors containing the uncertain constant parameters in $\mathbf{D_d}$, $\mathbf{K}$, and $\mathcal{B}$. The constants p_1 and $p_2 \in \mathbb{N}$ denote the number of uncertain parameters in the vectors θ_1 and θ_2, respectively.

To facilitate the subsequent Lyapunov-based adaptive control law development to compensate for the input-multiplicative uncertain matrix $\mathcal{B}$, an estimate $\widehat{\theta}_2(t) \in \mathbb{R}^{p_2}$ of the uncertain vector θ_2 is defined via the linear parameterization:

$$\mathbf{Y}_2\widehat{\theta}_2 = \widehat{\mathcal{B}}\mathbf{v}. \tag{26}$$

In Eq. (26), $\widehat{\mathcal{B}}(t) \in \mathbb{R}^{2\times 2}$ denotes a time-varying estimate of the uncertain constant matrix $\mathcal{B}$. By adding and subtracting the term $\widehat{\mathcal{B}}(t)\mathbf{v}(t)$ in Eq. (23) and using Eqs. (24) and (26), the open-loop error dynamics can be expressed as

$$\mathbf{M}\dot{\mathbf{r}} = \mathbf{Y}_1\theta_1 + \mathbf{h}(\eta, \dot{\eta}) + \mathbf{Y}_2\tilde{\theta}_2 + \widehat{\mathcal{B}}\mathbf{v} + \tau_d, \tag{27}$$

where $\tilde{\theta}_2(t) \in \mathbb{R}^{p_2}$ denotes the parameter estimate mismatch, which is defined as

$$\tilde{\theta}_2 \triangleq \theta_2 - \widehat{\theta}_2. \tag{28}$$

The error dynamics in Eq. (27) are now in a form amenable for the design of a robust and adaptive control law, which compensates for the parametric uncertainty and unmodeled nonlinearities present in the system dynamics.

Assumption 2 The unknown nonlinear term $\mathbf{h}(\eta, \dot{\eta})$ can be upper bounded as

$$\|\mathbf{h}(\eta, \dot{\eta})\| \le \rho(\|\mathbf{z}\|)\|\mathbf{z}\|, \tag{29}$$

where $\rho(\cdot) \in \mathbb{R}$ is a positive, globally invertible nondecreasing function and $\mathbf{z}(t) \in \mathbb{R}^4$ is defined as

$$\mathbf{z}(t) = \left[\eta^T(t)\ \ \mathbf{r}^T(t)\right]^T. \tag{30}$$

In Eq. (29), $\|\cdot\|$ denotes the standard Euclidean norm of the vector argument.

Assumption 2 is mild in the sense that inequality (29) is satisfied for a wide range of nonlinear function $\mathbf{h}(\eta,\dot{\eta})$.

4.2. Closed-loop error system

Based on the open-loop error system in Eq. (27), the control input $\mathbf{v}(t)$ is designed as

$$\mathbf{v} = \widehat{\mathcal{B}}^{-1}\left(-\mathbf{Y}_1\widehat{\theta}_1 - (k_s+1)\mathbf{r} - \beta\text{sgn}\,(\mathbf{r}) - \eta\right), \tag{31}$$

where $k_s \in \mathbb{R}$ denotes a positive, constant control gain and $\beta \in \mathbb{R}^{2\times 2}$ is a positive-definite, diagonal control gain matrix. In Eq. (31), sgn$(\cdot)$ denotes a vector form of the standard signum function.

After substituting the control input expression in Eq. (34) into the open-loop dynamics in Eq. (27), the closed-loop error system is obtained as

$$\mathbf{M}\dot{\mathbf{r}} = \mathbf{Y}_1\tilde{\theta}_1 + \mathbf{Y}_2\tilde{\theta}_2 + \mathbf{h}(\eta,\dot{\eta}) - \eta - (k_s+1)\mathbf{r} - \beta\text{sgn}\,(\mathbf{r}) + \tau_d, \tag{32}$$

where $\tilde{\theta}_1(t) \in \mathbb{R}^{p_1}$ is the parameter estimate mismatch defined as

$$\tilde{\theta}_1 \triangleq \theta_1 - \widehat{\theta}_1. \tag{33}$$

Based on Eq. (32) and the subsequent stability analysis, the parameter estimates $\widehat{\theta}_1(t)$ and $\widehat{\theta}_2(t)$ are generated online according to the adaptive laws:

$$\dot{\widehat{\theta}}_1 = \text{proj}(\Gamma_1\mathbf{Y}^T\mathbf{r}), \qquad \dot{\widehat{\theta}}_2 = \text{proj}(\Gamma_2\mathbf{Y}_2^T\mathbf{r}), \tag{34}$$

where $\Gamma_1 \in \mathbb{R}^{p_1\times p_1}$ and $\Gamma_2 \in \mathbb{R}^{p_2\times p_2}$ are positive-definite adaptation gains.

Remark 2 *The function* proj$(\cdot)$ *in* Eq. (34) *denotes a normal projection algorithm, which ensures that the following inequalities are satisfied:*

$$\underline{\theta}_1 \le \widehat{\theta}_1 \le \overline{\theta}_1, \qquad \underline{\theta}_2 \le \widehat{\theta}_2 \le \overline{\theta}_2, \tag{35}$$

where $\underline{\theta}_1, \overline{\theta}_1, \underline{\theta}_2$ and $\overline{\theta}_2 \in \mathbb{R}$ *represent known, constant lower and upper bounds of the elements of* $\widehat{\theta}_1(t)$ *and* $\widehat{\theta}_2(t)$*, respectively. In the current result, the use of the* proj$(\cdot)$ *function is primarily motivated by the desire to avoid singularities in the matrix estimate and facilitate the matrix inverse calculation in* Eq. (31).

To facilitate the following stability analysis, the control gain matrix β will be selected to satisfy the sufficient condition:

$$\lambda_{\min}\{\beta\} > \zeta, \tag{36}$$

where ζ is introduced in Eq. (20) and $\lambda_{\min}\{\cdot\}$ denotes the minimum eigenvalue of the argument.

5. Stability analysis

Theorem 1 *The control law in* Eq. (31) *with adaptive laws defined as in* Eq. (34) *ensures asymptotic regulation of the thermoacoustic modes* $\eta_1(t)$ *and* $\eta_2(t)$ *in the sense that*

$$\|\eta(t)\| \to 0 \qquad \text{as} \qquad t \to \infty \tag{37}$$

provided that k_s *is selected as sufficiently large (see the subsequent stability proof) and* β *is selected to satisfy inequality* (36).

Proof. Let $V\left(\eta, \mathbf{r}, \widehat{\theta}_1, \widehat{\theta}_2, t\right) \in \mathbb{R}$ be defined as the nonnegative function:

$$V(t) \triangleq \frac{1}{2}\eta^T\eta + \frac{1}{2}\mathbf{r}^T\mathbf{M}\mathbf{r} + \frac{1}{2}\tilde{\theta}_1^T\Gamma_1^{-1}\tilde{\theta}_1 + \frac{1}{2}\tilde{\theta}_2^T\Gamma_2^{-1}\tilde{\theta}_2. \tag{38}$$

After taking the time derivative of Eq. (38) and using Eq. (32), $\dot{V}(t)$ can be expressed as

$$\begin{aligned}\dot{V}(t) = \eta^T(\mathbf{r} - \alpha\eta) + \mathbf{r}^T\left(\mathbf{h}(\eta, \dot{\eta}) - \eta - (k_s + 1)\mathbf{r} - \beta\text{sgn}(\mathbf{r}) + \tau_d\right) \\ + \mathbf{r}^T\left(\mathbf{Y}_1\tilde{\theta}_1 + \mathbf{Y}_2\tilde{\theta}_2\right) - \tilde{\theta}_1^T\Gamma_1^{-1}\dot{\hat{\theta}}_1 - \tilde{\theta}_2^T\Gamma_2^{-1}\dot{\hat{\theta}}_2\end{aligned}, \tag{39}$$

where Eq. (22) was utilized. After substituting the adaptive laws in Eq. (34) and canceling common terms, $\dot{V}(t)$ can be expressed as

$$\dot{V}(t) = -\alpha\eta^T\eta + \mathbf{r}^T\left(\mathbf{h}(\eta, \dot{\eta}) - (k_s + 1)\mathbf{r} - \beta\text{sgn}(\mathbf{r}) + \tau_d\right). \tag{40}$$

By using inequalities of Eqs. (20) and (29), the expression in Eq. (40) can be upper bounded as

$$\dot{V}(t) \le -\alpha\|\eta\|^2 - \left(k_s\|\mathbf{r}\|^2 - \rho(\|\mathbf{z}\|)\|\mathbf{z}\|\|\mathbf{r}\|\right) - \|\mathbf{r}\|^2 - \beta\mathbf{r}^T\,\text{sgn}(\mathbf{r}) + \zeta\|\mathbf{r}\|. \tag{41}$$

After completing the squares for the parenthetic terms in Eq. (41), the upper bound on $\dot{V}(t)$ can be expressed as

$$\dot{V}(t) \le -\alpha\|\eta\|^2 - \|\mathbf{r}\|^2 - \beta|\mathbf{r}| + \zeta\|\mathbf{r}\| - k_s\left(\|\mathbf{r}\| - \frac{\rho(\|\mathbf{z}\|)}{2k_s}\|\mathbf{z}\|\right)^2 + \frac{\rho^2(\|\mathbf{z}\|)}{4k_s}\|\mathbf{z}\|^2, \tag{42}$$

where the fact that $\mathbf{r}^T\text{sgn}(\mathbf{r}) = |\mathbf{r}|$ was utilized. After using inequality (36), the upper bound in Eq. (42) can be expressed as

$$\dot{V}(t) \le -\left(\lambda_0 - \frac{\rho^2(\|\mathbf{z}\|)}{4k_s}\right)\|\mathbf{z}\|^2, \tag{43}$$

where $\lambda_0 \triangleq \min\{\alpha, 1\}$ and the triangle inequality (i.e., $|\mathbf{r}| \ge \|\mathbf{r}\|\ \forall\mathbf{r} \in \mathbb{R}^n$) was utilized. Based on Eq. (43), $\dot{V} \le -c\|\mathbf{z}\|^2$, for some positive constant c, inside the set $\mathcal{R}$, where $\mathcal{R}$ is defined as

$$\mathcal{R} \triangleq \left\{ \mathbf{z} | \mathbf{z} < \rho^{-1}\left(2\sqrt{\lambda_0 k_s}\right) \right\}. \tag{44}$$

The expressions in Eqs. (38) and (43) can be used to prove that $\eta(t)$, $\mathbf{r}(t)$, $\tilde{\theta}_1(t)$, and $\tilde{\theta}_2(t) \in \mathcal{L}_\infty$ in $\mathcal{R}$ and Eq. (22) can then be used to prove that $\dot{\eta}(t) \in \mathcal{L}_\infty$ in $\mathcal{R}$. Given that $\eta(t)$ and $\mathbf{r}(t) \in \mathcal{L}_\infty$ in $\mathcal{R}$, Eq. (31) can be used along with Eq. (35) to prove that $\mathbf{v}(t) \in \mathcal{L}_\infty$ in $\mathcal{R}$. Since $\eta(t)$, $\dot{\eta}(t)$, and $\mathbf{v}(t) \in \mathcal{L}_\infty$ in $\mathcal{R}$, $\mathbf{Y}_1(\eta, \dot{\eta})$ and $\mathbf{Y}_2(\mathbf{v}) \in \mathcal{L}_\infty$ in $\mathcal{R}$. Given that $\eta(t)$, $\mathbf{r}(t)$, $\dot{\eta}(t)$, $\mathbf{Y}_1(\eta, \dot{\eta})$, and $\mathbf{Y}_2(\mathbf{v}) \in \mathcal{L}_\infty$ in $\mathcal{R}$, Eq. (32) can be used along with Eq. (35) to show that $\dot{\mathbf{r}}(t) \in \mathcal{L}_\infty$ in $\mathcal{R}$. Since $\dot{\eta}(t)$ and $\dot{\mathbf{r}}(t) \in \mathcal{L}_\infty$ in $\mathcal{R}$, $\eta(t)$ and $\mathbf{r}(t)$ are uniformly continuous in $\mathcal{R}$. It then follows from Eq. (30) that $\mathbf{z}(t)$ is uniformly continuous in $\mathcal{R}$. Given that $\eta(t)$, $\mathbf{r}(t)$, $\tilde{\theta}_1(t)$, and $\tilde{\theta}_2(t) \in \mathcal{L}_\infty$ in $\mathcal{R}$, $V(t) \in \mathcal{L}_\infty$ in $\mathcal{R}$, and Eq. (43) can be integrated to prove that $\int_0^\infty \|\mathbf{z}(t)\|^2 dt \in \mathcal{L}_\infty$ in $\mathcal{R}$. Thus, $\mathbf{z}(t) \in \mathcal{L}_\infty \cap \mathcal{L}_2$ in $\mathcal{R}$. Barbalat's lemma can now be invoked to prove that $\|\mathbf{z}(t)\| \to 0$ as $t \to \infty$. Hence, $\|\eta(t)\| \to 0$ as $t \to \infty$ in $\mathcal{R}$, where the set $\mathcal{R}$ can be made arbitrarily large by increasing the control gain k_s—a semiglobal result.

6. Sliding-mode observer design

In practical thermoacoustic systems, the full state of the dynamic system is not directly measurable, and so it must be estimated through direct sensor measurements of velocity and pressure. This section presents an observer design, which is utilized to estimate the complete state of the system. The necessary observability condition can easily be satisfied through judicious sensor placement.

Let $\mathbf{x} = \left[\eta^T, \dot{\eta}^T \mathbf{M}\right]^T$ denote the state. For simplicity in the subsequent development, only one sensor is assumed to be available at location x_s. Then, the system can be rewritten as

$$\dot{\mathbf{x}}(t) = \mathbf{A}\mathbf{x}(t) + \mathbf{B}\mathbf{x}(t - \tau) + \mathbf{G}\mathbf{u}(t), \tag{45}$$

$$y = \mathbf{C}\mathbf{x}, \tag{46}$$

where

$$\mathbf{A} = \begin{bmatrix} \mathbf{0} & \mathbf{M}^{-1} \\ -\mathbf{M}^{-1} & -\mathbf{D} \end{bmatrix}, \quad \mathbf{B} = \begin{bmatrix} \mathbf{0} & \mathbf{0} \\ -\mathbf{W} & \mathbf{0} \end{bmatrix}, \quad \mathbf{G} = \begin{bmatrix} \mathbf{0} \\ \mathbf{I} \end{bmatrix}, \tag{47}$$

where

$$\mathbf{W} = \sqrt{3}(\gamma - 1)\mathcal{K}\mathbf{\Phi}_f \mathbf{\Psi}_f^T, \tag{48}$$

and the output matrix $\mathbf{C}$ is determined by the sensor choice and its location. The output equation for the velocity sensor case is given by

$$y = u(x_s, t) = \mathbf{\Psi}_s^T \eta, \tag{49}$$

and thus

$$\mathbf{C} = \begin{bmatrix} \mathbf{\Psi}_s^T & \mathbf{0}_{1\times 2} \end{bmatrix}. \tag{50}$$

For the pressure sensor case, we have

$$y = p(x_s, t) = -\mathbf{\Phi}_s^T \dot{\eta}, \tag{51}$$

and hence

$$\mathbf{C} = \begin{bmatrix} \mathbf{0}_{1\times 2} & -\mathbf{\Phi}_s^T \mathbf{M}. \end{bmatrix}. \tag{52}$$

It is assumed that the sensor location x_s is chosen such that the systems (45) and (46) subject to $\tau = 0$ are observable, i.e.,

$$\text{rank}\left[\mathbf{C}^T \ \ \mathbf{E}^T\mathbf{C}^T \ \ \left(\mathbf{E}^T\right)^2\mathbf{C}^T \ \ \left(\mathbf{E}^T\right)^3\mathbf{C}^T\right] = 4, \tag{53}$$

where $\mathbf{E} = \mathbf{A} + \mathbf{B}$.

It can be shown that there exists a coordinate transformation of the forms $\mathbf{z} = [z_1, \mathbf{z}_2]^T = \mathbf{T}\mathbf{x}$, $z_1 \in \mathbb{R}$, and $\mathbf{z}_2 \in \mathbb{R}^3$, such that the systems (45) and (46) in the new variables take the following form:

$$\dot{\mathbf{z}}(t) = \tilde{\mathbf{A}}\mathbf{z}(t) + \tilde{\mathbf{B}}\mathbf{z}(t-\tau) + \tilde{\mathbf{G}}\mathbf{u}(t), \tag{54}$$

$$y(t) = z_1(t), \tag{55}$$

where $\tilde{\mathbf{A}} = \mathbf{T}\mathbf{A}\mathbf{T}^{-1}$, $\tilde{\mathbf{B}} = \mathbf{T}\mathbf{B}\mathbf{T}^{-1}$, and $\tilde{\mathbf{G}} = \mathbf{T}\mathbf{G}$.

The estimate $\widehat{\mathbf{z}}$ of the the state $\mathbf{z}$ will now be generated via the observer equation

$$\dot{\widehat{\mathbf{z}}}(t) = \tilde{\mathbf{A}}\widehat{\mathbf{z}}(t) + \tilde{\mathbf{B}}\widehat{\mathbf{z}}(t-\tau) + \tilde{\mathbf{G}}\mathbf{u}(t) + \mathbf{L}\text{sgn}\left(y(t) - \widehat{z}_1(t)\right). \tag{56}$$

Then, the error dynamics $\overline{\mathbf{z}} = \mathbf{z} - \widehat{\mathbf{z}}$ can be written as

$$\dot{\overline{\mathbf{z}}}(t) = \tilde{\mathbf{A}}\overline{\mathbf{z}}(t) + \tilde{\mathbf{B}}\overline{\mathbf{z}}(t-\tau) - \mathbf{L}\text{sgn}\left(\overline{z}_1(t)\right). \tag{57}$$

Partition the system above as

$$\dot{\overline{z}}_1(t) = \tilde{A}_{11}\overline{z}_1(t) + \tilde{\mathbf{A}}_{12}\overline{\mathbf{z}}_2(t) + \tilde{B}_{11}\overline{z}_1(t-\tau) + \tilde{\mathbf{B}}_{12}\overline{\mathbf{z}}_2(t-\tau) - L_1\,\text{sgn}\left(\overline{z}_1(t)\right), \tag{58}$$

$$\dot{\overline{\mathbf{z}}}_2(t) = \tilde{\mathbf{A}}_{21}\overline{z}_1(t) + \tilde{\mathbf{A}}_{22}\overline{\mathbf{z}}_2(t) + \tilde{\mathbf{B}}_{21}\overline{z}_1(t-\tau) + \tilde{\mathbf{B}}_{22}\overline{\mathbf{z}}_2(t-\tau) - \mathbf{L}_2\,\text{sgn}\left(\overline{z}_1(t)\right). \tag{59}$$

The following result can now be stated:

Theorem 2 *Let* $\tilde{\mathbf{E}}_{12} = \tilde{\mathbf{A}}_{12} + \tilde{\mathbf{B}}_{12}$ *and* $\tilde{\mathbf{E}}_{22} = \tilde{\mathbf{A}}_{22} + \tilde{\mathbf{B}}_{22}$, *and assume that the pair* $\left[\tilde{\mathbf{E}}_{22}, \tilde{\mathbf{E}}_{12}\right]$ *is observable. Then, the observer gain* L_1 *can be chosen such that the system* (57) *is asymptotically stable for all* $\tau \in [0, \tau_{max}]$ *with* τ_{max} *defined by:*

$$\tau_{\max} = \frac{1}{2\sqrt{\lambda_{\max}\left(\mathbf{Q}_1^{-T}\mathbf{F}^T\mathbf{PHQ}^{-1}\mathbf{H}^T\mathbf{PFQ}_1^{-1}\right)}}, \tag{60}$$

where $\mathbf{F} = \tilde{\mathbf{E}}_{22} - \mathbf{L}_2\tilde{\mathbf{E}}_{12}/L_1$, $\mathbf{H} = \tilde{\mathbf{B}}_{22} - \mathbf{L}_2\tilde{\mathbf{B}}_{12}/L_1$, and $\mathbf{Q}$ is any symmetric, positive-definite matrix such that $\mathbf{P}$ is a symmetric, positive-definite matrix $\mathbf{P}$ solution of the Lyapunov matrix equation:

$$\mathbf{F}^T\mathbf{P} + \mathbf{PF} = -\mathbf{Q}, \tag{61}$$

and $\mathbf{Q}_1$ is the square root of the matrix $\mathbf{Q}$, i.e.,

$$\mathbf{Q}_1^T\mathbf{Q}_1 = \mathbf{Q}. \tag{62}$$

Proof. From Eq. (58), sliding mode exists in an area:

$$L_1 > |\tilde{A}_{11}\bar{z}_1(t) + \tilde{\mathbf{A}}_{12}\bar{\mathbf{z}}_2(t) + \tilde{B}_{11}\bar{z}_1(t-\tau) + \tilde{\mathbf{B}}_{12}\bar{\mathbf{z}}_2(t-\tau)|. \tag{63}$$

Condition (63) guarantees sliding in Eq. (58) along the manifold $\bar{z}_1 = 0$; thus $\hat{\eta} \to \eta$. According to the equivalent control method [32], the system in sliding mode behaves as if $L_1 \operatorname{sgn}(\bar{z}_1)$ is replaced by its equivalent value $(L_1 \operatorname{sgn}(\bar{z}_1))_{\text{eq}}$ which can be calculated from subsystem (58) assuming $\bar{z}_1 = 0$ and $\dot{\bar{z}}_1 = 0$. Hence,

$$(L_1 \operatorname{sgn}(\bar{z}_1))_{\text{eq}} = \tilde{\mathbf{A}}_{12}\bar{\mathbf{z}}_2(t) + \tilde{\mathbf{B}}_{12}\bar{\mathbf{z}}_2(t-\tau). \tag{64}$$

Substitution of Eq. (64) into Eq. (59) yields

$$\dot{\bar{\mathbf{z}}}_2 = \left(\tilde{\mathbf{A}}_{22} - \mathbf{L}_2\tilde{\mathbf{A}}_{12}/L_1\right)\bar{\mathbf{z}}_2(t) + \left(\tilde{\mathbf{B}}_{22} - \mathbf{L}_2\tilde{\mathbf{B}}_{12}/L_1\right)\bar{\mathbf{z}}_2(t-\tau). \tag{65}$$

Using the fact that the pair $\left[\tilde{\mathbf{E}}_{22}, \tilde{\mathbf{E}}_{12}/L_1\right]$ is observable, the observer gain $\mathbf{L}_2$ can be chosen such that the eigenvalues of the matrices $\tilde{\mathbf{E}}_{22} - \mathbf{L}_2\tilde{\mathbf{E}}_{12}/L_1$ have negative real parts. Thus, the subsystem (65) is asymptotically stable for $\tau \le \tau_{\max}$. This implies that $\bar{\mathbf{z}} \to 0$ and $\hat{\mathbf{z}}(t) = \mathbf{z}(t)$, and hence, $\hat{\mathbf{x}} = \mathbf{x}$.

7. Simulation results

A numerical simulation was created for two modes and two actuators (i.e., $N = K = 2$) to demonstrate the performance of the control law described by Eqs. (26), (31), and (34). The simulation utilizes the dynamics described in Eq. (16) for a system with two modes. The physical parameters used in the simulation are given in **Table 1**.

The initial conditions for the modes were selected as $\eta_1(0) = 0.07$ and $\eta_2(0)=0$. The initial conditions of the velocity and pressure parameters at the location of the heat source are $u(x_f, 0) = -0.1646$ and $p(x_f, 0)=0$ for $x_f = 0.7$. **Figure 4** shows the open-loop velocity and pressure perturbations at the heat source in the absence of control actuation.

In closed-loop operation, the adaptation gain matrices used in the simulation were selected as $\mathbf{\Gamma}_1=0.1\mathbf{I}_8$ and $\mathbf{\Gamma}_2=0.001\mathbf{I}_4$. The value for the gain α was selected as 1, and the control gain matrices β and $\mathbf{k}_s$ were selected as $\beta=\text{diag}\ \{2.43, 2.1\}$, $\mathbf{k}_s=\text{diag}\ \{4.5, 2.4\}$. The results of 20 Monte Carlo-type simulations for the closed-loop operation are shown in **Figures 5–10. Figure 5** shows the time evolution of the velocity and pressure values at the heat source location during

ρ	1.025 kg/m^3	λ	0.0328 W/m K
c_v	719 J/kg K	γ	1.4
L_0	1 m	L_w	2.5 m
c	344 m/s	u_0	0.3 m/s
T_0	295 K	T_w	1680 K
d_w	0.5×10^{-3} m	S	1.56×10^{-3} m
P_0	8.69×10^4 Pa	$\alpha_{a_1}=\alpha_{a_2}$	0.01
ζ_1	0.0440	ζ_2	0.1657

Table 1. Physical parameters.

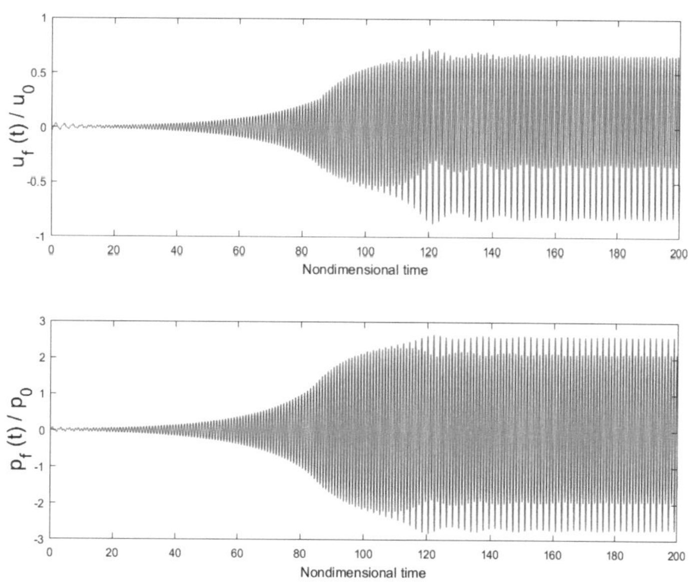

Figure 4. Time response of the velocity $u(t)$ and pressure $p(t)$ at the heat source location during open-loop (uncontrolled) operation.

closed-loop controller operation. **Figure 6** shows the time history of the modes $\eta_1(t)$ and $\eta_2(t)$ during closed-loop operation. The results clearly show the capability of the proposed robust and adaptive control law to drive the states to zero. The commanded control signals are shown in **Figure 7**. The control actuation remains within the reasonable limits throughout closed-loop operation. **Figures 8–10** show the time responses of the elements of the parameter estimate vectors $\widehat{\theta}_1(t)$ and $\widehat{\theta}_2(t)$ during closed-loop controller operation.

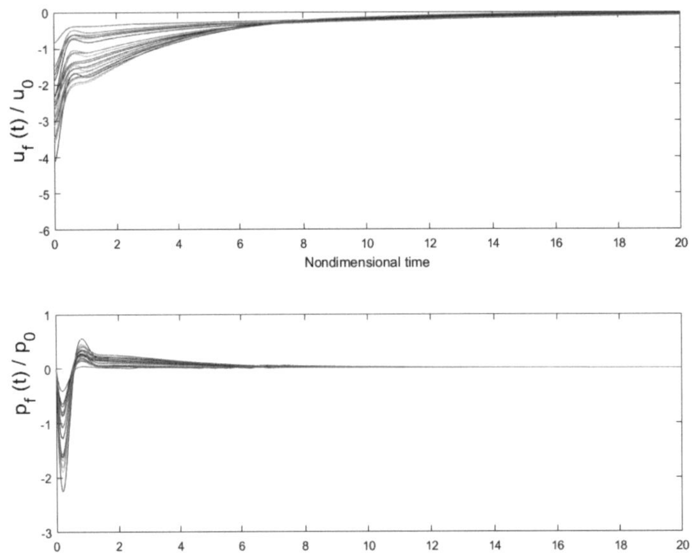

Figure 5. Time response of the velocity $u(t)$ and pressure $p(t)$ during closed-loop controller operation at the heat source location.

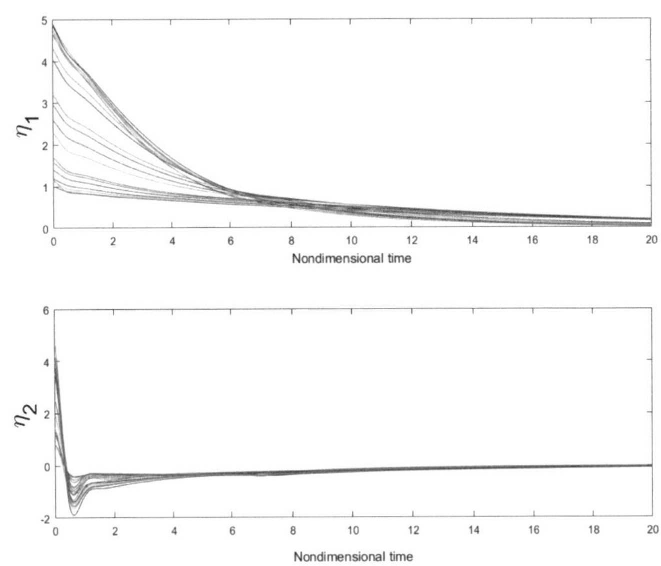

Figure 6. Time response of the oscillation modes $\eta_1(t)$ and $\eta_2(t)$ during closed-loop controller operation.

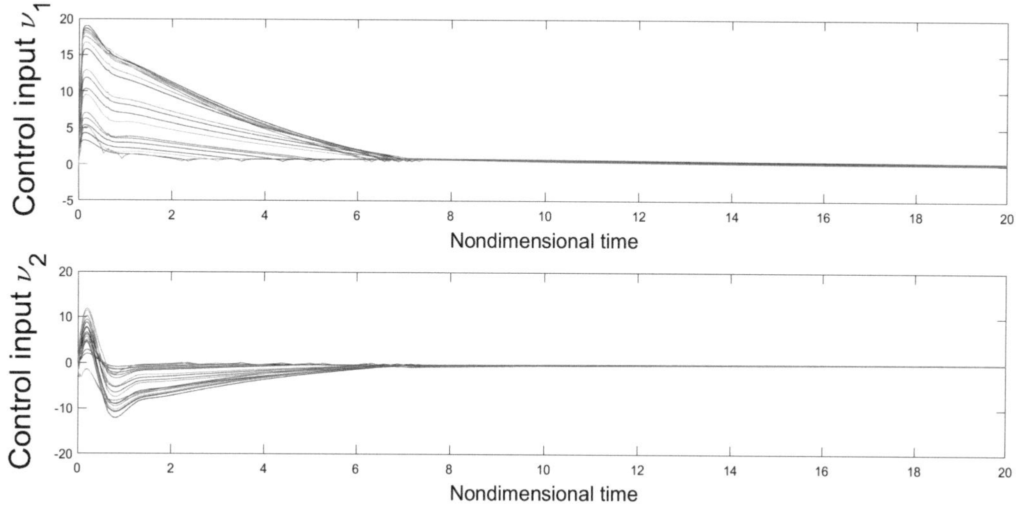

Figure 7. Commanded control inputs $v_1(t)$ and $v_2(t)$ during closed-loop controller operation.

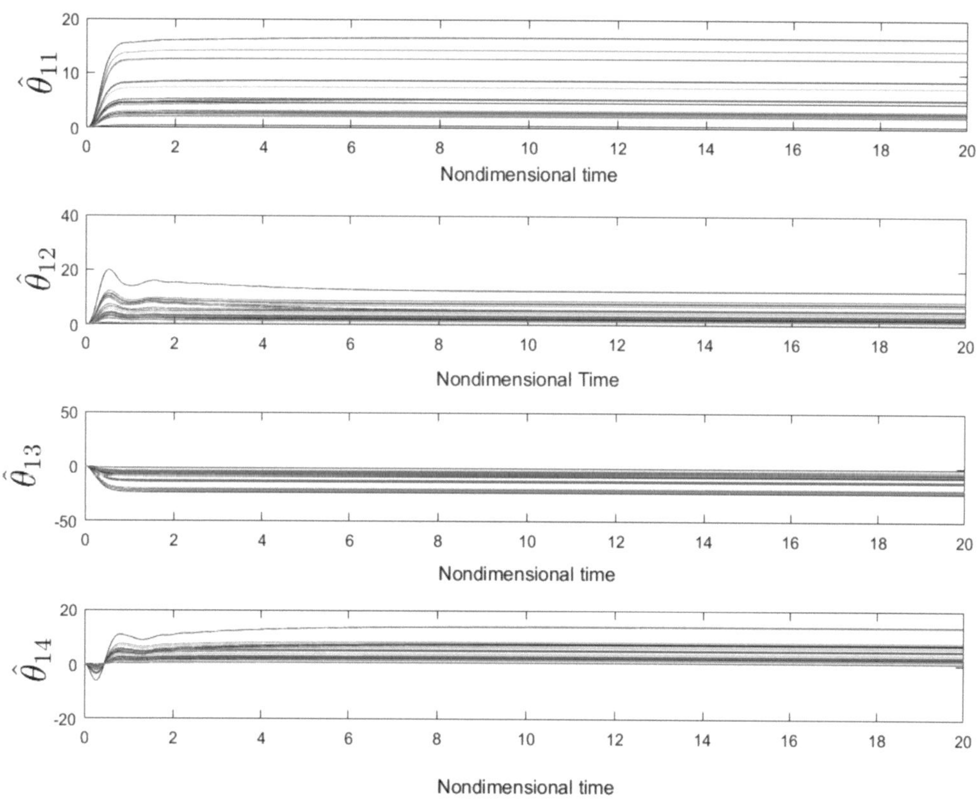

Figure 8. Time response of the adaptive parameter estimates $\widehat{\theta}_{11}(t)$, $\widehat{\theta}_{12}(t)$, $\widehat{\theta}_{13}(t)$, and $\widehat{\theta}_{14}(t)$ during closed-loop controller operation.

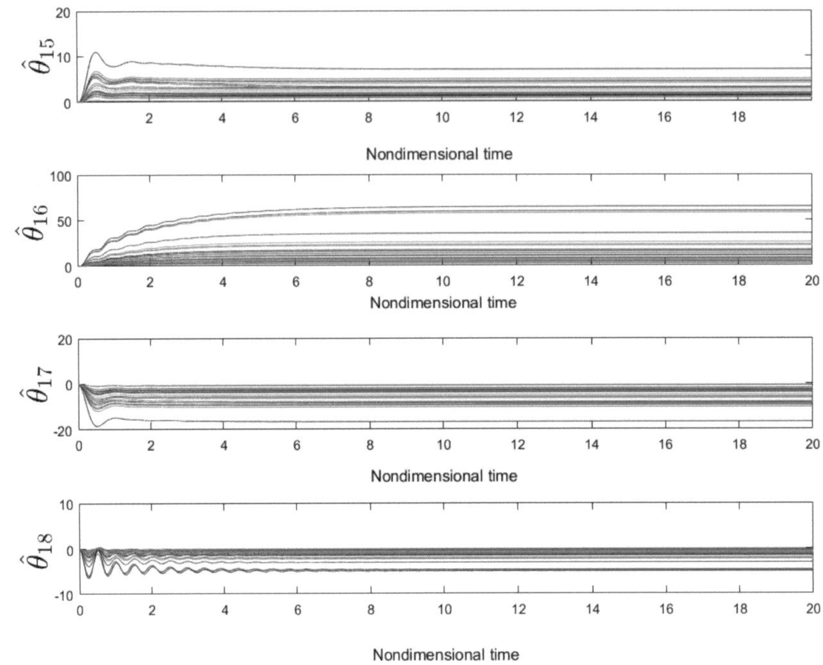

Figure 9. Time response of the adaptive parameter estimates $\widehat{\theta}_{15}(t)$, $\widehat{\theta}_{16}(t)$, $\widehat{\theta}_{17}(t)$, and $\widehat{\theta}_{18}(t)$ during closed-loop controller operation.

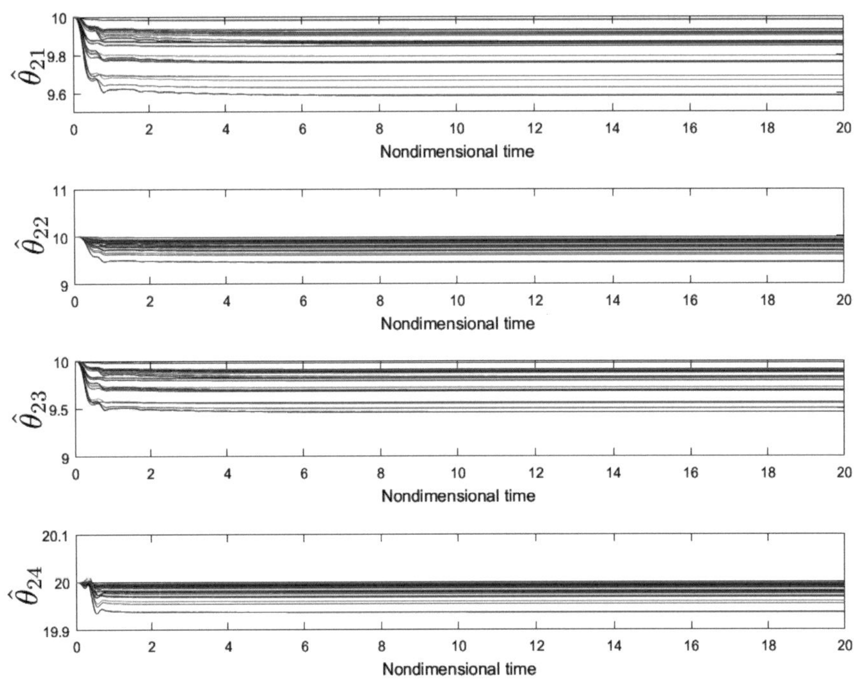

Figure 10. Time response of the adaptive parameter estimates $\widehat{\theta}_{21}(t)$, $\widehat{\theta}_{22}(t)$, $\widehat{\theta}_{23}(t)$, and $\widehat{\theta}_{24}(t)$ during closed-loop controller operation.

8. Conclusion

A robust and adaptive nonlinear control method is presented, which asymptotically regulates thermoacoustic oscillations in a Rijke-type system in the presence of dynamic model uncertainty and unknown disturbances. To demonstrate the methodology, a well-accepted thermoacoustic dynamic model is introduced, which includes arrays of sensors and monopole-like actuators. To facilitate the derivation of the adaptive control law, the dynamic model is recast as a set of nonlinear ordinary differential equations, which are amenable to control design. To compensate for the unmodeled disturbances in the dynamic model, a robust nonlinear feedback term is included in the control law. One of the primary challenges in the control design is the presence of input-multiplicative parametric uncertainty in the dynamic model for the control actuator. This challenge is mitigated through innovative algebraic manipulation in the regulation error system derivation along with a Lyapunov-based adaptive control law. To address practical implementation considerations, where sensor measurements of the complete state are not available for feedback, a detailed analysis is provided to demonstrate that system observability can be ensured through judicious placement of pressure (and/or velocity) sensors. A sliding-mode observer design is developed, which is shown to estimate the unmeasurable states using only the available sensor measurements. A detailed Lyapunov-based stability analysis is provided to prove that the proposed closed-loop active thermoacoustic control system achieves asymptotic (zero steady-state error) regulation of multiple thermoacoustic modes in the presence of the aforementioned model uncertainty. Numerical Monte Carlo-type simulation results are also provided, which demonstrate the performance of the proposed closed-loop control system under 20 different sets of operating conditions.

Author details

William MacKunis*, Mahmut Reyhanoglu and Krishna Bhavithavya Kidambi

*Address all correspondence to: mackuniw@erau.edu

Physical Sciences Department, Embry-Riddle Aeronautical University, Daytona Beach, FL, USA

References

[1] Epperlein J, Bamieh B, Astrom K. Thermoacoustics and the Rijke tube: Experiments, identification, and modeling. IEEE Control Systems. 2015;**35**(2):57-77

[2] McManus KR, Poinsot T, Candel SM. A review of active control of combustion instabilities. Progress in Energy and Combustion Science. 1993;**19**(1):1-29

[3] Dowling AP. A kinematic model of a ducted flame. Journal of Fluid Mechanics. 1999;**394**:51-72

[4] Dowling AP, Morgans AS. Feedback control of combustion oscillations. Annual Review of Fluid Mechanics. 2005;**37**:151-182

[5] Huang Y, Yang V. Dynamics and stability of lean-premixed swirl-stabilized combustion. Progress in Energy and Combustion Science. 2009;**35**(4):293-364

[6] Kim KT, Hochgreb S. Measurements of triggering and transient growth in a model lean-premixed gas turbine combustor. Combustion and Flame. 2012;**159**(3):1215-1227

[7] Langhorne PJ. Reheat buzz: An acoustically coupled combustion instability. Part 1. Experiment. Journal of Fluid Mechanics. 1988;**193**:417-443

[8] Lieuwen TC, Yang V. Combustion instabilities in gas turbine engines (operational experience, fundamental mechanisms and modeling). Progress in Astronautics and Aeronautics. 2005;**210**:8-13

[9] Palies P, Durox D, Schuller T, Candel S. Nonlinear combustion instability analysis based on the flame describing function applied to turbulent premixed swirling flames. Combustion and Flame. 2011;**158**(10):1980-1991

[10] Yang V, Yoon MW, Wicker JM. Acoustic Waves in Baffled Liquid-Propellant Rocket Engines. Technical Report AD-A267-260, Air Force Office of Scientific Research, Bolling Air Force Base, 1993

[11] Ken HY, Trouv A, Daily JW. Low-frequency pressure oscillations in a model ramjet combustor. Journal of Fluid Mechanics. 1991;**232**:47-72

[12] Eldredge JD, Dowling AP. The absorption of axial acoustic waves by a perforated liner with bias flow. Journal of Fluid Mechanics. 2003;**485**:307-335

[13] Gysling DL, Copeland GS, McCormick DC, Proscia WM. Combustion system damping augmentation with Helmholtz resonators. In: ASME 1998 International Gas Turbine and Aeroengine Congress and Exhibition

[14] Richards GA, Straub DL, Robey EH. Passive control of combustion dynamics in stationary gas turbines. Journal of Propulsion and Power. 2003;**19**(5):795-810

[15] Zhao D, Morgans AS. Tuned passive control of combustion instabilities using multiple Helmholtz resonators. Journal of Sound and Vibration. 2009;**320**(4):744-757

[16] Zhao D. Transient growth of flow disturbances in triggering a Rijke tube combustion instability. Combustion and Flame. 2012;**159**(6):2126-2137

[17] Zhao D, Li J. Feedback control of combustion instabilities using a Helmholtz resonator with an oscillating volume. Combustion Science and Technology. 2012;**184**(5):694-716

[18] Zhong Z, Zhao D. Time-domain characterization of the acoustic damping of a perforated liner with bias flow. The Journal of the Acoustical Society of America. 2012;**132**(1):271-281

[19] Heckl MA. Active control of the noise from a Rijke tube. Journal of Sound and Vibration. 1988;**124**(1):117-133

[20] Sattinger SS, Neumeier Y, Nabi A, Zinn BT, Amos DJ, Darling DD. Sub-scale demonstration of the active feedback control of gas-turbine combustion instabilities. In: ASME 1998 International Gas Turbine and Aeroengine Congress and Exhibition

[21] Seume JR, Vortmeyer N, Krause W, Hermann J, Hantschk CC, Zangl P, Gleis S, Vortmeyer D, Orthmann A. Application of active combustion instability control to a heavy duty gas turbine. In: ASME 1997 Turbo Asia Conference

[22] Hervas JR, Zhao D, Reyhanoglu M. Nonlinear feedback control of thermoacoustic oscillations in a Rijke tube. In: 2014 IEEE 23rd International Symposium on Industrial Electronics (ISIE); 2014. pp. 173-177

[23] Hervas JR, Zhao D, Reyhanoglu M. Observer-based control of Rijke-type combustion instability. In: Sivasundaram S, editor. AIP Conference Proceedings. 2014;**1637**(1):899-906

[24] Zhao D, Reyhanoglu M. Feedback control of acoustic disturbance transient growth in triggering thermoacoustic instability. Journal of Sound and Vibration. 2014;**333**(16):3639-3656

[25] Annaswamy AM, Fleifil M, Hathout JP, Ghoniem AF. Impact of linear coupling on the design of active controllers for the thermoacoustic instability. Combustion Science and Technology. 1997;**128**(1–6):131-180

[26] Hervas JR, Reyhanoglu M, MacKunis W. Sliding mode control of Rijke-type thermoacoustic systems. In: 2015 International Workshop on Recent Advances in Sliding Modes (RASM); 2015. pp. 1-6

[27] Rubio-Hervas J, Zhao D, Reyhanoglu M. Nonlinear feedback control of self-sustained thermoacoustic oscillations. Aerospace Science and Technology. 2015;**41**:209-215

[28] Fleifil M, Hathout JP, Annaswamy AM, Ghoniem AF. The origin of secondary peaks with active control of thermoacoustic instability. Combustion Science and Technology. 1998; **133**(4–6):227-265

[29] Juniper MP. Triggering in the horizontal Rijke tube: Non-normality, transient growth and bypass transition. Journal of Fluid Mechanics. 2011;**667**:272-308

[30] Matveev KI, Culick FE. A model for combustion instability involving vortex shedding. Combustion Science and Technology. 2003;**175**(6):1059-1083

[31] Balasubramanian K, Sujith RI. Thermoacoustic instability in a Rijke tube: Non-normality and nonlinearity. Physics of Fluids. 2008;**20**(4):044103

[32] Drakunov SV, Reyhanoglu M. Hierarchical sliding mode observers for distributed parameter systems. Journal of Vibration and Control. 2011;**17**(10):1441-1453

Section 3

Other Robust Adaptive Control Systems

Chapter 10

Adaptive Gain Robust Control Strategies for Uncertain Dynamical Systems

Shunya Nagai, Hidetoshi Oya, Tsuyoshi Matsuki and Yoshikatsu Hoshi

Additional information is available at the end of the chapter

http://dx.doi.org/10.5772/intechopen.71733

Abstract

In this chapter, adaptive gain robust control strategies for uncertain dynamical systems are presented. Firstly, synthesis of centralized adaptive gain robust controllers for a class of uncertain linear systems is shown. The design problem of the centralized controller is reduced to the constrained convex optimization problem, and allowable perturbation regions of unknown parameters are discussed. Next, the result for the centralized robust controller is extended to uncertain large-scale interconnected systems, that is, an LMI-based design approach for decentralized adaptive gain robust controllers is suggested.

Keywords: adaptive gain robust control, adjustable time-varying parameter, allowable perturbation regions of unknown parameters, LMIs

1. Introduction

It is well known that control systems can be found in abundance in all sectors of industry such as robotics, power systems, transportation systems space technologies, and many others, and thus control theory has been well studied. In order to design control systems, designers have to derive mathematical models for dynamical systems, and there are mainly two types of representations for mathematical models, that is, transfer functions and state equations. In other words, control theory is divided into "classical control" and "modern control" (e.g., see [12]).

Classical control means an analytical theory based on transfer function representations and frequency responses, and for classical control theory, we can find a large number of useful and typical results such as Routh-Hurwitz stability criterion [20] based on characteristic equations in the nineteenth century, Nyquist criterion [28] in the 1930s, and so on. Moreover, by using classical control ideas, some design methods of controllers such as proportional, derivative, and integral (PID) controllers and phase lead-lag compensators have also been presented [21]. In classical control, controlled systems are mainly linear and time-invariant and have a single input and a single output only. Furthermore, it is well known that design approaches based on classical control theory need experiences and trial and error. On the other hand, in the 1960s, state variables and state equations (i.e., state-space representations) have been introduced by Kalman as system representations, and he has proposed an optimal regulator theory [14–16] and an optimal filtering one [17]. Namely, controlled systems are represented by state equations, and controller design problems are reduced to optimization problems based on the concept of state variables. Such controller design approach based on the state-space representation has been established as "modern control theory." Modern control is a theory of time domain, and whereas the transfer function and the frequency response are of limited applicability to nonlinear systems, state equations and state variables are equally appropriate to linear multi-input and multi-output systems or nonlinear one. Therefore, many existing results based on the state-space representation for controller design problems have been suggested (e.g., [7, 43]).

Now, as mentioned above, in order to design control systems, the derivation of a mathematical model for controlled system based on state-space representation is needed. If the mathematical model describes the controlled system with sufficient accuracy, a satisfactory control performance is achievable by using various controller design methods. However, there inevitably exists some gaps between the controlled system and its mathematical model, and the gaps are referred to as "uncertainties." The uncertainties in the mathematical model may cause deterioration of control performance or instability of the control system. From this viewpoint, robust control for dynamical systems with uncertainties has been well studied, and a large number of existing results for robust stability analysis and robust stabilization have been obtained [34, 36, 47, 48]. One can see that quadratic stabilization based on Lyapunov stability criterion and $\mathcal{H}^\infty$ control is a typical robust controller (e.g., [1, 6]). Furthermore, some researchers investigated quadratic stabilizing control with an achievable performance level in Ref. to such as a quadratic cost function [23, 28, 35, 37], robust $\mathcal{H}^2$ control [18, 39], and robust $\mathcal{H}^\infty$-type disturbance attenuation [46]. However, these approaches result in worst-case design, and, therefore, these controllers with a fixed feedback gain which is designed by considering the worst-case variations of uncertainties/unknown parameters become cautious when the perturbation region of uncertainties has been estimated larger than the proper region. In contrast with the conventional robust control with fixed gains, several design methods of some robust controllers with time-varying adjustable parameters have also been proposed (e.g., [3, 24, 36]). In the work of Maki and Hagino [25], by introducing time-varying adjustable parameters, adaptation mechanisms for improving transient behavior have been suggested. Moreover, robust controllers with adaptive compensation inputs have also been shown [29–31]. In particular, for linear systems with matched uncertainties, Oya and Hagino [29] have introduced an adaptive compensation input which is determined so as to reduce the effect of unknown parameters. Furthermore, a design method of a variable gain robust controller based on LQ optimal control for a class of uncertain linear system has also been shown [32]. These robust controllers have fixed gains and variable ones tuned by updating laws and are more flexible and adaptive compared with the conventional robust controllers with fixed gains only, and one can easily see that these robust controllers with adjustable parameters differ from gain-scheduling control techniques [22, 41, 42]. Additionally, these robust controllers with time-varying adjustable parameters may also be referred to as "variable gain robust controller" or "adaptive gain robust controller."

In recent years, a great number of control systems are brought about by present technologies and environmental and societal processes which are highly complex and large in dimension, and such systems are referred to as "large-scale complex systems" or "large-scale interconnected systems." Namely, large-scale and complex systems are progressing due to the rapid development of industry, and large-scale interconnected systems can be seen in diverse fields such as economic systems, electrical systems, and so on. For such large-scale interconnected systems, it is difficult to apply centralized control strategies because of calculation amount, physical communication constraints, and so on. Namely, a notable characteristic of the most large-scale interconnected systems is that centrality fails to hold due to either the lack of centralized computing capability of or centralized information. Moreover, large-scale interconnected systems are controlled by more than one controller or decision-maker involving decentralized computation. In the decentralized control strategy, large-scale interconnected systems are divided into several subsystems, and various types of decentralized control problems have been widely studied [13, 38, 44]. The major problem of large-scale interconnected systems is how to deal with the interactions among subsystems. A large number of results in decentralized control systems can be seen in the work of Šijjak [38]. Moreover, a framework for decentralized fault-tolerant control has also been studied [44]. Additionally, decentralized robust control strategies for uncertain large-scale interconnected systems have also attracted the attention of many researchers (e.g., [3–5, 11]). Moreover, in the work of Mao and Lin [24], for large-scale interconnected systems with unmodeled interaction, the aggregative derivation is tracked by using a model following the technique with online improvement, and a sufficient condition for which the overall system when controlled by the completely decentralized control is asymptotically stable has been established. Furthermore, decentralized guaranteed cost controllers for uncertain large-scale interconnected systems have also been suggested [26, 27].

In this chapter, for a class of uncertain linear systems, we show LMI-based design strategies for adaptive gain robust controllers for a class of uncertain dynamical systems. The adaptive gain robust controllers consist of fixed gains and adaptive gains which are tuned by time-varying adjustable parameters. The proposed adaptive gain robust controller can achieve asymptotical stability but also improving transient behavior of the resulting closed-loop system. Moreover, by adjusting design parameters, the excessive control input is avoided [32]. In this chapter, firstly, a design method of the centralized adaptive gain robust stabilizing controllers for a class of uncertain linear systems has been shown, and the maximum allowable perturbation region of uncertainties is discussed. Namely, the proposed adaptive gain robust controllers can achieve robustness for the derived perturbation regions for unknown parameters. Additionally, the result for the centralized adaptive gain robust stabilizing controllers is extended to the design problem of decentralized robust control systems.

The contents of this chapter are as follows, where the item numbers in the list accord with the section numbers:

2. Synthesis of centralized adaptive gain robust controllers.

3. Synthesis of decentralized adaptive gain robust controllers.

4. Conclusions and future works.

The basic symbols are listed below.

$\mathbb{R}$	The set of the real number
$\mathbb{R}^n$	The set of n-dimensional vectors
$\mathbb{R}^{n\times m}$	The set of $n \times m$-dimensional matrices
$\mathbb{C}$	The set of complex numbers

Other than the above, we use the following notation and terms: For a matrix $\mathcal{A}$, the transpose of matrix A and the inverse of one are denoted by $\mathcal{A}^T$ and $\mathcal{A}^{-1}$, respectively. The notations $H_e\{\mathcal{A}\}$ and $\mathrm{diag}(\mathcal{A}_1, \cdots, \mathcal{A}_\mathcal{N})$ represent $\mathcal{A} + \mathcal{A}^T$ and a block diagonal matrix composed of matrices $\mathcal{A}_i$ for $i = 1, \cdots, \mathcal{N}$. The n-dimensional identity matrix and $n \times m$-dimensional zero matrix are described by I_n and $0_{n\times m}$, and for real symmetric matrices $\mathcal{A}$ and $\mathcal{B}$, $\mathcal{A} > \mathcal{B}$ $(\text{resp. } \mathcal{A} \geq \mathcal{B})$ means that $\mathcal{A} - \mathcal{B}$ is a positive (resp. nonnegative) definite matrix. For a vector $\alpha \in \mathbb{R}^n$, $\|\alpha\|$ denotes standard Euclidian norm, and for a matrix $\mathcal{A}$, $\|\mathcal{A}\|$ represents its induced norm. The real part of a complex number s (i.e., $s \in \mathbb{C}$) is denoted by $\mathbb{R}_e\{s\}$, and the symbols "$\triangleq$" and "$\star$" mean equality by definition and symmetric blocks in matrix inequalities, respectively.

Furthermore, the following useful lemmas are used in this chapter.

Lemma 1.1. *For arbitrary vectors λ and ξ and the matrices $\mathcal{G}$ and $\mathcal{H}$ which have appropriate dimensions, the following relation holds:*

$$2\lambda^T \mathcal{G}\Delta(t)\mathcal{H}\xi \leq 2\|\mathcal{G}^T\lambda\|\|\mathcal{H}\xi\|,$$

where $\Delta(t) \in \mathbb{R}^{p\times q}$ is a time-varying unknown matrix satisfying $\|\Delta(t)\| \leq 1$.

Proof. The above relation can be easily obtained by Schwartz's inequality (see [9]).

Lemma 1.2. *(Schur complement) For a given constant real symmetric matrix Ξ, the following arguments are equivalent:*

(i) $\Xi = \begin{pmatrix} \Xi_{11} & \Xi_{12} \\ \Xi_{12}^T & \Xi_{22} \end{pmatrix} > 0.$

(ii) $\Xi_{11} > 0$ *and* $\Xi_{22} - \Xi_{12}^T \Xi_{11}^{-1} \Xi_{12} > 0.$

(iii) $\Xi_{22} > 0$ *and* $\Xi_{11} - \Xi_{12}\Xi_{22}^{-1}\Xi_{12}^T > 0$.

Proof. See Boyd et al. [2].

2. Synthesis of centralized adaptive gain robust controllers

A centralized adaptive gain robust state feedback control scheme for a class of uncertain linear systems is proposed in this section. The adaptive gain robust controller under consideration is composed of a state feedback with a fixed gain matrix and a time-varying adjustable parameter. In this section, we show an LMI-based design method of the adaptive gain robust state feedback controller, and the allowable perturbation region of unknown parameters is discussed.

2.1. Problem statement

Consider the uncertain linear system described by the following state-space representation:

$$\frac{d}{dt}\mathrm{x}(t) = (A + \Delta(t))\mathrm{x}(t) + Bu(t), \tag{1}$$

where $\mathrm{x}(t) \in \mathbb{R}^{\mathrm{n}}$ and $u(t) \in \mathbb{R}^{\mathrm{m}}$ are the vectors of the state (assumed to be available for feedback) and the control input, respectively. In Eq. (1) the constant matrices A and B mean the nominal values of the system, and (A, B) is stabilizable pair. Moreover, the matrix $\Delta(t) \in \mathbb{R}^{\mathrm{n}\times\mathrm{n}}$ represents unknown time-varying parameters which satisfy $\Delta^T(t)\Delta(t) \leq \delta^\star I_n$, and the elements of $\Delta(t) \in \mathbb{R}^{\mathrm{n}\times\mathrm{n}}$ are Lebesgue measurable [1, 34]. Namely, the unknown time-varying matrix $\Delta(t) \in \mathbb{R}^{\mathrm{n}\times\mathrm{n}}$ is bounded, and the parameter $\delta^\star$ denotes the upper bound of the perturbation region for the unknown parameter $\Delta(t) \in \mathbb{R}^{\mathrm{n}\times\mathrm{n}}$. Additionally, we suppose that the nominal system which can be obtained by ignoring the unknown parameter $\Delta(t)$ in Eq. (1) is given by

$$\frac{d}{dt}\overline{\mathrm{x}}(t) = A\overline{\mathrm{x}}(t) + B\overline{u}(t). \tag{2}$$

In Eq. (2), $\overline{\mathrm{x}}(t) \in \mathbb{R}^{\mathrm{n}}$ and $\overline{u}(t) \in \mathbb{R}^{\mathrm{m}}$ are the vectors of the state and the control input for the nominal system, respectively.

First of all, we design the state feedback control for the nominal system of Eq. (2) so as to generate the desirable transient behavior in time response for the uncertain linear system of Eq. (1). Namely, the nominal control input is given as

$$\overline{u}(t) = K\overline{\mathrm{x}}(t), \tag{3}$$

and thus the following nominal closed-loop system is obtained:

$$\frac{d}{dt}\overline{\mathrm{x}}(t) = A_K\overline{\mathrm{x}}(t), \tag{4}$$

where A_K is a matrix given by $A_K \triangleq A + BK$. Note that the standard LQ control theory for the nominal system of Eq. (2) for designing the fixed feedback gain $K \in \mathbb{R}^{m \times n}$ is adopted in the existing result [32]. In this section, for the nominal system of Eq. (2), we derive a state feedback controller with pole placement constraints [8]. Note that for simplicity the sector constraints are introduced only in this chapter, and of course, one can adopt some other design constraints or another controller design approach for designing the fixed gain matrix $K \in \mathbb{R}^{m \times n}$. Therefore, we consider the matrix inequality condition:

$$(A_K + \alpha I_n)^T \mathcal{P} + \mathcal{P}(A_K + \alpha I_n) + \mathcal{Q} < 0, \tag{5}$$

where $\mathcal{P} \in \mathbb{R}^{n \times n}$ and $\mathcal{Q} \in \mathbb{R}^{n \times n}$ are a symmetric positive definite matrix and a symmetric semi-positive definite matrix, respectively, and the matrix $\mathcal{Q} \in \mathbb{R}^{n \times n}$ is selected by designers. If the symmetric positive definite matrix $\mathcal{P} \in \mathbb{R}^{n \times n}$ satisfying the matrix inequality of Eq. (5) exists, then poles for the nominal closed-loop system of Eq. (4) are located into the subspace $\mathcal{S}_\alpha = \{s | \mathbb{R}_e\{s\} \leq -\alpha\}$ in the complex plane. Namely, the nominal closed-loop system of Eq. (4) is asymptotically stable, and the quadratic function $\mathcal{V}(\overline{x}, t) \triangleq \overline{x}^T(t)\mathcal{P}\overline{x}(t)$ becomes a Lyapunov function for the nominal closed-loop system of Eq. (4), because the time derivative of the quadratic function $\mathcal{V}(\overline{x}, t)$ can be expressed as

$$\begin{aligned} \frac{d}{dt}\mathcal{V}(\overline{x}, t) &< -\overline{x}^T(t)(\mathcal{Q} + 2\alpha\mathcal{P})\overline{x}(t) \\ &< 0, \quad \forall\, \overline{x}(t) \neq 0. \end{aligned} \tag{6}$$

Now, we introduce complementary matrices $\mathcal{Y} \in \mathbb{R}^{n \times n}$ and $\mathcal{W} \in \mathbb{R}^{m \times m}$ which satisfy the relations $\mathcal{Y} \triangleq \mathcal{P}^{-1}$, $K = -\mathcal{W}B^T\mathcal{P}$, and $\mathcal{W} = \mathcal{W}^T > 0$, respectively. Then, some algebraic manipulations gives

$$\mathcal{Y}A^T + A\mathcal{Y} - B\mathcal{W}^T B^T - B\mathcal{W}B^T + 2\alpha\mathcal{Y} + \mathcal{Y}\mathcal{Q}\mathcal{Y} < 0. \tag{7}$$

Additionally, applying **Lemma 1.2** (Schur complement) to Eq. (7), one can easily see that the matrix inequality condition of Eq. (7) is equivalent to

$$\begin{pmatrix} \mathcal{Y}A^T + A\mathcal{Y} - B\mathcal{W}^T B^T - B\mathcal{W}B^T + 2\alpha\mathcal{Y} & \mathcal{Y} \\ \star & -\mathcal{Q}^{-1} \end{pmatrix} < 0. \tag{8}$$

Thus, the control gain matrix $K \in \mathbb{R}^{m \times n}$ is determined as $K = -\mathcal{W}B^T\mathcal{P} = -\mathcal{W}B^T\mathcal{Y}^{-1}$.

Now, for the uncertain linear system of Eq. (1), we define the following control input [37]:

$$u(t) \triangleq (1 + \theta(x, t))Kx(t), \tag{9}$$

where $\theta(x, t)$: $\mathbb{R}^n \times \mathbb{R} \to \mathbb{R}$ is an adjustable time-varying parameter [32] which plays the important role for correcting the effect of uncertainties, that is, the control input $u(t) \in \mathbb{R}^m$

consists of a fixed gain matrix $K \in \mathbb{R}^{m \times n}$ and $\theta(x, t) \in \mathbb{R}$. Note that, the robust control input of the form of Eq. (9) is called "adaptive gain robust control" in this chapter. Thus, from Eqs. (1) and (9), the uncertain closed-loop system can be written as

$$\frac{d}{dt}x(t) = A_K x(t) + \Delta(t)x(t) + \theta(x, t)BKx(t). \tag{10}$$

From the above, the control objective in this section is to design the adaptive gain robust control which achieves satisfactory transient behavior. Namely, the control problem is to derive the adjustable time-varying parameter $\theta(x, t) \in \mathbb{R}$ such that the closed-loop system of Eq. (10) can achieve the desired transient response. In addition, we evaluate the allowable perturbation region of the unknown parameter $\Delta(t) \in \mathbb{R}^{n \times n}$.

2.2. Synthesis of centralized adaptive gain robust state feedback controllers

In this subsection, we deal with design problems for the adjustable time-varying parameter $\theta(x, t) \in \mathbb{R}$ so that the satisfactory transient response for the uncertain linear system of Eq. (1) can be achieved. For the proposed adaptive gain robust control, the following theorem gives an LMI-based design synthesis.

Theorem 1: *Consider the uncertain linear system of Eq. (1) and the adaptive gain robust control of Eq. (9) with the adjustable time-varying parameter $\theta(x, t) \in \mathbb{R}$.*

For a given design parameter $\vartheta > 0$ and the known upper bound $\delta^\star$ for the unknown parameter $\Delta(t) \in \mathbb{R}^{n \times n}$, if the scalar parameter $\gamma > 0$ exists satisfying

$$\begin{pmatrix} A_K^T \mathcal{P} + \mathcal{P} A_K + \gamma \mathcal{P}^2 & I_n \\ \star & -\dfrac{\gamma}{\delta^\star} I_n \end{pmatrix} < 0, \tag{11}$$

the adjustable time-varying parameter $\theta(x, t) \in \mathbb{R}$ is determined as

$$\theta(x, t) = \begin{cases} \dfrac{\sqrt{\delta^\star} \| \mathcal{P}x(t) \| \| x(t) \|}{\| \mathcal{W}^{1/2} B^T \mathcal{P} x(t) \|^2} & \text{if} \quad x^T(t)\mathcal{P}B\mathcal{W}B^T\mathcal{P}x(t) \geq \vartheta x^T(t)x(t), \\ \\ \dfrac{\sqrt{\delta^\star} \| \mathcal{P}x(t) \| \| x(t) \|}{\vartheta x^T(t)x(t)} & \text{if} \quad x^T(t)\mathcal{P}B\mathcal{W}B^T\mathcal{P}x(t) < \vartheta x^T(t)x(t). \end{cases} \tag{12}$$

Then, the uncertain closed-loop system of Eq. (10) is asymptotically stable.

Proof. In order to prove Theorem 1, by using symmetric positive definite matrix $\mathcal{P} \in \mathbb{R}^{n \times n}$ which satisfies the standard Riccati equation of Eq. (4), we introduce the quadratic function

$$\mathcal{V}(x, t) \triangleq x^T(t)\mathcal{P}x(t), \tag{13}$$

as a Lyapunov function candidate. Let $x(t)$ be the solution of the uncertain closed-loop system of Eq. (10) for $t \geq t_0$, and then the time derivative of the quadratic function $V(x, t)$ along the trajectory of the uncertain closed-loop system of Eq. (10) can be written as

$$\begin{aligned}\frac{d}{dt}\mathcal{V}(x,t) &= x^T(t)\left(A_K^T\mathcal{P} + \mathcal{P}A_K\right)x(t)\\ &+2x^T(t)\mathcal{P}\Delta(t)x(t) + 2\theta(x,t)x^T(t)\mathcal{P}BKx(t).\end{aligned} \tag{14}$$

Firstly, the case of $x^T(t)\mathcal{P}B\mathcal{W}B^T\mathcal{P}x(t) \geq \vartheta x^T(t)x(t)$ is considered. In this case, one can see from the relation $\| \Delta(t) \| \leq \sqrt{\delta^\star}$, Eq. (14), and **Lemma 1.1** that the following inequality holds:

$$\begin{aligned}\frac{d}{dt}\mathcal{V}(x,t) &\leq x^T(t)\left(A_K^T\mathcal{P} + \mathcal{P}A_K\right)x(t) + 2\sqrt{\delta^\star}\| \mathcal{P}x(t) \|\| x(t) \|\\ &+2\theta(x,t)x^T(t)\mathcal{P}BKx(t).\end{aligned} \tag{15}$$

Moreover, since the relation $K = -\mathcal{W}B^T\mathcal{P}$ holds, the inequality of Eq. (15) can be rewritten as

$$\begin{aligned}\frac{d}{dt}\mathcal{V}(x,t) &\leq x^T(t)\left(A_K^T\mathcal{P} + \mathcal{P}A_K\right)x(t) + 2\sqrt{\delta^\star}\| \mathcal{P}x(t) \|\| x(t) \|\\ &-2\theta(x,t)x^T(t)\mathcal{P}B\mathcal{W}B^T\mathcal{P}x(t).\end{aligned} \tag{16}$$

Substituting the adjustable time-varying parameter $\theta(x, t)$ of Eq. (12) into Eq. (16) gives

$$\begin{aligned}\frac{d}{dt}\mathcal{V}(x,t) &\leq x^T(t)\left(A_K^T\mathcal{P} + \mathcal{P}A_K\right)x(t) + 2\sqrt{\delta^\star}\| \mathcal{P}x(t) \|\| x(t) \|\\ &-2x^T(t)\mathcal{P}\left(\frac{\sqrt{\delta^\star}\| \mathcal{P}x(t) \|\| x(t) \|}{\| \mathcal{W}^{1/2}B^T\mathcal{P}x(t) \|^2}\right)B\mathcal{W}B^T\mathcal{P}x(t)\\ &\leq x^T(t)\left(A_K^T\mathcal{P} + \mathcal{P}A_K\right)x(t).\end{aligned} \tag{17}$$

If the solution of the LMI of Eq. (11) exists, then the inequality

$$A_K^T\mathcal{P} + \mathcal{P}A_K < 0 \tag{18}$$

is satisfied. Thus, one can see that the following relation holds:

$$\frac{d}{dt}\mathcal{V}(x,t) < 0, \quad \forall x(t) \neq 0. \tag{19}$$

Next, we consider the case of $x^T(t)\mathcal{P}B\mathcal{W}B^T\mathcal{P}x(t) < \vartheta x^T(t)x(t)$. By using the well-known inequality for any vectors α and β with appropriate dimensions and a positive scalar ζ

$$2\alpha^T\beta \leq \zeta\alpha^T\alpha + \frac{1}{\zeta}\beta^T\beta, \tag{20}$$

we see from Eq. (14) that some algebraic manipulations give

$$\begin{aligned}\frac{d}{dt}\mathcal{V}(\mathrm{x},t) &\leq \mathrm{x}^T(t)\left(A_K^T\mathcal{P}+\mathcal{P}A_K\right)\mathrm{x}(t)+\gamma\mathrm{x}^T(t)\mathcal{P}^2\mathrm{x}(t)+\frac{1}{\gamma}\mathrm{x}^T(t)\Delta^T(t)\Delta(t)\mathrm{x}(t)\\ &\quad+2\theta(\mathrm{x},t)\mathrm{x}^T(t)\mathcal{P}B K\mathrm{x}(t)\\ &\leq \mathrm{x}^T(t)\left(A_K^T\mathcal{P}+\mathcal{P}A_K+\gamma\mathcal{P}^2+\frac{\delta^\star}{\gamma}I_n\right)\mathrm{x}(t)+2\theta(\mathrm{x},t)\mathrm{x}^T(t)\mathcal{P}BK\mathrm{x}(t)\end{aligned} \tag{21}$$

where γ is a positive constant.

Let us consider the last term of the right-hand side of Eq. (21). We see from Eq. (12) and the relation $K=-\mathcal{W}B^T\mathcal{P}$ that the last term of the right-hand side of Eq. (21) is nonpositive. Thus, if the scalar parameter γ exists satisfying

$$A_K^T\mathcal{P}+\mathcal{P}A_K+\gamma\mathcal{P}^2+\frac{\delta^\star}{\gamma}I_n<0, \tag{22}$$

then the following relation for the quadratic function $\mathcal{V}(\mathrm{x},t)$ holds:

$$\frac{d}{dt}\mathcal{V}(\mathrm{x},t)<0, \quad \forall\mathrm{x}(t)\neq 0. \tag{23}$$

Furthermore, applying **Lemma 1.2** (Schur complement) to Eq. (22), we find that the matrix inequality condition of Eq. (22) can be transformed into the LMI of Eq. (11). Namely, the quadratic function $\mathcal{V}(\mathrm{x},t)$ of Eq. (13) becomes a Lyapunov function of the uncertain closed-loop system of Eq. (10) with the adjustable time-varying parameter of Eq. (12), that is, asymptotical stability of the uncertain closed-loop system of Eq. (10) is ensured. It follows that the result of this theorem is true.

From the above, we show an LMI-based design strategy for the proposed adaptive gain robust control. Namely, the design problem of the proposed adaptive gain robust controller can be reduced to the feasibility of the LMI of Eq. (11). Note that the LMI of Eq. (11) defines a convex solution set of γ, and therefore one can easily see that various efficient convex optimization algorithms can be used to test whether the LMI is solvable and to generate particular solution. Furthermore, the LMI of Eq. (11) can also be exploited to design the proposed adaptive gain robust controller with some additional requirements. Thus, in this paper, we consider the allowable region of the unknown parameter $\Delta(t)\in\mathbb{R}^{n\times n}$ and introduce the additional constraints $\gamma=\delta^\star$ and

$$\gamma-\frac{1}{\varepsilon}>0, \tag{24}$$

where ε is a positive constant. From the relation of Eq. (24), we find that the minimization of the parameter ε means the maximization of the upper bound $\delta^\star$. Then, by using Lemma 2 (Schur complement), we find that the LMI of Eq. (11) is equivalent to

$$A_K^T\mathcal{P}+\mathcal{P}A_K+\gamma\mathcal{P}^2+I_n<0, \tag{25}$$

and the constraint of Eq. (24) can be transformed into

$$\begin{pmatrix} \gamma & 1.0 \\ \star & \varepsilon \end{pmatrix} > 0. \tag{26}$$

From the above, we consider the following constrained optimization problem:

$$\underset{\gamma>0,\, \varepsilon}{\text{Minimize}} \left[\varepsilon\right] \text{ subject to (25) and (26).} \tag{27}$$

If the optimal solution of the constrained optimization problem of Eq. (27) exists, in which are denoted by $\gamma^{\star}$ and $\varepsilon^{\star}$, the proposed adaptive gain robust controller can be done, and the allowable upper bound of the unknown parameter $\Delta(t) \in \mathbb{R}^{n \times n}$ is given by

$$\delta^{\star} = \gamma^{\star}. \tag{28}$$

Consequently, the following theorem for the proposed adaptive gain robust control with guaranteed allowable region of unknown parameter $\Delta(t) \in \mathbb{R}^{n \times n}$ is developed.

Theorem 2: *Consider the uncertain linear system of Eq. (1) and the adaptive gain robust control of Eq. (8) with the adjustable time-varying parameter* $\theta(x, t) \in \mathbb{R}$.

If the optimal solution $\gamma^{\star}$ *of the constrained optimization problem of Eq. (27) exists, then the adjustable time-varying parameter* $\theta(x, t) \in \mathbb{R}$ *is designed as Eq. (12), and asymptotical stability of the uncertain closed-loop system of Eq. (10) is ensured. Moreover, the upper bound* $\delta^{\star}$ *for the unknown parameter* $\Delta(t) \in \mathbb{R}^{n \times n}$ *is given by Eq. (28).*

Remark 1: *In this section, the uncertain linear dynamical system of Eq. (1) is considered, and the centralized adaptive gain robust controller has been proposed. Although the uncertain linear system of Eq. (1) has uncertainties in the state matrix only, the proposed adaptive gain robust controller can also be applied to the case that the uncertainties are included in both the system matrix and the input one. Namely, by introducing additional actuator dynamics and constituting an augmented system, unknown parameters in the input matrix are embedded in the system matrix of the augmented system [45]. As a result, the proposed controller design procedure can be applied to such case.*

Remark 2: *In* Theorem 1, *the design problem of the proposed adaptive gain robust controller can be reduced to the feasibility of the LMI of Eq. (11). Namely, in order to design the proposed robust control system, designers have to solve the LMI of Eq. (11). If the LMI of Eq. (11) is feasible for* $\exists \delta^{\star} > 0$, *then one can easily see that the LMI of Eq. (11) is always satisfied for the positive scalar* $\forall \delta^{-} < \delta^{\star}$. *Moreover, if a positive scalar* γ *exists satisfying the LMI of Eq. (11) for* $\exists \delta^{+} > \delta^{\star}$, *then the proposed adaptive gain robust controller can also be designed, and note that the adaptive gain robust controller for* $\delta^{\star} > 0$ *coincides exactly with the one for* $\delta^{+} > \delta^{\star} > 0$. *Furthermore, one can see from* Theorem 2 *that the resultant adaptive gain robust controller derived by solving the constrained convex optimization problem of Eq. (27) is same, because the solution of LMI of Eq. (8) or one of the constrained convex optimization problem of Eq. (27) cannot be reflected the resultant controller. Note that in the general controller design strategies for the conventional fixed gain robust control, the solution of the some constraints can be applied to the resultant robust controller. This is a fascinating fact for the proposed controller design strategy.*

Remark 3: *The proposed adaptive gain robust controller with the adjustable time-varying parameter has some advantages as follows: the proposed controller design approach is very simple, and by selecting the design parameter, the proposed adaptive gain robust control system can achieve good transient performance which is close to the nominal one or avoid the excessive control input (see [32]). Besides, the structure of the proposed control system is also simple compared with the existing results for robust controllers with adjustable parameters* (e.g., *[29, 30]). However, the online adjustment strategy for the design parameter ϑ has not been established, and this problem is one of our future research subjects.*

Remark 4: *In this section, firstly the nominal control input is designed by adopting pole placement constraints, and the fixed gain $K \in \mathbb{R}^{m \times n}$ can be derived by using the solution of the LMI of Eq. (8). Note that the quadratic function $\mathcal{V}(x, t)$ is a Lyapunov function for both the uncertain linear system of Eq. (1) and the nominal system of Eq. (2), that is, the Lyapunov function for the uncertain linear system of Eq. (1) and one for the nominal system of Eq. (2) have same level set. Therefore, by selecting the design parameter $\vartheta > 0$, the proposed adaptive gain robust control system can achieve good transient performance which is close to the nominal one or avoid the excessive control input.*

On the other hand, if the design problem for a state feedback control $u(t) = K_s x(t)$ is considered, the quadratic function $\mathcal{V}(x, t)$ is replaced as $\mathcal{V}_s(x, t) = x^T(t)\mathcal{P}_s x(t)$ where $\mathcal{P}_s \in \mathbb{R}^{n \times n}$ is a Lyapunov matrix. Moreover, $\mathcal{P}_s \in \mathbb{R}^{n \times n}$ becomes a variable for resultant LMI conditions, and the standard techniques for the quadratic stabilization can also be used.

2.3. Illustrative examples

In order to demonstrate the efficiency of the proposed control strategy, we have run a simple example.

Consider the following linear system with unknown parameter $\Delta(t) \in \mathbb{R}^{2 \times 2}$:

$$\frac{d}{dt}x(t) = \begin{pmatrix} 1.0 & 4.0 \\ 0.0 & -1.0 \end{pmatrix} x(t) + \Delta(t)x(t) + \begin{pmatrix} 0.0 \\ 1.0 \end{pmatrix} u(t). \tag{29}$$

Firstly, we design the nominal control input $\overline{u}(t) = K\overline{x}(t)$. By selecting the design parameters α and Q in Eq. (5) such as $\alpha = 3.0$ and $Q = 1.0 \times I_2$ and solving the LMI of Eq. (8), we obtain the following solution:

$$\begin{aligned} \mathcal{Y} &= \begin{pmatrix} 1.0855 & -1.5356 \\ \star & 4.5318 \end{pmatrix}, \\ \mathcal{W} &= 2.1708 \times 10^1. \end{aligned} \tag{30}$$

Thus, the following fixed gain matrix can be computed:

$$K = \begin{pmatrix} -1.3017 \times 10^1 & -9.2008 \end{pmatrix}. \tag{31}$$

Next, we solve the constrained optimization problem of Eq. (27), then the solutions

$$\begin{aligned}\gamma &= 3.1612,\\ \varepsilon &= 3.1633 \times 10^{-1},\end{aligned} \tag{32}$$

can be derived, and therefore the allowable upper bound of unknown parameter is given as

$$\delta^{\star} = 3.1612. \tag{33}$$

In this example, we consider the following two cases for the unknown parameter $\Delta(t) \in \mathbb{R}^{2\times 2}$:

- Case 1) $\Delta(t) = \delta^{\star} \times \begin{pmatrix} 0.0 & 0.0 \\ -7.2289 & 6.8530 \end{pmatrix} \times 10^{-1}$.
- Case 2) $\Delta(t) = \delta^{\star} \times \begin{pmatrix} \sin(5.0 \times \pi \times t) & -\cos(5.0 \times \pi \times t) \\ \star & -\sin(5.0 \times \pi \times t) \end{pmatrix}$.

Note that the unknown parameter of Case 1 satisfies the matching condition [45]. In addition, for the design parameter ϑ, the numerical simulation for two cases such as $\vartheta = 1.0 \times 10^{2}$ and $\vartheta = 5.0 \times 10^{-1}$ is run. Moreover, the initial values of the uncertain system of Eq. (29) and the nominal system are selected as $\mathrm{x}(0) = \overline{\mathrm{x}}(0) = (1.0\ -2.0)^{T}$. The results of the simulation of this example are shown in **Figures 1–4** and **Table 1**. In these figures, "Case 1)" and "Case 2)" represent the time histories of the state variables $\mathrm{x}_1(t)$ and $\mathrm{x}_2(t)$ and the control input $u(t)$ and Lyapunov function $\mathcal{V}(\mathrm{x}.t)$ for the proposed adaptive gain robust control, and "nominal" means the desired time response and the desired control input and Lyapunov function $\mathcal{V}(\overline{\mathrm{x}}, t)$ for the nominal system. In **Table 1**, $\mathcal{J}_e$ means

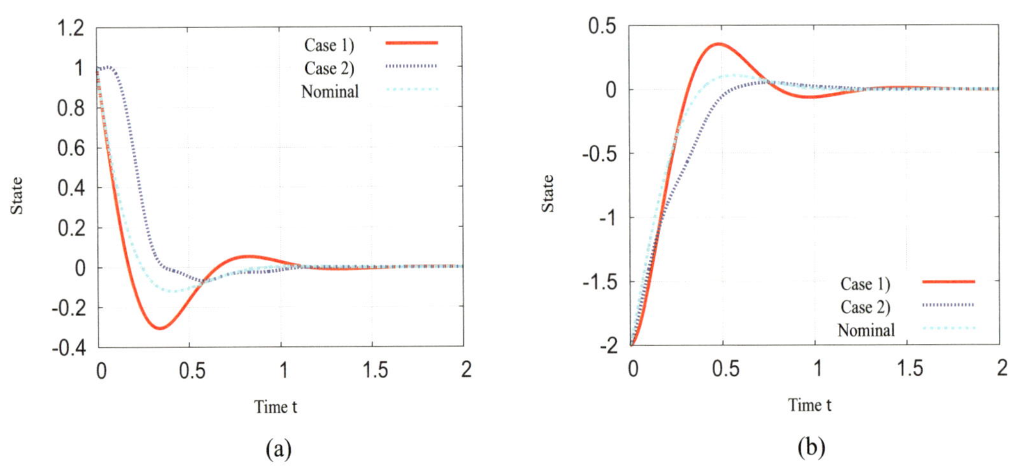

Figure 1. Time histories of the states for $\vartheta = 1.0 \times 10^{2}$. (a) The time histories of $x_1(t)$, (b) The time histories of $x_2(t)$.

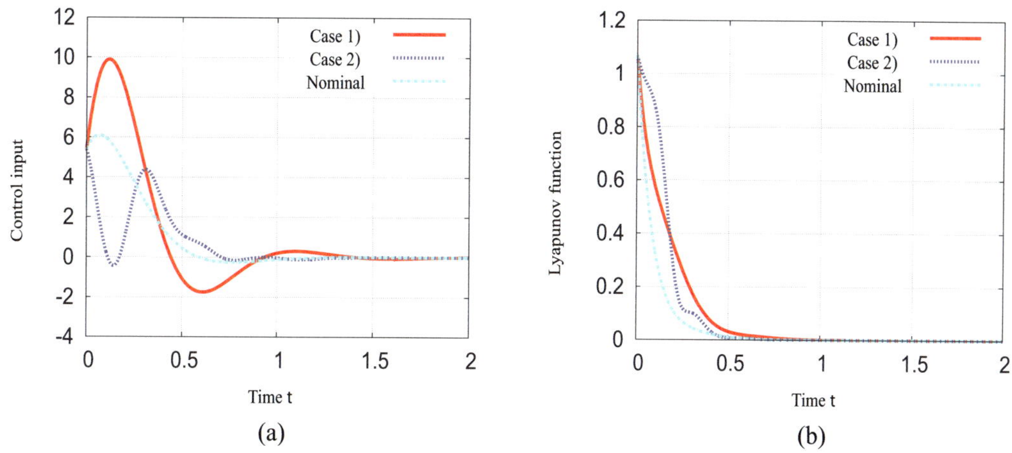

Figure 2. Time histories of the control inputs and the Lyapunov function for $\vartheta = 1.0 \times 10^2$. (a) The time histories of $u(t)$ and $\overline{u}(t)$, (b) The time histories of $\mathcal{V}(x, t)$ and $\mathcal{V}(\overline{x}, t)$.

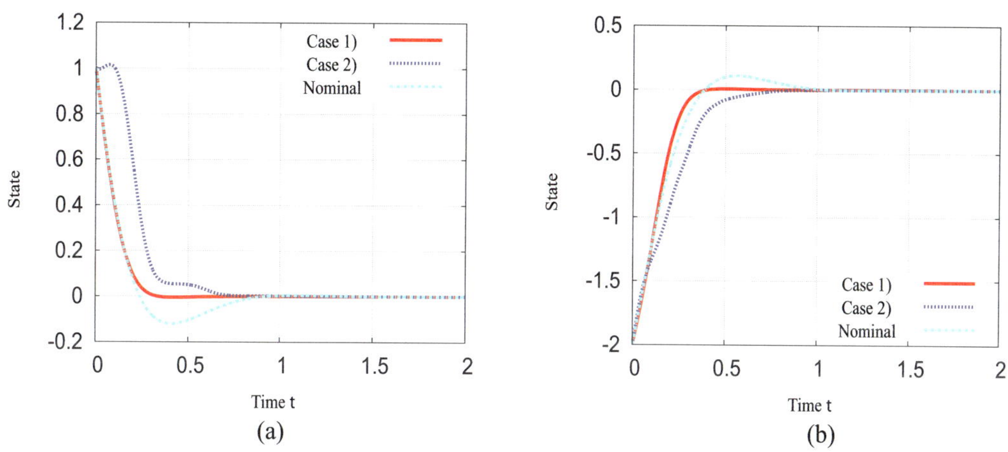

Figure 3. Time histories of the states for $\vartheta = 5.0 \times 10^{-1}$. (a) The time histories of $x_1(t)$, (b) The time histories of $x_2(t)$.

$$\mathcal{J}_e \overset{\Delta}{=} \int_0^{\infty} e^T(t)e(t)dt, \tag{34}$$

where $e(t)$ is an error vector between the time response and the desired one generated by the nominal system, that is, $e(t) \overset{\Delta}{=} x(t) - \overline{x}(t)$. Namely, $\mathcal{J}_e$ of Eq. (34) is a performance index so as to evaluate the transient performance.

From **Figures 1–4** the proposed adaptive gain robust state feedback controller stabilizes the uncertain linear system of Eq. (29) in spite of uncertainties. Furthermore, we also find that the proposed adaptive gain robust controller achieves the good transient performance close to the nominal system.

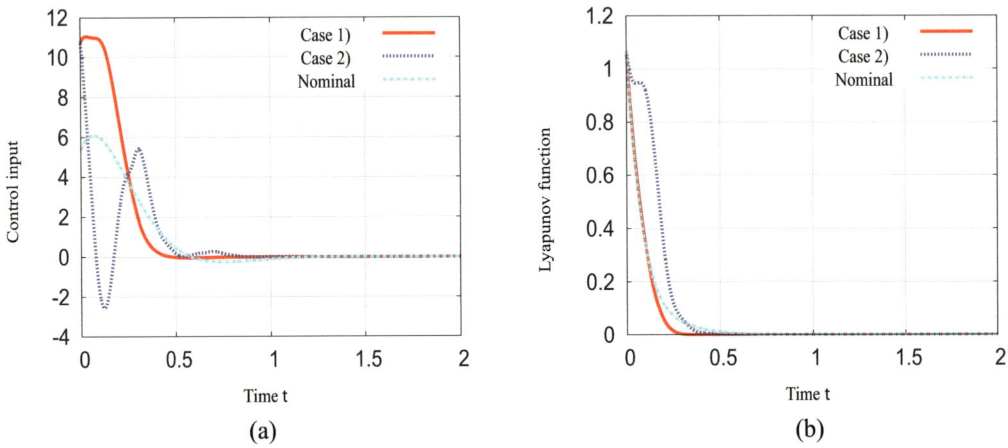

Figure 4. Time histories of the control inputs and the Lyapunov function for $\vartheta = 5.0 \times 10^{-1}$. (a) The time histories of $u(t)$ and $\overline{u}(t)$, (b) The time histories of $\mathcal{V}(x, t)$ and $\mathcal{V}(\overline{x}, t)$.

	$\boldsymbol{\vartheta = 1.0 \times 10^2}$	$\boldsymbol{\vartheta = 5.0 \times 10^{-1}}$
Case 1)	4.2584×10^{-2}	1.0160×10^{-2}
Case 2)	9.7403×10^{-2}	1.0038×10^{-1}

Table 1. The performance index $\mathcal{J}_e$.

For Case 1 in this example, one can see from **Table 1** that the adaptive gain robust controller for $\vartheta = 5.0 \times 10^{-1}$ is more desirable comparing with one for $\vartheta = 1.0 \times 10^2$, that is, the error between the time response and the desired one generated by the nominal system ("nominal" in figures) is small. But for the result of Case 2), we find that the robust controller with the parameter $\vartheta = 1.0 \times 10^2$ achieves more desirable performance. Additionally, one can see from **Figures 2(a)** and **4(a)** that by selecting the design parameter ϑ the proposed adaptive gain robust controller can adjust the magnitude of the control input. In this example, the magnitude of the control input for $\vartheta = 1.0 \times 10^2$ is suppressed comparing with one for $\vartheta = 5.0 \times 10^{-1}$. However, the online adjustment way of the design parameter ϑ for the purpose of improving transient behavior and avoiding excessive control input cannot to developed, and thus it is an important problem of our research subjects.

Therefore, the effectiveness of the proposed adaptive gain robust controller is shown.

2.4. Summary

In this section, an LMI-based design scheme of the centralized adaptive gain robust state feedback controller for a class of uncertain linear systems has been proposed, and by simple numerical simulations, the effectiveness of the proposed robust control strategy has been presented. Since the proposed adaptive gain robust controller can easily be obtained by solving the constrained convex optimization problem, the proposed design approach is simple. Moreover, by selecting the design parameter, the proposed adaptive gain robust controller can achieve good transient performance and/or avoid excessive control input. Note that there are trade-offs between achieving good transient performance and avoiding excessive control input.

The future research subject is the extension of proposed robust control scheme to such a broad class of systems as linear systems with state delays, uncertain systems with some constraints, and so on. Additionally, we will discuss the online adjustment for the design parameter ϑ and the design problem for output feedback control systems.

3. Synthesis of decentralized adaptive gain robust controllers

In this section, on the basis of the result derived in Section 2, an LMI-based design method of decentralized adaptive gain robust state feedback controllers for a class of uncertain large-scale interconnected systems is suggested. The design problem of the decentralized adaptive gain robust controller under consideration can also be reduced to the feasibility of LMIs, and the allowable perturbation region of uncertainties is also discussed.

3.1. Problem statement

Consider the uncertain large-scale interconnected system composed of $\mathcal{N}$ subsystems described as

$$\frac{d}{dt}\mathrm{x}_i(t) = A_{ii}(t)\mathrm{x}_i(t) + \sum_{\substack{j=1 \\ j\neq i}}^{\mathcal{N}} A_{ij}(t)\mathrm{x}_j(t) + B_i u_i(t), \tag{35}$$

where $\mathrm{x}_i(t) \in \mathbb{R}^{\mathrm{n}_i}$ and $u_i(t) \in \mathbb{R}^{\mathrm{m}_i}$ $(i = 1, \cdots, \mathcal{N})$ are the vectors of the state and the control input for the ith subsystem, respectively, and $\mathrm{x}(t) = \left(\mathrm{x}_1^T(t), \cdots, \mathrm{x}_{\mathcal{N}}^T(t)\right)^T$ is the state of the overall system. The matrices $A_{ii}(t) \in \mathbb{R}^{\mathrm{n}_i\times\mathrm{n}_i}$ and $A_{ij}(t) \in \mathbb{R}^{\mathrm{n}_i\times\mathrm{n}_j}$ in Eq. (35) are given by

$$\begin{aligned} A_{ii}(t) &= A_{ii} + \Delta_{ii}(t), \\ A_{ij}(t) &= A_{ij} + \Delta_{ij}(t). \end{aligned} \tag{36}$$

In Eqs. (35) and (36), the matrices $A_{ii} \in \mathbb{R}^{\mathrm{n}_i\times\mathrm{n}_i}$, $A_{ij} \in \mathbb{R}^{\mathrm{n}_i\times\mathrm{n}_j}$, and $B_i \in \mathbb{R}^{\mathrm{n}_i\times\mathrm{m}_i}$ denote the nominal values of the system, and matrices $\Delta_{ii}(t) \in \mathbb{R}^{\mathrm{n}_i\times\mathrm{n}_i}$ and $\Delta_{ij}(t) \in \mathbb{R}^{\mathrm{n}_i\times\mathrm{n}_j}$ show unknown parameters which satisfy $\Delta_{ii}^T(t)\Delta_{ii}(t) \le \rho_{ii}^\star I_{n_i}$ and $\Delta_{ij}^T(t)\Delta_{ij}(t) \le \rho_{ij}^\star I_{n_j}$, respectively. Note that the elements of these unknown parameters are Lebesgue measurable [1, 34]. For Eq. (35), the nominal subsystem, ignoring the unknown parameters, is given by

$$\frac{d}{dt}\overline{x}_i(t) = A_{ii}\overline{x}_i(t) + \sum_{\substack{j=1\\ j\neq i}}^{\mathcal{N}} A_{ij}\overline{x}_j(t) + B_i\overline{u}_i(t), \tag{37}$$

where $\overline{x}_i(t) \in \mathbb{R}^{n_i}$ and $\overline{u}_i(t) \in \mathbb{R}^{m_i}$ are the vectors of the state and the control input for the ith nominal subsystem, respectively. Furthermore, the control input for the nominal subsystem of Eq. (37) is determined as

$$\overline{u}_i(t) = -K_i\overline{x}_i(t), \tag{38}$$

where $K_i \in \mathbb{R}^{m_i \times n_i}$ is a fixed gain matrix. From Eqs. (37) and (38), the following nominal closed-loop subsystem is obtained:

$$\frac{d}{dt}\overline{x}_i(t) = A_{K_i}\overline{x}_i(t) + \sum_{\substack{j=1\\ j\neq i}}^{\mathcal{N}} A_{ij}\overline{x}_j(t), \tag{39}$$

where $A_{K_i} \stackrel{\Delta}{=} A_{ii} - B_iK_i$.

Now, by using symmetric positive definite matrices $\mathcal{P}_i \in \mathbb{R}^{n_i \times n_i}$, we consider the quadratic function

$$\mathcal{V}(\overline{x}, t) \stackrel{\Delta}{=} \sum_{i=1}^{\mathcal{N}} \mathcal{V}_i(\overline{x}_i, t), \tag{40}$$

$$\mathcal{V}_i(\overline{x}_i, t) \stackrel{\Delta}{=} \overline{x}_i^T(t)\mathcal{P}_i\overline{x}_i(t), \tag{41}$$

as a Lyapunov function candidate. For the quadratic function $\mathcal{V}_i(\overline{x}_i, t)$ of Eq. (41), its time derivative along the trajectory of the nominal closed-loop subsystem of Eq. (39) is given by

$$\frac{d}{dt}\mathcal{V}_i(\overline{x}_i, t) = \overline{x}_i^T(t)\left(A_{K_i}^T\mathcal{P}_i + \mathcal{P}_iA_{K_i}\right)\overline{x}_i(t) + \sum_{\substack{j=1\\ j\neq i}}^{\mathcal{N}} 2\overline{x}_i^T(t)\mathcal{P}_iA_{ij}\overline{x}_j(t). \tag{42}$$

For the second term on the right side of Eq. (42), by using the well-known relation of Eq. (20), we can obtain the following relation:

$$\frac{d}{dt}\mathcal{V}_i(\overline{x}_i, t) \leq \overline{x}_i^T(t)\left(A_{K_i}^T\mathcal{P}_i + \mathcal{P}_iA_{K_i}\right)\overline{x}_i(t) + \sum_{\substack{j=1\\ j\neq i}}^{\mathcal{N}} \mu_{ij}\overline{x}_i^T(t)\mathcal{P}_iA_{ij}A_{ij}^T\mathcal{P}\overline{x}_i(t) + \sum_{\substack{j=1\\ j\neq i}}^{\mathcal{N}} \frac{1}{\mu_{ij}}\overline{x}_j^T(t)\overline{x}_j(t). \tag{43}$$

From Eqs. (40) and (43), we have

$$\frac{d}{dt}\mathcal{V}(\overline{x},t)\leq\sum_{i=1}^{\mathcal{N}}\overline{x}_i^T(t)\left(A_{K_i}^T\mathcal{P}_i+\mathcal{P}_iA_{K_i}\right)\overline{x}_i(t)+\sum_{i=1}^{\mathcal{N}}\sum_{\substack{j=1\\ j\neq i}}^{\mathcal{N}}\mu_i\overline{x}_i^T(t)\mathcal{P}_iA_{ij}A_{ij}^T\mathcal{P}_i\overline{x}_i(t)$$
$$+\sum_{i=1}^{\mathcal{N}}\sum_{\substack{j=1\\ j\neq i}}^{\mathcal{N}}\frac{1}{\mu_{ij}}\overline{x}_j^T(t)\overline{x}_j(t). \tag{44}$$

The inequality of Eq. (44) can also be rewritten as

$$\frac{d}{dt}\mathcal{V}(\overline{x},t)\leq\sum_{i=1}^{\mathcal{N}}\overline{x}_i^T(t)\left(A_{K_i}^T\mathcal{P}_i+\mathcal{P}_iA_{K_i}+\sum_{\substack{j=1\\ j\neq i}}^{\mathcal{N}}\mu_i\mathcal{P}_iA_{ij}A_{ij}^T\mathcal{P}_i+\sum_{\substack{j=1\\ j\neq i}}^{\mathcal{N}}\frac{1}{\mu_{ji}}I_n\right)\overline{x}_i(t). \tag{45}$$

Therefore, if the matrix inequality

$$A_{K_i}^T\mathcal{P}_i+\mathcal{P}_iA_{K_i}+\sum_{\substack{j=1\\ j\neq i}}^{\mathcal{N}}\mu_i\mathcal{P}_iA_{ij}A_{ij}^T\mathcal{P}_i+\sum_{\substack{j=1\\ j\neq i}}^{\mathcal{N}}\frac{1}{\mu_{ji}}I_n<0 \tag{46}$$

holds, then the following relation for the time derivative of $\mathcal{V}(\overline{x},t)$ is satisfied:

$$\frac{d}{dt}\mathcal{V}(\overline{x},t)<0,\quad\forall\overline{x}(t)\neq 0. \tag{47}$$

Now, as with Section 2, we derive a decentralized controller with pole placement constraints for the nominal subsystem of Eq. (37). Namely, from Eq. (46), the matrix inequality

$$\left(A_{K_i}+\alpha_iI_n\right)^T\mathcal{P}_i+\mathcal{P}_i\left(A_{K_i}+\alpha_iI_n\right)+\sum_{\substack{j=1\\ j\neq i}}^{\mathcal{N}}\mu_i\mathcal{P}_iA_{ij}A_{ij}^T\mathcal{P}_i+\sum_{\substack{j=1\\ j\neq i}}^{\mathcal{N}}\frac{1}{\mu_{ji}}I_n+\mathcal{Q}_i<0, \tag{48}$$

is considered. In Eq. (48), $\alpha_i\in\mathbb{R}$ is a positive scalar and is selected by designers.

We introduce symmetric positive definite matrices $\mathcal{Y}_i\overset{\Delta}{=}\mathcal{P}_i^{-1}$ and $\mathcal{W}_i\in\mathbb{R}^{m_i\times m_i}$ and define the fixed gain K_i as $K_i\overset{\Delta}{=}\mathcal{W}_iB_i^T\mathcal{P}_i$. Then for the matrix inequality of Eq. (48), by pre- and post-multiplying both sides of the matrix inequality of Eq. (48) by $\mathcal{Y}_i$, it can be obtained that

$$A_{ii}\mathcal{Y}_i-B_i\mathcal{W}_iB_i^T+\mathcal{Y}_iA_{ii}^T-B_i\mathcal{W}_i^TB_i^T+2\alpha_i\mathcal{Y}_i+\sum_{\substack{j=1\\ j\neq i}}^{\mathcal{N}}\mu_{ij}A_{ij}A_{ij}^T+\sum_{\substack{j=1\\ j\neq i}}^{\mathcal{N}}\frac{1}{\mu_{ji}}\mathcal{Y}_i\mathcal{Y}_i+\mathcal{Y}_i\mathcal{Q}_i\mathcal{Y}_i<0. \tag{49}$$

Thus, by applying **Lemma 1.2** (Schur complement) to Eq. (49), we find that the matrix inequality of Eq. (49) is equivalent to the following LMI:

$$\begin{pmatrix} \Lambda_i\left(\mathcal{Y}_i, \mathcal{W}_i, \mu_{ij}\right) & \Theta_i(\mathcal{Y}_i) \\ \star & -\Gamma_i\left(\mu_{ij}\right) \end{pmatrix} < 0. \tag{50}$$

In Eq. (50), matrices $\Lambda_i\left(\mathcal{Y}_i, \mathcal{W}_i, \mu_{ij}\right) \in \mathbb{R}^{n_i \times n_i}$, $\Theta_i(\mathcal{Y}_i) \in \mathbb{R}^{n_i \times \mathcal{N} n_i}$, and $\Gamma_i\left(\mu_{ij}\right) \in \mathbb{R}^{\mathcal{N} n_i \times \mathcal{N} n_i}$ are given by

$$\Lambda_i\left(\mathcal{Y}_i, \mathcal{W}_i, \mu_{ij}\right) \stackrel{\Delta}{=} A_{ii}\mathcal{Y}_i - B_i\mathcal{W}_i B_i^T + \mathcal{Y}_i A_{ii}^T - B_i \mathcal{W}_i^T B_i^T + 2\alpha_i \mathcal{Y}_i + \sum_{\substack{j=1 \\ j \neq i}}^{\mathcal{N}} \mu_{ij} A_{ij} A_{ij}^T,$$

$$\Theta_i(\mathcal{Y}_i) \stackrel{\Delta}{=} \left(\overbrace{\mathcal{Y}_i \; \mathcal{Y}_i \; \cdots \; \mathcal{Y}_i}^{\mathcal{N}}\right),$$

$$\Gamma_i\left(\mu_{ij}\right) \stackrel{\Delta}{=} \mathrm{diag}\left(\mathcal{Q}_i^{-1}, \mu_{1i} I_n, \mu_{2i} I_n, \cdots, \mu_{i-1i} I_n, \mu_{i+1i} I_n, \cdots, \mu_{\mathcal{N}i} I_n\right). \tag{51}$$

Therefore, if matrices $\mathcal{Y}_i \in \mathbb{R}^{n_i \times n_i}$ and $\mathcal{W}_i \in \mathbb{R}^{m_i \times m_i}$ and positive scalars μ_{ij} exist, the nominal closed-loop subsystem is asymptotically stable, and the fixed gain matrix K_i is determined as $K_i = \mathcal{W}_i B_i^T \mathcal{Y}_i^{-1}$.

Now, by using the fixed gain matrix $K_i \in \mathbb{R}^{m_i \times n_i}$ which is designed for the nominal subsystem, we define the control input

$$u_i(t) \stackrel{\Delta}{=} -(1 + \theta_i(t)) K_i x_i(t), \tag{52}$$

where $\theta_i(t) \in \mathbb{R}^1$ is an adjustable time-varying parameter. From Eqs. (35) and (52), the uncertain closed-loop subsystem can be obtained as

$$\frac{d}{dt} x_i(t) = A_{K_i} x_i(t) + \Delta_{ii}(t) x_i(t) + \sum_{i=1}^{\mathcal{N}} \left(A_{ij} + \Delta_{ij}(t)\right) x_j(t) - \theta_i(t) B_i K_i x_i(t). \tag{53}$$

From the above discussion, the designed objective in this section is to determine the decentralized robust control of Eq. (52) such that the resultant overall system achieves robust stability. That is to design the adjustable time-varying parameter $\theta_i(t) \in \mathbb{R}^1$ such that asymptotical stability of the overall system composed of $\mathcal{N}$ subsystems of Eq. (53) is guaranteed.

3.2. Decentralized variable gain controllers

The following theorem shows sufficient conditions for the existence of the proposed decentralized adaptive gain robust control system.

Theorem 3: *Consider the uncertain large-scale interconnected system of Eq. (35) and the control input of Eq. (52).*

For a given positive constant ϑ_i, if positive constants ξ_{ii}, σ_{ij}, and ε_{ij} exist which satisfy the LMIs

$$\begin{pmatrix} \Pi_i\left(\xi_{ii}, \varepsilon_{ij}, \sigma_{ij}\right) & \Xi_i \\ \star & -\Omega_i\left(\xi_{ii}, \varepsilon_{ij}, \sigma_{ij}\right) \end{pmatrix} < 0, \tag{54}$$

the time-varying adjustable parameters $\theta_i(t) \in \mathbb{R}$ are determined as

$$\theta_i(t) \triangleq \begin{cases} \dfrac{\sqrt{\rho_{ii}^\star}\|\mathcal{P}_i x_i(t)\|\|x_i(t)\|}{x_i^T(t)\mathcal{P}_i B_i \mathcal{W}_i B_i^T \mathcal{P}_i x_i(t)} & \text{if} \quad x_i^T(t)\mathcal{P}_i B_i \mathcal{W}_i B_i^T \mathcal{P}_i x_i(t) \geq \vartheta_i x_i^T(t) x_i(t), \\ \dfrac{\sqrt{\rho_{ii}^\star}\|\mathcal{P}_i x_i(t)\|\|x_i(t)\|}{\vartheta_i x_i^T(t) x_i(t)} & \text{if} \quad x_i^T(t)\mathcal{P}_i B_i \mathcal{W}_i B_i^T \mathcal{P}_i x_i(t) < \vartheta_i x_i^T(t) x_i(t), \end{cases} \tag{55}$$

where matrices $\Pi_i\left(\xi_i, \varepsilon_{ij}, \sigma_{ij}\right) \in \mathbb{R}^{n_i \times n_i}$, $\Xi_i \in \mathbb{R}^{n_i \times (2\mathcal{N}-1)n_i}$, and $\Omega_i\left(\xi_{ii}, \varepsilon_{ij}, \sigma_{ij}\right) \in \mathbb{R}^{(2\mathcal{N}-1)n_i \times (2\mathcal{N}-1)n_i}$ are given by

$$\Pi_i\left(\xi_{ii}, \varepsilon_{ij}, \sigma_{ij}\right) \triangleq \left(A_{K_i}^T \mathcal{P}_i + \mathcal{P}_i A_{K_i}\right) + \xi_{ii}\mathcal{P}_i\mathcal{P}_i + \sum_{\substack{j=1 \\ j\neq i}}^{\mathcal{N}} \varepsilon_{ij}\mathcal{P}_i A_{ij} A_{ij}^T \mathcal{P}_i + \sum_{\substack{j=1 \\ j\neq i}}^{\mathcal{N}} \sigma_{ij}\mathcal{P}_i\mathcal{P}_i,$$

$$\Xi_i \triangleq \left(\overbrace{I_n \ I_n \ \cdots \ I_n}^{2\mathcal{N}-1}\right),$$

$$\begin{aligned} \Omega_i\left(\xi_{ii}, \varepsilon_{ij}, \sigma_{ij}\right) \triangleq \ & \mathrm{diag}\left(\xi_{ii}\rho_{ii}^\star I_n, \varepsilon_{1i} I_n, \varepsilon_{2i} I_n, \cdots, \varepsilon_{i-1i} I_n, \varepsilon_{i+1i} I_n, \cdots, \varepsilon_{\mathcal{N}i} I_n, \sigma_{1i}\rho_{1i}^\star I_n, \sigma_{2i}\rho_{2i}^\star I_n, \right. \\ & \left. \cdots, \sigma_{i-1i}\rho_{i-1i}^\star I_n, \sigma_{i+1i}\rho_{i+1i}^\star I_n, \cdots, \sigma_{\mathcal{N}i}\rho_{\mathcal{N}i}^\star I_n\right). \end{aligned} \tag{56}$$

Then, the overall close-loop system composed of $\mathcal{N}$ closed-loop subsystems is asymptotically stable.

Proof. In order to prove Theorem 3, the following Lyapunov function candidate is introduced by using symmetric positive definite matrices $\mathcal{P}_i \in \mathbb{R}^{n_i \times n_i}$ which satisfy the LMIs of (50):

$$\mathcal{V}(x, t) \triangleq \sum_{i=1}^{\mathcal{N}} \mathcal{V}_i(x_i, t), \tag{57}$$

where $\mathcal{V}_i(x_i, t)$ is a quadratic function given by

$$\mathcal{V}_i(x_i, t) \triangleq x_i^T(t)\mathcal{P}_i x_i(t). \tag{58}$$

We can obtain the following relation for the time derivative of the quadratic function $\mathcal{V}_i(x_i, t)$ of Eq. (58):

$$\begin{aligned} \frac{d}{dt}\mathcal{V}_i(x_i, t) = \ & x_i^T(t)\left(A_{K_i}^T \mathcal{P}_i + \mathcal{P}_i A_{K_i}\right)x_i(t) + 2x_i(t)\mathcal{P}_i\Delta_{ii}(t)x_i(t) \\ & + 2x_i^T(t)\mathcal{P}_i\sum_{\substack{j=1 \\ j\neq i}}^{\mathcal{N}}\left(A_{ij} + \Delta_{ij}(t)\right)x_j(t) - 2\theta_i(t)x_i^T(t)\mathcal{P}_i B_i K_i x_i(t). \end{aligned} \tag{59}$$

Firstly, we consider the case of $x_i^T(t)\mathcal{P}_iB_iW_iB_i^T\mathcal{P}_ix_i(t) \geq \vartheta_i x_i^T(t)x_i(t)$. In this case, one can see from the relations $\Delta_{ii}^T(t)\Delta_{ii}(t) \leq \rho_{ii}^\star I_{n_i}$ and $\Delta_{ij}^T(t)\Delta_{ij}(t) \leq \rho_{ij}^\star I_{n_j}$, the well-known inequality of Eq. (20), and **Lemma 1.1** that the following relation for the quadratic function $\mathcal{V}_i(x_i,t)$ of Eq. (58) can be obtained:

$$\begin{aligned}\frac{d}{dt}\mathcal{V}_i(x_i,t) \leq{}& x_i^T(t)\left(A_{K_i}^T\mathcal{P}_i + \mathcal{P}_iA_{K_i}\right)x_i(t) + 2\sqrt{\rho_{ii}^\star}\|\mathcal{P}_ix_i(t)\|\|x_i(t)\| \\ &+ \sum_{\substack{j=1\\ j\neq i}}^{\mathcal{N}} \varepsilon_{ij}x_i^T(t)\mathcal{P}_iA_{ij}A_{ij}^T\mathcal{P}_ix_i(t) + \sum_{\substack{j=1\\ j\neq i}}^{\mathcal{N}} \frac{1}{\varepsilon_{ij}}x_j^T(t)x_j(t) + \sum_{\substack{j=1\\ j\neq i}}^{\mathcal{N}} \sigma_{ij}x_i^T(t)\mathcal{P}_i\mathcal{P}_ix_i(t) \\ &+ \sum_{\substack{j=1\\ j\neq i}}^{\mathcal{N}} \frac{\rho_{ij}^\star}{\sigma_{ij}}x_j^T(t)x_j(t) - 2\theta_i(t)x_i^T(t)\mathcal{P}_iB_iK_ix_i(t).\end{aligned} \tag{60}$$

Substituting the adjustable time-varying parameter $\theta_i(t)$ of Eq. (55) into Eq. (60) gives

$$\begin{aligned}\frac{d}{dt}\mathcal{V}_i(x_i,t) \leq{}& x_i^T(t)\left(A_{K_i}^T\mathcal{P}_i + \mathcal{P}_iA_{K_i}\right)x_i(t) + 2\sqrt{\rho_{ii}^\star}\|\mathcal{P}_ix_i(t)\|\|x_i(t)\| \\ &+ \sum_{\substack{j=1\\ j\neq i}}^{\mathcal{N}} \varepsilon_{ij}x_i^T(t)\mathcal{P}_iA_{ij}A_{ij}^T\mathcal{P}_ix_i(t) + \sum_{\substack{j=1\\ j\neq i}}^{\mathcal{N}} \frac{1}{\varepsilon_{ij}}x_j^T(t)x_j(t) + \sum_{\substack{j=1\\ j\neq i}}^{\mathcal{N}} \sigma_{ij}x_i^T(t)\mathcal{P}_i\mathcal{P}_ix_i(t) \\ &+ \sum_{\substack{j=1\\ j\neq i}}^{\mathcal{N}} \frac{\rho_{ij}^\star}{\sigma_{ij}}x_j^T(t)x_j(t) - 2\left(\frac{\sqrt{\rho_{ii}^\star}\|\mathcal{P}_ix_i(t)\|\|x_i(t)\|}{x_i^T(t)\mathcal{P}_iB_i\mathcal{W}_iB_i^T\mathcal{P}_ix_i(t)}\right)x_i^T(t)\mathcal{P}_iB_iK_ix_i(t) \\ ={}& x_i^T(t)\left(A_{K_i}^T\mathcal{P}_i + \mathcal{P}_iA_{K_i}\right)x_i(t) + \sum_{\substack{j=1\\ j\neq i}}^{\mathcal{N}} \varepsilon_{ij}x_i^T(t)\mathcal{P}_iA_{ij}A_{ij}^T\mathcal{P}_ix_i(t) + \sum_{\substack{j=1\\ j\neq i}}^{\mathcal{N}} \frac{1}{\varepsilon_{ij}}x_j^T(t)x_j(t) \\ &+ \sum_{\substack{j=1\\ j\neq i}}^{\mathcal{N}} \sigma_{ij}x_i^T(t)\mathcal{P}_i\mathcal{P}_ix_i(t) + \sum_{\substack{j=1\\ j\neq i}}^{\mathcal{N}} \frac{\rho_{ij}^\star}{\sigma_{ij}}x_j^T(t)x_j(t),\end{aligned} \tag{61}$$

and, thus, we have the following inequality for the function $\mathcal{V}(x,t)$ of Eq. (57):

$$\begin{aligned}\frac{d}{dt}\mathcal{V}(x,t) \leq{}& \sum_{i=1}^{\mathcal{N}} x_i^T(t)\left(A_{K_i}^T\mathcal{P}_i + \mathcal{P}_iA_{K_i}\right)x_i(t) + \sum_{i=1}^{\mathcal{N}}\sum_{\substack{j=1\\ j\neq i}}^{\mathcal{N}} \varepsilon_{ij}x_i(t)\mathcal{P}_iA_{ij}A_{ij}^T\mathcal{P}_ix_i(t) \\ &+ \sum_{i=1}^{\mathcal{N}}\sum_{\substack{j=1\\ j\neq i}}^{\mathcal{N}} \frac{1}{\varepsilon_{ij}}x_j^T(t)x_j(t) + \sum_{i=1}^{\mathcal{N}}\sum_{\substack{j=1\\ j\neq i}}^{\mathcal{N}} \sigma_{ij}x_i^T(t)\mathcal{P}_i\mathcal{P}_ix_i(t) + \sum_{i=1}^{\mathcal{N}}\sum_{\substack{j=1\\ j\neq i}}^{\mathcal{N}} \frac{\rho_{ij}^\star}{\sigma_{ij}}x_j^T(t)x_j(t).\end{aligned} \tag{62}$$

Furthermore, the inequality of Eq. (62) can be rewritten as

$$\frac{d}{dt}\mathcal{V}(\mathrm{x},t) \le \sum_{i=1}^{\mathcal{N}} \mathrm{x}_i^T(t)\left(A_{K_i}^T\mathcal{P}_i + \mathcal{P}_i A_{K_i} + \sum_{\substack{j=1\\ j\neq i}}^{\mathcal{N}} \varepsilon_{ij}\mathcal{P}_i A_{ij}A_{ij}^T\mathcal{P}_i + \sum_{\substack{j=1\\ j\neq i}}^{\mathcal{N}} \frac{1}{\varepsilon_{ji}} I_{\mathrm{n}} + \sum_{\substack{j=1\\ j\neq i}}^{\mathcal{N}} \sigma_{ij}\mathcal{P}_i\mathcal{P}_i + \sum_{\substack{j=1\\ j\neq i}}^{\mathcal{N}} \frac{\rho_{ji}^\star}{\sigma_{ji}} I_{\mathrm{n}} \right)\mathrm{x}_i(t). \quad (63)$$

Therefore, if the matrix inequality

$$A_{K_i}^T\mathcal{P}_i + \mathcal{P}_i A_{K_i} + \sum_{\substack{j=1\\ j\neq i}}^{\mathcal{N}} \varepsilon_{ij}\mathcal{P}_i A_{ij}A_{ij}^T\mathcal{P}_i + \sum_{\substack{j=1\\ j\neq i}}^{\mathcal{N}} \frac{1}{\varepsilon_{ji}} I_{\mathrm{n}} + \sum_{\substack{j=1\\ j\neq i}}^{\mathcal{N}} \sigma_{ij}\mathcal{P}_i\mathcal{P}_i + \sum_{\substack{j=1\\ j\neq i}}^{\mathcal{N}} \frac{\rho_{ij}^\star}{\sigma_{ji}} I_{\mathrm{n}} < 0 \quad (64)$$

holds, then the following relation for the time derivative of $\mathcal{V}(\mathrm{x},t)$ is satisfied:

$$\frac{d}{dt}\mathcal{V}(\mathrm{x},t) < 0, \quad \forall \mathrm{x}(t) \neq 0. \quad (65)$$

Next, we consider the case of $\mathrm{x}_i^T(t)\mathcal{P}_i B_i \mathcal{W}_i B_i^T\mathcal{P}_i\mathrm{x}_i(t) < \vartheta_i \mathrm{x}_i^T(t)\mathrm{x}_i(t)$. In this case, by using the relations $\Delta_{ii}^T(t)\Delta_{ii}(t) \le \rho_{ii}^\star I_{n_i}$ and $\Delta_{ij}^T(t)\Delta_{ij}(t) \le \rho_{ij}^\star I_{n_j}$, and Eq. (20) and substituting the adjustable time-varying parameter $\theta_i(t)$ of Eq. (55) into Eq. (59), we have

$$\begin{aligned}\frac{d}{dt}\mathcal{V}_i(\mathrm{x}_i,t) \le\ & \mathrm{x}_i^T(t)\left(A_{K_i}^T\mathcal{P}_i + \mathcal{P}_i A_{K_i}\right)\mathrm{x}_i(t) + \xi_{ii}\mathrm{x}_i^T(t)\mathcal{P}_i\mathcal{P}_i\mathrm{x}_i(t) + \frac{\rho_{ii}^\star}{\xi_{ii}}\mathrm{x}_i^T(t)\mathrm{x}_i(t)\\ & + \sum_{\substack{j=1\\ j\neq i}}^{\mathcal{N}} \varepsilon_{ij}\mathrm{x}_i(t)\mathcal{P}_i A_{ij}A_{ij}^T\mathcal{P}_i\mathrm{x}_i(t) + \sum_{\substack{j=1\\ j\neq i}}^{\mathcal{N}} \frac{1}{\varepsilon_{ij}}\mathrm{x}_j^T(t)\mathrm{x}_j(t) + \sum_{\substack{j=1\\ j\neq i}}^{\mathcal{N}} \sigma_{ij}\mathrm{x}_i^T(t)\mathcal{P}_i\mathcal{P}_i\mathrm{x}_i(t)\\ & + \sum_{\substack{j=1\\ j\neq i}}^{\mathcal{N}} \frac{\rho_{ij}^\star}{\sigma_{ij}}\mathrm{x}_j^T(t)\mathrm{x}_j(t) - 2\left(\frac{\sqrt{\rho_{ii}^\star}\|\mathcal{P}_i\mathrm{x}_i(t)\|\|\mathrm{x}_i(t)\|}{\vartheta_i\mathrm{x}_i^T(t)\mathrm{x}_i(t)}\right)\mathrm{x}_i^T(t)\mathcal{P}_i B_i K_i\mathrm{x}_i(t).\end{aligned} \quad (66)$$

The last term on the right side of Eq. (66) is less than 0 because the matrix $K_i \in \mathbb{R}^{\mathrm{m}_i\times\mathrm{n}_i}$ is defined as $K_i = \mathcal{W}_i B_i^T\mathcal{P}_i$ and $\theta_i(t)$ is a positive scalar function. Therefore, we find that the following relation for the quadratic function $\mathcal{V}_i(\mathrm{x}_i,t)$ is satisfied:

$$\begin{aligned}\frac{d}{dt}\mathcal{V}_i(\mathrm{x}_i,t) \le\ & \mathrm{x}_i^T(t)\left(A_{K_i}^T\mathcal{P}_i + \mathcal{P}_i A_{K_i}\right)\mathrm{x}_i(t) + \xi_{ii}\mathrm{x}_i^T(t)\mathcal{P}_i\mathcal{P}_i\mathrm{x}_i(t) + \frac{\rho_{ii}^\star}{\xi_{ii}}\mathrm{x}_i^T(t)\mathrm{x}_i(t)\\ & + \sum_{\substack{j=1\\ j\neq i}}^{\mathcal{N}} \varepsilon_{ij}\mathrm{x}_i(t)\mathcal{P}_i A_{ij}A_{ij}^T\mathcal{P}_i\mathrm{x}_i(t) + \sum_{\substack{j=1\\ j\neq i}}^{\mathcal{N}} \frac{1}{\varepsilon_{ij}}\mathrm{x}_j^T(t)\mathrm{x}_j(t) + \sum_{\substack{j=1\\ j\neq i}}^{\mathcal{N}} \sigma_{ij}\mathrm{x}_i^T(t)\mathcal{P}_i\mathcal{P}_i\mathrm{x}_i(t)\\ & + \sum_{\substack{j=1\\ j\neq i}}^{\mathcal{N}} \frac{\rho_{ij}^\star}{\sigma_{ij}}\mathrm{x}_j^T(t)\mathrm{x}_j(t).\end{aligned} \quad (67)$$

Therefore, we see from Eqs. (57) and (67) that the following inequality:

$$\frac{d}{dt}\mathcal{V}(\mathrm{x},t) \leq \sum_{i=1}^{\mathcal{N}} \mathrm{x}_i^T(t)\left(A_{K_i}^T\mathcal{P}_i + \mathcal{P}_iA_{K_i}\right)\mathrm{x}_i(t) + \sum_{i=1}^{\mathcal{N}} \xi_{ii}\mathrm{x}_i^T(t)\mathcal{P}_i\mathcal{P}_i\mathrm{x}_i(t) + \sum_{i=1}^{\mathcal{N}} \frac{\rho_{ii}^\star}{\xi_{ii}}\mathrm{x}_i^T(t)\mathrm{x}_i(t)$$
$$+ \sum_{i=1}^{\mathcal{N}} \sum_{\substack{j=1 \\ j\neq i}}^{\mathcal{N}} \varepsilon_{ij}\mathrm{x}_i(t)P_iA_{ij}A_{ij}^TP_i\mathrm{x}_i(t) + \sum_{i=1}^{\mathcal{N}} \sum_{\substack{j=1 \\ j\neq i}}^{\mathcal{N}} \frac{1}{\varepsilon_{ij}}\mathrm{x}_j^T(t)\mathrm{x}_j(t) \tag{68}$$
$$+ \sum_{i=1}^{\mathcal{N}} \sum_{\substack{j=1 \\ j\neq i}}^{\mathcal{N}} \sigma_{ij}\mathrm{x}_i^T(t)\mathcal{P}_i\mathcal{P}_i\mathrm{x}_i(t) + \sum_{i=1}^{\mathcal{N}} \sum_{\substack{j=1 \\ j\neq i}}^{\mathcal{N}} \frac{\rho_{ij}^\star}{\sigma_{ij}}\mathrm{x}_j^T(t)\mathrm{x}_j(t)$$

can be derived. Moreover, one can easily see that the inequality of Eq. (68) can be rewritten as

$$\frac{d}{dt}\mathcal{V}(\mathrm{x},t) \leq \sum_{i=1}^{\mathcal{N}} \mathrm{x}_i^T(t)\left(A_{K_i}^T\mathcal{P}_i + \mathcal{P}_iA_{K_i} + \xi_{ii}\mathcal{P}_i\mathcal{P}_i + \frac{\rho_{ii}^\star}{\xi_{ii}}I_{\mathrm{n}} + \sum_{\substack{j=1 \\ j\neq i}}^{\mathcal{N}} \varepsilon_{ij}\mathcal{P}_iA_{ij}A_{ij}^T\mathcal{P}_i\right.$$
$$\left. + \sum_{\substack{j=1 \\ j\neq i}}^{\mathcal{N}} \frac{1}{\varepsilon_{ji}}I_{\mathrm{n}} + \sum_{\substack{j=1 \\ j\neq i}}^{\mathcal{N}} \sigma_{ij}\mathcal{P}_i\mathcal{P}_i + \sum_{\substack{j=1 \\ j\neq i}}^{\mathcal{N}} \frac{\rho_{ji}^\star}{\sigma_{ji}}I_{\mathrm{n}}\right)x_i(t). \tag{69}$$

Therefore, if the matrix inequality

$$A_{K_i}^T\mathcal{P}_i + \mathcal{P}_iA_{K_i} + \xi_{ii}\mathcal{P}_i\mathcal{P}_i + \frac{\rho_{ii}^\star}{\xi_{ii}}I_{\mathrm{n}} + \sum_{\substack{j=1 \\ j\neq i}}^{\mathcal{N}} \varepsilon_{ij}\mathcal{P}_iA_{ij}A_{ij}^T\mathcal{P}_i + \sum_{\substack{j=1 \\ j\neq i}}^{\mathcal{N}} \frac{1}{\varepsilon_{ji}}I_{\mathrm{n}} + \sum_{\substack{j=1 \\ j\neq i}}^{\mathcal{N}} \sigma_{ij}\mathcal{P}_i\mathcal{P}_i$$
$$+ \sum_{\substack{j=1 \\ j\neq i}}^{\mathcal{N}} \frac{\rho_{ji}^\star}{\sigma_{ji}}I_{\mathrm{n}} < 0 \tag{70}$$

holds, then the relation of Eq. (65) for the time derivative of the function $\mathcal{V}(\mathrm{x},t)$ of Eq. (57) is satisfied. Due to the 3rd and 4th terms on the left side of Eq. (70) which are positive definite, if the inequality of Eq. (70) is satisfied, then the inequality of Eq. (64) is also constantly satisfied.

For the matrix inequality of Eq. (70), by applying **Lemma 1.2** (Schur complement), one can find that the matrix inequalities of Eq. (70) are equivalent to the LMIs of Eq. (54). Therefore, by solving the LMIs of Eq. (54), the adjustable time-varying parameter is given by Eq. (55), and proposed control input of Eq. (52) stabilizes the overall system of Eq. (35). Thus, the proof of Theorem 3 is completed.

Next, as mentioned in Section 2, we discuss the allowable region of the unknown parameters $\Delta_{ii}(t) \in \mathbb{R}^{\mathrm{n}_i \times \mathrm{n}_i}$ and $\Delta_{ij}(t) \in \mathbb{R}^{\mathrm{n}_i \times \mathrm{n}_j}$. Thus, the following additional constraints are introduced:

$$\begin{aligned}\rho_{ii}^{\star} &= \xi_{ii},\\ \rho_{ij}^{\star} &= \sigma_{ij}.\end{aligned} \tag{71}$$

From the relations of Eq. (71), one can find that the maximization of ξ_{ii} and σ_{ij} is equivalent to the maximization of $\rho_{ii}^{\star}$ and $\rho_{ij}^{\star}$. Then, the LMIs of Eq. (54) can be rewritten as

$$\begin{pmatrix} \Pi_{i'}(\xi_{ii}, \varepsilon_{ij}, \sigma_{ij}) & \Xi_{i'} \\ \star & -\Omega_{i'}(\varepsilon_{ij}) \end{pmatrix} < 0, \tag{72}$$

$$\begin{aligned}&\Pi_{i'}(\xi_{ii}, \varepsilon_{ij}, \sigma_{ij}) \stackrel{\Delta}{=} A_{K_i}^T \mathcal{P}_i + \mathcal{P}_i A_{K_i} + \xi_{ii}\mathcal{P}_i\mathcal{P}_i + \mathcal{N} I_n + \sum_{\substack{j=1\\ j\neq i}}^{\mathcal{N}} \varepsilon_{ij}\mathcal{P}_i A_{ij} A_{ij}^T \mathcal{P}_i + \sum_{\substack{j=1\\ j\neq i}}^{\mathcal{N}} \sigma_{ij}\mathcal{P}_i\mathcal{P}_i,\\ &\Xi_{i'} \stackrel{\Delta}{=} \Bigg(\overbrace{I_n \quad I_n \quad \cdots \quad I_n}^{\mathcal{N}-1}\Bigg),\\ &\Omega_{i'}(\varepsilon_{ij}) \stackrel{\Delta}{=} \mathrm{diag}(\varepsilon_{1i}I_n, \varepsilon_{2i}I_n, \cdots, \varepsilon_{i-1i}I_n, \varepsilon_{i+1i}I_n, \cdots, \varepsilon_{\mathcal{N}i}I_n).\end{aligned} \tag{73}$$

Furthermore, we introduce a positive scalar λ and a complementary matrix $\Gamma \in \mathbb{R}^{\mathcal{N}^2 \times \mathcal{N}^2}$ defined as

$$\Gamma \stackrel{\Delta}{=} \mathrm{diag}(\xi_{11}, \xi_{22}, \cdots, \xi_{\mathcal{N}\mathcal{N}}, \sigma_{12}, \sigma_{13}, \cdots, \sigma_{1\mathcal{N}}, \sigma_{21}, \sigma_{23}, \cdots, \sigma_{\mathcal{N}\mathcal{N}-1}), \tag{74}$$

and consider the following additional condition:

$$\Gamma - \frac{1}{\lambda} I_{\mathcal{N}^2} > 0. \tag{75}$$

Namely, we can replace the maximization problem of ξ_{ii} and σ_{ij} with the minimization problem of λ. From Eq. (75) and **Lemma 1.2** (Schur complement), one can easily see that the constraint of Eq. (75) can be transformed into

$$\begin{pmatrix} \Gamma & I_{\mathcal{N}^2} \\ \star & \lambda I_{\mathcal{N}^2} \end{pmatrix} > 0. \tag{76}$$

Thus, in order to design the proposed decentralized adaptive gain robust controller, the constrained convex optimization problem

$$\underset{\xi_{ii}>0,\ \varepsilon_{ij}>0,\ \sigma_{ij}>0}{\mathrm{Minimize}} \quad [\lambda] \text{ subject to } (72) \text{ and } (76) \tag{77}$$

should be solved.

As a result, the following theorem can be obtained:

Theorem 4: *Consider the uncertain large-scale interconnected system of Eq. (35) and the control input of Eq. (52).*

If positive constants ξ_{ii}, ε_{ij}, σ_{ij}, and λ exist which satisfy the constrained convex optimization problem of Eq. (77), the adjustable time-varying parameter $\theta_i(t)$ is designed as Eq. (55). Then, the overall *uncertain closed-loop system of Eq. (53) is asymptotically stable. Furthermore, by using the optimal solution* $\xi_{ii}^{\star}$ *and* $\sigma_{ij}^{\star}$ *for Eq. (77), the upper bound of unknown parameters* $\Delta_{ii}(t) \in \mathbb{R}^{n_i \times n_i}$ *and* $\Delta_{ij}(t) \in \mathbb{R}^{n_i \times n_j}$ *is given by*

$$\begin{aligned} \rho_{ii}^{\star} &= \xi_{ii}^{\star}, \\ \rho_{ij}^{\star} &= \sigma_{ij}^{\star}. \end{aligned} \tag{78}$$

3.3. Illustrative examples

To demonstrate the efficiency of the proposed decentralized robust controller, an illustrative example is provided. In this example, we consider the uncertain large-scale interconnected system consisting of three two-dimensional subsystems, that is, $\mathcal{N} = 3$. The system parameters are given as follows:

$$\begin{aligned} &\mathcal{A}_{11} = \begin{pmatrix} -1.0 & 1.0 \\ 0.0 & 1.0 \end{pmatrix}, \mathcal{A}_{22} = \begin{pmatrix} 0.0 & 1.0 \\ -1.0 & -1.0 \end{pmatrix}, \mathcal{A}_{33} = \begin{pmatrix} 1.0 & 0.0 \\ 1.0 & -3.0 \end{pmatrix}, \\ &\mathcal{B}_1 = \begin{pmatrix} 0.0 \\ 1.0 \end{pmatrix}, \mathcal{B}_2 = \begin{pmatrix} 1.0 \\ 1.0 \end{pmatrix}, \mathcal{B}_3 = \begin{pmatrix} 1.0 \\ 0.0 \end{pmatrix}, \\ &\mathcal{A}_{12} = \begin{pmatrix} 0.5 & 0.0 \\ 0.0 & 1.0 \end{pmatrix}, \mathcal{A}_{13} = \begin{pmatrix} 0.0 & 0.5 \\ 0.0 & 0.0 \end{pmatrix}, \mathcal{A}_{21} = \begin{pmatrix} 0.0 & 0.0 \\ 0.0 & 0.5 \end{pmatrix}, \\ &\mathcal{A}_{23} = \begin{pmatrix} 0.0 & 0.5 \\ 1.0 & 0.0 \end{pmatrix}, \mathcal{A}_{31} = \begin{pmatrix} 0.5 & 0.0 \\ 0.0 & 0.0 \end{pmatrix}, \mathcal{A}_{32} = \begin{pmatrix} 0.0 & 0.5 \\ 0.0 & 0.5 \end{pmatrix}. \end{aligned} \tag{79}$$

Firstly, by selecting the design parameters $\alpha_i \in \mathbb{R}^1$ and $\mathcal{Q}_i \in \mathbb{R}^{2\times 2}$ $(i = 1, 2, 3)$ as $\alpha_1 = \alpha_2 = \alpha_3 = 1.0$ and $\mathcal{Q}_1 = \mathcal{Q}_2 = \mathcal{Q}_3 = 2.0 \times I_2$ and solving LMIs of Eq. (50), we have the symmetric positive definite matrices $\mathcal{Y}_i \in \mathbb{R}^{2\times 2}$ and $\mathcal{W}_i \in \mathbb{R}^{1\times 1}$, and positive scalars μ_{ij} can be obtained:

$$\begin{aligned} &\mathcal{Y}_1 = \begin{pmatrix} 1.8972 & -2.1976 \\ \star & 8.1021 \end{pmatrix} \times 10^{-1}, \qquad \mathcal{W}_1 = 3.9298, \\ &\mathcal{Y}_2 = \begin{pmatrix} 3.4941 & 4.7825 \\ \star & 8.8702 \end{pmatrix} \times 10^{-1}, \qquad \mathcal{W}_2 = 2.2200, \\ &\mathcal{Y}_3 = \begin{pmatrix} 4.0414 \times 10^{-1} & 3.2732 \times 10^{-2} \\ \star & 3.2709 \times 10^{-1} \end{pmatrix}, \quad \mathcal{W}_3 = 3.2166, \\ &\mu_{12} = 7.0526 \times 10^{-1}, \ \mu_{13} = 4.5522 \times 10^{-1}, \ \mu_{21} = 1.3986, \\ &\mu_{23} = 3.2285 \times 10^{-1}, \ \mu_{31} = 3.4477, \ \mu_{32} = 2.0763. \end{aligned} \tag{80}$$

Thus, the symmetric positive definite matrices $\mathcal{P}_i = \mathcal{Y}_i^{-1}$ and the fixed gain matrices $K_i = \mathcal{W}_i B_i^T \mathcal{Y}_i^{-1}$ can be calculated as

$$\begin{aligned}
\mathcal{P}_1 &= \begin{pmatrix} 7.6854 & 2.0845 \\ \star & 1.7996 \end{pmatrix}, & K_1 &= \begin{pmatrix} 8.1918 & 7.0723 \end{pmatrix}, \\
\mathcal{P}_2 &= \begin{pmatrix} 1.0923 \times 10^1 & -5.8891 \\ \star & 4.3025 \end{pmatrix}, & K_2 &= \begin{pmatrix} 1.1174 \times 10^1 & -3.5221 \end{pmatrix}, \\
\mathcal{P}_3 &= \begin{pmatrix} 2.4946 & -2.4964 \times 10^{-1} \\ \star & 3.0823 \end{pmatrix}, & K_3 &= \begin{pmatrix} 8.0240 & -8.0297 \times 10^{-1} \end{pmatrix}.
\end{aligned} \tag{81}$$

Next, by solving the constrained convex optimization problem of Eq. (77), the following solution can be obtained:

$$\begin{aligned}
&\xi_{11} = 3.4167 \times 10^{-2}, && \xi_{22} = 3.5524 \times 10^{-2}, && \xi_{33} = 1.5590 \times 10^{-1}, \\
&\varepsilon_{12} = 8.5122 \times 10^{-1}, && \varepsilon_{13} = 5.9622 \times 10^{-1}, && \varepsilon_{21} = 1.4174, \\
&\varepsilon_{23} = 3.1440 \times 10^{-1}, && \varepsilon_{31} = 9.9709, && \varepsilon_{32} = 1.9446, \\
&\sigma_{12} = 3.4167 \times 10^{-2}, && \sigma_{13} = 3.4167 \times 10^{-2}, && \sigma_{21} = 3.5524 \times 10^{-2}, \\
&\sigma_{23} = 3.5524 \times 10^{-2}, && \sigma_{31} = 1.5590 \times 10^{-1}, && \sigma_{32} = 1.5590 \times 10^{-1}, \\
&\lambda = 1.0001.
\end{aligned} \tag{82}$$

Therefore, the allowable upper bound of unknown parameters is given as

$$\begin{aligned}
&\rho_{11}^{\star} = 3.4167 \times 10^{-2}, && \rho_{22}^{\star} = 3.5524 \times 10^{-2}, && \rho_{33}^{\star} = 1.5590 \times 10^{-1}, \\
&\rho_{12}^{\star} = 3.4167 \times 10^{-2}, && \rho_{13}^{\star} = 3.4167 \times 10^{-2}, && \rho_{21}^{\star} = 3.5524 \times 10^{-2}, \\
&\rho_{23}^{\star} = 3.5524 \times 10^{-2}, && \rho_{31}^{\star} = 1.5590 \times 10^{-1}, && \rho_{32}^{\star} = 1.5590 \times 10^{-1}.
\end{aligned} \tag{83}$$

In this example, unknown parameters $\Delta_{ii}(t) \in \mathbb{R}^{2\times2}$ and $\Delta_{ij}(t) \in \mathbb{R}^{2\times2}$ are chosen as

$$\begin{aligned}
\Delta_{ii}(t) &= \rho_{ii}^{\star} \times \begin{pmatrix} \sin(5.0 \times \pi \times t) & -\cos(2.0 \times \pi \times t) \\ \star & \cos(5.0 \times \pi \times t) \end{pmatrix}, \\
\Delta_{ij}(t) &= \rho_{ij}^{\star} \times \begin{pmatrix} -\cos(\pi \times t) & \sin(3.0 \times \pi \times t) \\ \star & \sin(\pi \times t) \end{pmatrix}.
\end{aligned} \tag{84}$$

Moreover, the design parameters $\vartheta_i (i = 1, 2, 3)$, the initial value of the uncertain large-scale system with system parameters of Eq. (79), and one of the nominal systems are selected as $\vartheta_1 = \vartheta_2 = \vartheta_3 = 1.0 \times 10^{-1}$ and $\mathrm{x}(0) = \overline{\mathrm{x}}(0) = \left(1.5\ –1.0\ –1.0\ 5.0 \times 10^{-1}\ 2.0\ –1.0\right)^T$.

The result of this example is shown in **Figures 5** and **6**. In these figures, $\mathrm{x}_i^{(l)}(t)$, $u_i(t)$, $\overline{\mathrm{x}}_i^{(l)}(t)$, and $\overline{u}_i(t)$ denote the lth element $(l = 1, 2)$ of the state $\mathrm{x}_i(t)$ and the control input $u_i(t)$ for ith subsystem and one of the states $\overline{\mathrm{x}}_i(t)$ and the control input $\overline{u}_i(t)$ for ith nominal subsystem.

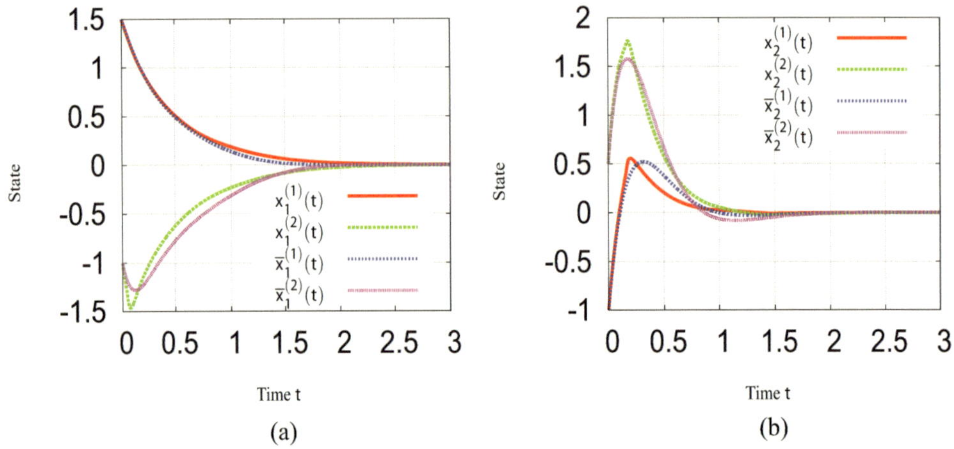

Figure 5. Time histories of $x_i(t)$ and $\overline{x}_i(t)$ $(i = 1, 2)$. (a) The time histories of $x_1(t)$ and $\overline{x}_1(t)$, (b) Time histories of $x_2(t)$ and $\overline{x}_2(t)$.

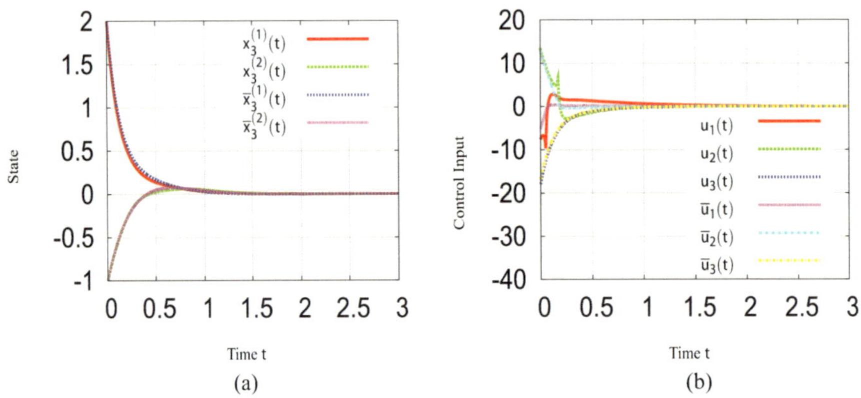

Figure 6. Time histories of $x_3(t)$, $\overline{x}_3(t)$, $u(t)$ and $\overline{u}(t)$. (a) Time histories of $x_3(t)$ and $\overline{x}_3(t)$, (b) Time histories of $u(t)$ and $\overline{u}(t)$.

From these figures, the proposed decentralized adaptive gain robust controller stabilizes the uncertain large-scale interconnected system with system parameters of Eq. (79). Furthermore, one can see that each subsystem achieves good transient behavior close to nominal subsystems by the proposed decentralized robust controller. Thus, the effectiveness of the proposed robust control strategy is shown.

3.4. Summary

In this section, on the basis of the result of Section 2, we have suggested the decentralized adaptive gain robust controller for the large-scale interconnected system with uncertainties.

Furthermore, the effectiveness of the proposed controller has been shown via an illustrative example. The proposed adaptive gain robust controller can be easily designed by solving a constrained convex optimization problem and adjust the magnitude of the control input for each subsystem. Therefore, we find that the proposed decentralized robust controller design method is very useful.

Future research subjects include analysis of conservatism for the proposed controller design approach and extension of the proposed adaptive gain robust control strategies to uncertain systems with time delay, decentralized output/observer-based control systems, and so on.

4. Conclusions and future works

In this chapter, firstly the centralized adaptive gain robust controller for a class of uncertain linear systems has been proposed, and through a simple numerical example, we have shown the effectiveness/usefulness for the proposed adaptive gain robust control strategy. Next, for a class of uncertain large-scale interconnected systems, we have presented an LMI-based design method of decentralized adaptive gain robust controllers. In the proposed controller robust synthesis, advantages are as follows: the proposed adaptive gain robust controller can achieve satisfactory transient behavior and/or avoid the excessive control input, that is, the proposed robust controller with adjustable time-varying parameters is more flexible and adaptive than the conventional robust controller with a fixed gain which is derived by the worst-case design for the unknown parameter variations. Moreover, in this chapter we have derived the allowable perturbation region of unknown parameters, and the proposed robust controller can be obtained by solving constrained convex optimization problems. Although the solution of the some matrix inequalities can be applied to the resultant robust controller in the general controller design strategies for the conventional fixed gain robust control, the solutions of the constrained convex optimization problem derived in this chapter cannot be reflected to the resultant robust controller. Note that the proposed controller design strategy includes this fascinating fact.

In Section 2 for a class of uncertain linear systems, we have dealt with a design problem of centralized adaptive gain robust state feedback controllers. Although the standard LQ regulator theory for the purpose of generating the desired response is adopted in the existing result [32], the nominal control input is designed by using pole placement constraints. By using the controller gain for the nominal system, the proposed robust control with adjustable time-varying parameter has been designed by solving LMIs. Additionally, based on the derived LMI-based conditions, the constrained convex optimization problem has been obtained for the purpose of the maximization of the allowable perturbation region of uncertainties included in the controlled system. Section 3 extends the result for the centralized adaptive gain robust state feedback controller given in Section 2 to decentralized adaptive gain robust state feedback controllers for a class of uncertain large-scale interconnected systems. In this section, an LMI-based controller synthesis of decentralized adaptive gain robust state feedback control has also been presented. Furthermore, in order to maximize the allowable region of uncertainties, the design problem of the decentralized adaptive gain robust controller for the uncertain large-scale interconnected system has been reduced to the constrained convex optimization problem.

In the future research, an extension of the proposed adaptive gain robust state feedback controller to output feedback control systems or observer-based control ones is considered. Moreover, the problem for the extension to such a broad class of systems as uncertain time-delay systems, uncertain discrete-time systems, and so on should be tackled. Furthermore, we will examine the conservativeness of the proposed adaptive gain robust control strategy and online adjustment way of the design parameter which plays important roles such as avoiding the excessive control input.

On the other hand, it is well known that the design of control systems is often complicated by the presence of physical constraints: temperatures, pressures, saturating actuators, within safety margins, and so on. If such constraints are violated, serious consequences may ensue. For example, physical components will suffer damage from violating some constraints, or saturations for state/input constraints may cause a loss of closed-loop stability. In particular, input saturation is a common feature of control systems, and the stabilization problems of linear systems with control input saturation have been studied (e.g., [33, 40]). Additionally, some researchers have investigated analysis of constrained systems and reference managing for linear systems subject to input and state constraints (e.g., [10, 19]). Therefore, the future research subjects include the constrained robust controller design reducing the effect of unknown parameters.

Acknowledgements

The authors would like to thank the associate editor for his valuable and helpful comments that greatly contributed to this chapter.

Author details

Shunya Nagai[1], Hidetoshi Oya[2*], Tsuyoshi Matsuki[3] and Yoshikatsu Hoshi[2]

*Address all correspondence to: hide@tcu.ac.jp

1 Tokushima University, Japan

2 Tokyo City University, Japan

3 National Institute of Technology, Niihama College, Japan

References

[1] Bermish BR, Corless M, Leitmann G. A new class of stabilizing controllers for uncertain dynamical systems. SIAM Journal on Control and Optimization. 1983;**21**(2):246-255

[2] Boyd S, El Ghaoui L, Feron E, Balakrishnan V. Linear matrix inequalities in system and control theory. SIAM Studies in Applied Mathmatics. 1994

[3] Chen YH. Decentralized robust control system Design for Large-Scale Uncertain Systems. International Journal of Control. 1988;**47**(5):1195-1205

[4] Choi YK, Chung MJ, Bien Z. An adaptive control scheme for robot manipulators. International Journal of Control. 1986;**44**(4):1185-1191

[5] Davison EJ, Tripathi NK. The optimal decentralized control of a large power system: Load and frequency control. IEEE Transactions on Automatic Control. 1978;**23**(2):312-325

[6] Doyle JC, Glover K, Khargonekar PP, Francis BA. State-space solutions to standard $\mathcal{H}^2$ and $\mathcal{H}^\infty$ control problems. IEEE Transactions on Automatic Control. 1989;**34**(8):831-847

[7] Faub PL, Wolovich WA. Decoupling in the design and synthesis of multivariable control systems. IEEE Transactions on Automatic Control. 1967;**12**(6):651-659

[8] Chilali M, Gahinet P, Apkarian P. Robust pole placement in LMI regions. IEEE Transactions on Automatic Control. 1999;**44**(12):2257-2270

[9] Gantmacher FR. The Theory of Matrices. Vol. 1. New York: Chelsea Publishing Company; 1960

[10] Gilbert EG, Kolmanovsky I. Nonlinear tracking control in the presence of state and control constraints: A generalized reference governor. Automatica. 2002;**38**(12):2071-2077

[11] Gong Z. Decentralized robust control of uncertain interconnected systems with prescribed degree of exponential convergence. IEEE Transactions on Automatic Control. 1995;**40**(4):704-707

[12] Goodwin GC, Graebe SF, Saogado ME. Control System Design. NJ, USA: Prentice-Hall Inc.; 2001

[13] Hua C, Guan X, Shi P. Decentralized robust model reference adaptive control for interconnected time-delay systems. Journal of Computational and Applied Mathematics. 2006;**193**:383-396

[14] Khalil HK. Nonlinear Systems. 3rd ed. New Jersey: Prentice-Hall Inc.; 2002

[15] Kalman RE. Contributions to the theory of optimal control. Boletín de la Sociedad Matemática Mexicana. 1960;(5):102-199

[16] Kalman RE. When is a linear control system optimal? Transactions of ASME Series D, Journal of Basic Engineering. 1964;**86**(1):51-60

[17] Kalman RE, Bucy RS. A new approach to linear filtering and prediction problems. Transactions of ASME Series D, Journal of Basic Engineering. 1960;**82**(1):35-48

[18] Khargonekar PP, Rotea MA. Mixed $\mathcal{H}_2/\mathcal{H}_\infty$ Control, A Convex Optimization Approach. IEEE Transactions on Automatic Control. 1991;**36**(7):824-837

[19] Kogiso K, Hirata K. Reference governor for constrained systems with time-varying references. Journal of Robotics and Autonomous System. 2009;**57**(3):289-295

[20] Krishnamurthi V. Correlation between Routh's stability criterion and relative stability of linear systems. IEEE Transactions on Automatic Control. 1972;**17**(1):144-145

[21] Kuo BC. Automatic Control Systems. 4th ed. New Jersey: Prentice-Hall Inc.; 1982

[22] Leith DJ, Leithead WE. Survey of gain-scheduling analysis and design. International Journal of Control. 2010;**73**(11):1001-1025

[23] Luo JS, Johnson A, Van Den Bosch PPJ. Minimax guaranteed cost control for linear continuous-time systems with large parameter uncertainty. Automatica. 1994;**30**(4):719-722

[24] Mao CJ, Lin WS. Decentralized control of interconnected systems with unmodelled nonlinearity and interaction. Automatica. 1990;**26**(2):263-268

[25] Maki M, Hagino K. Robust control with adaptation mechanism for improving transient behaviour. International Journal of Control. 1999;**72**(13):1218-1226

[26] Mukaidani H, Takato Y, Tanaka Y, Mizukami K. The guaranteed cost control for uncertain large-scale interconnected systems. The Preprints of the 15th IFAC World Congress, Barcelona, SPAIN; 2002

[27] Mukaidani H, Kimoto M, Yamamoto T. Decentralized guaranteed cost control for discrete-time uncertain large-scale systems using fuzzy control. The Proceedings of 2006 IEEE World Congress on Computational Intelligence,Vancouver, CANADA; 2006. pp. 3099-3105

[28] Nyquist H. Regeneration theory. Bell System Technical Journal. 1932;**11**(1):126-147

[29] Oya H, Hagino K. Robust control with adaptive compensation input for linear uncertain systems. IEICE Transactions on Fundamentals of Electronics, Communications and Computer Sciences. 2003;**E86-A**(6):1517-1524

[30] Oya H, Hagino K. Adaptive robust control scheme for linear systems with structured uncertainties. IEICE Transactions on Fundamentals of Electronics, Communications and Computer Sciences. 2004;**E87-A**(8):2168-2173

[31] Oya H, Hagino K. A new adaptive robust controller avoiding chattering phenomenon for a class of uncertain linear systems. Proceedings of the 28th IASTED International Conference on Modeling, Identification and Control. Innsbruck; 2009. pp. 236-241

[32] Oya H, Uehara Y. Synthesis of variable gain controllers based on LQ optimal control for a class of uncertain linear systems. Proceedings of the UKACC International Conference on CONTROL 2012, Cardiff; 2012. pp. 87-91

[33] Oba K, Oya H, Kubo T, Matsuki T. Synthesis of adaptive robust controllers giving consideration to input saturations for a class of nonlinear systems. Mathematical Problem in Engineering. 2017;**2017**. Article ID:5345812, 9 pages

[34] Petersen IR. A Riccati equation approach to the design of stabilizing controllers and observers for a class of uncertain linear systems. IEEE Transactions on Automatic Control. 1985;**30**(9):904-907

[35] Petersen IR, McFarlane DC. Optimal guaranteed cost control and filtering for uncertain linear systems. IEEE Transactions on Automatic Control. 1994;**39**(9):1971-1977

[36] Petersen IR, Hollot CC. A Riccati equation approach to the stabilization of uncertain linear systems. Automatica. 1986;**22**(4):397-411

[37] Reza Moheimani SO, Petersen IR. Optimal guaranteed cost control of uncertain systems via static and dynamic output feedback. Automatica. 1996;**32**(4):575-579

[38] Šijjak DD. Decentralized Control of Complex Systems. New York: Academic Press; 1991

[39] Stoorvogel AA. The robust $\mathcal{H}_2$ control problem, a worst-case design. IEEE Transactions on Automatic Control. 1993;**38**(9):1358-1370

[40] Turner MC, Postlethwaite I, Walker DJ. Non-linear tracking control for multivariable constrained input linear systems. International Journal of Control. 2000;**73**(12):1160-1172

[41] Ushida S, Yamamoto S, Kimura H. Quadratic stabilization by $\mathcal{H}^\infty$ state feedback controllers with adjustable parameters. Proceedings of the 35th IEEE Conference Decision and Control. Kobe, JAPAN, 1996. pp. 1003-1008

[42] Wei G, Wang Z, Li W, Ma L. A survey on gain-scheduled control and filtering for parameter-varying systems. Mathematical Problem in Eng. 2014;**2014**. Article ID:105815, 10 pages

[43] Wonham WM. On pole assignment in multi-input controllable linear systems. IEEE Transactions on Automatic Control. 1967;**12**(6):660-665

[44] Xu D, Jiang B, Liu H, Shi P. Decentralized asymptotic fault tolerant control of near space vehicle with high order actuator dynamics. Journal of Franklin Institute. 2013;**350**:2519-2534

[45] Zhou K, Khargonekar PP. Robust stabilization of linear systems with norm-bounded time-varying uncertainty. Systems & Control Letters. 1988;**10**(1):17-20

[46] Zhou K, Khargonekar PP, Stoustrup J, Niemann HH. Robust performance of systems with structured uncertainties in state space. Automatica. 1995;**31**(2):249-255

[47] Zhou K, Doyle JC, Glover K. Robust and Optimal Control. New Jersey: Prentice-Hall Inc.; 1996

[48] Zhou K. Essentials of Robust Control. New Jersey, USA: Prentice Hall Inc.; 1998

Chapter 11

Robust Control Applications to a Wind Turbine-Simulated System

Silvio Simani and Paolo Castaldi

Additional information is available at the end of the chapter

http://dx.doi.org/10.5772/intechopen.71526

Abstract

Wind turbine plants are complex dynamic and uncertain processes driven by stochastic inputs and disturbances, as well as different loads represented by gyroscopic, centrifugal and gravitational forces. Moreover, as their aerodynamic models are nonlinear, both modelling and control become challenging problems. On one hand, high-fidelity simulators should contain different parameters and variables in order to accurately describe the main dynamic system behaviour. Therefore, the development of modelling and control for wind turbine systems should consider these complexity aspects. On the other hand, these control solutions have to include the main wind turbine dynamic characteristics without becoming too complicated. The main point of this chapter is thus to provide two practical examples of development of robust control strategies when applied to a simulated wind turbine plant. Experiments with the wind turbine simulator represent the instruments for assessing the main aspects of the developed control methodologies.

Keywords: wind turbine simulator, data-driven and model-based approaches, fuzzy identification, online estimation, robustness and reliability

1. Introduction

Wind turbine plants represent complex and nonlinear dynamic systems usually driven by stochastic inputs and different disturbances describing gravitational, centrifugal and gyroscopic loads. Moreover, their aerodynamic models are uncertain and nonlinear, while wind turbine rotors are subject to complex turbulent wind fields, especially in large systems, thus yielding to extreme fatigue loading conditions. In this way, the development of viable, robust and reliable control solutions for wind turbines can become a challenging issue [1].

Usually, a model-based control design requires an accurate description of the system under investigation, which has to include different parameters and variables in order to model the most important nonlinear and dynamic aspects. Moreover, the wind turbine working conditions can produce further problems to the design of the control method. In general, commercial codes are not able to adequately describe the wind turbine overall dynamic behaviour; usually, special simulation software solutions are used. On the other hand, control schemes have to manage the most important turbine dynamics, without being too complex and unwieldy. Control methods for wind turbines usually rely on the signals from sensors and actuators, with a system that connects these elements together. Hardware or software modules elaborate these signals to generate the output signals for actuators. The main task of the control law consists of maintaining safe and reliable working conditions of the wind turbine, while achieving prescribed control performances and allowing for optimal energy conversion, as shown e.g. in recent works applied to the same wind turbine model considered in this chapter [2].

Today's wind turbines can implement several control strategies to allow for the required performances. Some turbines use passive control methods, such as in fixed-pitch, stall control machines. In this case, the system is designed so that the power is limited above rated wind speed through the blade stall. Therefore, the control of the blades is not required [1]. In this case, the rotational speed control is proposed, thus avoiding the inaccuracy of measuring the wind speed. Rotors with pitch regulation are usually used for constant-speed plants to provide a power control that works better than the blade stall solution. In these machines, the blade pitching is controlled in order to provide optimal power conversion with respect to modelling errors, wind gusts and disturbance. However, when the system works at constant speed and below rated wind speed, the optimal conversion rate cannot be obtained. Therefore, in order to maximise the power conversion rate, the rotational speed of the turbine must vary with wind speed. Blade pitch control is thus used also above the rated wind speed [1]. A different control method can introduce the yaw regulation to orient the machine into the wind field. A yaw error reference from a nacelle-mounted wind direction sensor system must be included in order to calculate this reference signal [3].

Regarding the regulation strategies proposed in this chapter, two control design examples are described and applied to a wind turbine system. The wind turbine model exploited in this chapter is freely available for the Matlab® and Simulink® environments and already proposed as benchmark for an international competition regarding the validation of fault diagnosis and fault-tolerant control approaches [2].

In particular, a first *data-driven* method relying on a fuzzy identification approach to the control design is considered. In fact, since the wind turbine mathematical model is nonlinear with uncertain inputs, fuzzy modelling represents an alternative tool for obtaining the mathematical description of the controlled process. In contrast to purely nonlinear identification schemes, see, e.g. [4], fuzzy modelling and identification methods are able to directly provide nonlinear models from the measured input-output signals. Therefore, this chapter suggests to model the wind turbine plant via Takagi-Sugeno (TS) fuzzy prototypes [5], whose parameters are obtained by identification procedures. This approach is also motivated by previous works by the same authors [6].

Regarding the second *model-based* strategy presented in this chapter, it relies on an adaptive control scheme [7]. Again, with respect to pure nonlinear control methods [8], it does not require a detailed knowledge about the model structure. Therefore, this chapter suggests the implementation of controllers based on adaptive schemes, used for the recursive derivation of

http://dx.doi.org/10.5772/intechopen.71526

the controller model. In particular, a recursive Frisch scheme extended to the adaptive case for control design is considered in this study, as proposed, e.g. in Simani and Castaldi [9], which makes use of exponential forgetting laws. This allows the online application of the Frisch scheme to derive the parameters of a time-varying controller.

Finally, the chapter is organised as follows. Section 2 recalls the wind turbine model considered for control design purposes. Section 3 addresses the data-driven scheme exploited for the derivation of the fuzzy controller, proposed in Section 3.1. On the other hand, the model-based control design is considered in Section 3.2, based on its mathematical derivation also described in Section 3. The achieved results and comparisons with different control strategies are outlined in Section 4.

2. Wind turbine simulator model

This section outlines the wind turbine model, whose sampled inputs and outputs will be used for the proposed control designs, as shown in Section 3.

The wind turbine system exploited in this chapter uses a nonlinear dynamic model representing the wind acting on the wind turbine blades, thus producing the movement of the low-speed rotor shaft. The higher speed required by the electric converter is produced by means of a gear box. The simulator is described in more detail, e.g. in Odgaard et al. [10]. A block scheme of the wind turbine simulator considered in this chapter is represented in **Figure 1**.

Both the generator speed and the generator power are controller by means of the two control inputs representing the generator torque $\tau_g(t)$ and the blade pitch angle $\beta(t)$. Several signals can be acquired from the wind turbine simulator. In particular, the signal $\omega_r(t)$ represents the rotor speed measurement, whereas $\omega_g(t)$ represents the converter velocity. Concerning the electric generator, $\tau_g(t)$ refers to its required torque, which is controlled by the converter. Therefore, this signal represents the measurement of the torque setpoint, $\tau_r(t)$. The aerodynamic model defining the aerodynamic torque provides the $\tau_{aero}(t)$ signal, which is a nonlinear function of the wind speed $v(t)$. This measurement is very difficult to be acquired correctly, as described in Odgaard et al. [10].

The aerodynamic model reported in **Figure 1** is described as follows:

$$\tau_{aero}(t) = C_p(\beta(t), \lambda(t)) \frac{\rho A v^3(t)}{2\omega_r(t)} \tag{1}$$

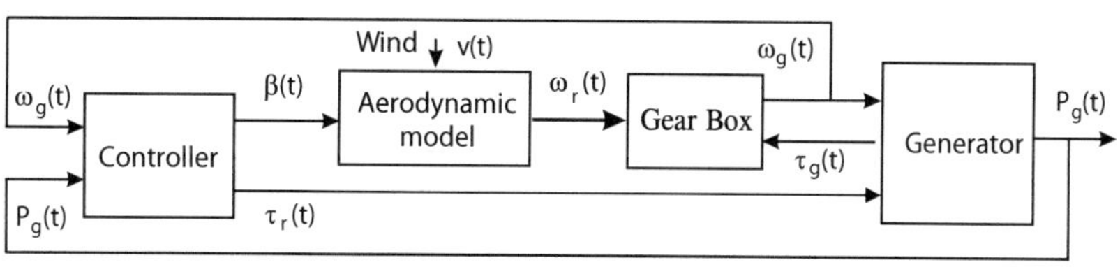

Figure 1. Scheme of the wind turbine process.

where the variable ρ represents the air density and A is the effective rotor area. Another important variable is represented by the so-called tip-speed ratio, which is defined as

$$\lambda(t) = \frac{\omega_r(t) R}{v(t)} \tag{2}$$

with R the rotor radius. $C_p(\cdot)$ represents the power coefficient that is normally represented via a two-dimensional map [10]. The expression of Eq. (1) allows the computation of the signal $\tau_{aero}(t)$, by means of the estimated wind speed $v(t)$, and the measured $\beta(t)$ and $\omega_r(t)$. Due to the uncertainty of the wind speed, the estimate of $\tau_{aero}(t)$ is considered affected by an unknown measurement error, which justifies the robust approaches described in Section 3. Moreover, the nonlinearity represented by the expressions of Eqs. (1) and (2) motivates the required reliable and robust control approaches suggested in this chapter.

A two-mass model is exploited to describe the drive-train system, while the hydraulic pitch system is modelled as second-order transfer function [10]. Moreover, the generator dynamics are described as a first-order transfer function. More details regarding the considered simulator are in Odgaard et al. [10]. Under these assumptions, the complete state-space description of the wind turbine model has the form of Eq. (3):

$$\begin{cases} \dot{x}_c(t) = f_c(x_c(t), u(t)) \\ y(t) = x_c(t) \end{cases} \tag{3}$$

where $u(t) = [\beta(t), \tau_g(t)]^T$ and $y(t) = x_c(t) = [P_g(t), \omega_g(t)]^T$ are the control inputs and the monitored output measurements, respectively, as shown in **Figure 1**. $P_g(t)$ is the generator power measurement, whereas $f_c(\cdot)$ represents the continuous-time nonlinear function that will be approximated via discrete-time models from N sampled data u_k and y_k, with the sample index $k = 1, 2, \ldots N$, as presented in Section 3. Finally, the model parameters and the map $C_p(\beta, \lambda)$ are chosen in order to represent a realistic turbine [10].

As described in Odgaard et al. [10], the baseline controller developed for this wind turbine system works in two normal operating conditions, namely, the region 1 corresponding to the power optimisation (partial load) and the region 2 of constant power production (full load). The partial load working condition (also known as working region 1), the optimal wind-power conversion is achieved without any pitching of the blades, which are fixed to $0°$. In this case, λ is constant at its optimal value λ_{opt} that is defined by the maximal value of the power coefficient map C_p when $\beta = 0$. Therefore, this working condition is completely defined by setting $\tau_g = \tau_r$ (i.e. the generator torque is equal to the reference one) with pitch angle $\beta = 0$.

The reference torque τ_r shown in **Figure 1** can be written as

$$\tau_r = K_{opt}\, \omega_r^2 \tag{4}$$

where:

$$K_{opt} = \frac{1}{2} \rho A R^3 \frac{C_{p_{\max}}}{\lambda_{opt}^3} \tag{5}$$

with $C_{p_{\max}}$ the maximal value of C_p, related the to λ_{opt}, i.e. the optimal tip-speed ratio.

When the power reference is achieved and the wind speed increases, the controller can be switched to the control region 2 (full load condition). In this zone, the control objective consists of tracking the power reference P_r, obtained by regulating β, such that the C_p is decreased. In a traditional industrial control scheme, usually a PI controller is used to keep ω_r at the prescribed value by changing β; the second input of the controlled is τ_g.

The baseline controller considered in this chapter was implemented with a sample frequency at 100 Hz, i.e. $T_s = 0.01$ s. In full load conditions, i.e. in region 2, the actuated input β is controlled via the relations of Eq. 6 [10]:

$$\begin{cases} \beta_k = \beta_{k-1} + k_p e_k + \left(k_i T_s - k_p\right) e_{k-1} \\ e_k = \omega_{gk} - \omega_{nom} \end{cases} \tag{6}$$

with the sample index $k = 1, 2, \dots, N$. The parameters for this PI speed controller are $k_i = 0.5$ and $k_p = 3$, with sampling time $T_s = 0.01$ s, as reported in [10].

To control the further input τ_g shown in **Figure 1**, a second PI regulator is used, in the form of Eq. (7):

$$\begin{cases} \tau_{rk} = \tau_{rk-1} + k_p e_k + \left(k_i T_s - k_p\right) e_{k-1} \\ e_k = P_{gk} - P_r \end{cases} \tag{7}$$

The parameters for this second PI power controller are $k_i = 0.014$ and $k_p = 447 \times 10^{-6}$ [10].

Finally, note that in region 1 (partial load, below the rated wind speed), the wind turbine is regulated only by means of the torque input $\tau_g(t)$. In this situation, the blade pitching system is not exploited to achieve the optimal power conversion. On the other hand, in region 2 (full load, above the rated wind speed), the wind turbine control regulates both the blade pitch angle $\beta(t)$ and the control torque $\tau_g(t)$. The wind turbine Simulink® simulator considered in this work includes also saturation blocks limiting the values of these control signals and their rates.

3. Data-driven and model-based designs

This section describes the two approaches considered in this chapter for obtaining the control laws by using data-driven and model-based methodologies. Once a suitable mathematical description of the monitored process is provided, the derivation of the controller structure is sketched in Section 3.1 for the fuzzy approach, whereas Section 3.2 proposes a different method relying on an adaptive technique.

The first method proposed in this chapter for the derivation of the wind turbine controller is based on a fuzzy clustering technique to partition the available data into subsets characterised by linear behaviours. The integration between clusters and linear regression is exploited, thus allowing for the combination of fuzzy logic techniques with system identification methodologies. These tools are already available and implemented in the Matlab® Fuzzy Modelling and

Identification (FMID) toolbox recalled below [5]. This study proposes the use of TS fuzzy prototypes since they are able to model nonlinear dynamic systems with arbitrary accuracy [5]. The switching between the local affine submodels is achieved through a smooth function of the system state defined exploiting the fuzzy set theory and its tools.

In more detail, the fuzzy estimation scheme relies on a two-step algorithm, in which the working regions are first defined by exploiting the data fuzzy clustering tool, i.e. the Gustafson-Kessel (GK) method [5]. On the other hand, the second step performs the identification of the controller structure and its parameters using the estimation method proposed by the same authors in Simani et al. [6]. This estimation approach can be considered as a generalisation of the general least-squares method for hybrid models.

Under these assumptions, the TS fuzzy prototypes have the form of the model of Eq. (8):

$$y_{k+1} = \frac{\sum_{i=1}^{M} \mu_i(x_k) y_i}{\sum_{i=1}^{M} \mu_i(x_k)} \tag{8}$$

where $y_i = a_i^T x + b_i$, with a_i the parameter vector (regressand) and b_i is the scalar offset. $x = x_k$ represents the regressor vector, which contains delayed samples of the signals u_k and y_k.

The antecedent fuzzy sets μ_i that determine the switching among the different submodels i are estimated from the data clusters [5]. The consequent parameters a_i and b_i are identified from the data by means of the methodology proposed in Simani et al. [6]. This identification scheme exploited for the estimation of the TS model parameters has been integrated into the FMID toolbox for Matlab® by the authors. This approach is preferable when the TS model of Eq. (8) is used as predictor, since it derives the consequent parameters via the so-called Frisch scheme, developed for the errors-in-variables (EIV) structures [6].

Once the description of the monitored process is obtained in the form of Eq. (8), the data-driven approach for the design of the fuzzy controller proposed in this chapter is presented in Section 3.1.

The second approach exploited for obtaining the mathematical description of the wind turbine system under investigation is based on a recursive methodology, which will be used for the design of the second control strategy presented in Section 3.2. An online version of the batch Frisch scheme estimation methodology summarised above is recalled in the remainder of this section for estimating the parameters of dynamic EIV models. For the derivation of the adaptation law, an online bias-compensating algorithm is also implemented. Thus, the online Frisch scheme estimation is generalised to enhance its applicability to real-time implementations. Moreover, by means of an exponential forgetting factor included in the adaptation law, the algorithm is able to deal with linear parameter-varying (LPV) structures that are exploited in connection with the model-based design of the adaptive control scheme, presented in Section 3.2.

Thus, the considered scheme is proposed for the online identification of the process modelled by the following transfer function $G(z)$:

$$G(z) = \frac{A(z^{-1})}{B(z^{-1})} = \frac{b_1 z^{-1} + \ldots + b_{n_b} z^{-n_b}}{1 + a_1 z^{-1} + \ldots + a_{n_a} z^{-n_a}} \tag{9}$$

where a_i, b_i, n_a and n_b represent the unknown parameters and the structure of the model, defining the polynomials $A(z^{-1})$ and $B(z^{-1})$, while z is the discrete-time complex variable.

The parameter vector describing the linear relationship is given by

$$\theta = \left[a_1 \ldots a_{n_a}\, b_1 \ldots b_{n_b}\right]^T \tag{10}$$

whose extended version is defined as in Eq. (11):

$$\overline{\theta} = \left[1\, \theta^T\right]^T \tag{11}$$

An equivalent expression of the considered relations is obtained by using vector and matrix notations, in the form of Eq. (12):

$$\psi_k^T \overline{\theta} = 0 \tag{12}$$

where the regressor vector ψ_k is defined as

$$\psi_k = \left[-y_k - y_{k-1} \cdots - y_{k-n_a}\; u_{k-1} \ldots u_{k-n_b}\right]^T \tag{13}$$

where the subscript k denotes the sample index.

The Frisch scheme provides the estimates of the measurement errors affecting the input and output signals u_k and y_k, i.e. σ_u and σ_y, and θ for a linear time-invariant dynamic system. Note that the polynomial orders n_a and n_b in the relation of Eq. 9 are assumed to be fixed in advance.

From the Frisch scheme method, the following expression is considered:

$$\left(\Sigma_\psi - \Sigma_{\tilde{\psi}}\right) \overline{\theta} = 0 \tag{14}$$

where the noise covariance matrix is given by

$$\Sigma_{\tilde{\psi}} - \begin{bmatrix} \sigma_y I_{n_a+1} & 0 \\ 0 & \sigma_u I_{n_b} \end{bmatrix} \tag{15}$$

which are approximated by the sample covariance matrix over N samples:

$$\Sigma_{\tilde{\psi}} \approx \frac{1}{N} \sum_{k=1}^{N} \psi_k \psi_k^T \tag{16}$$

Thus, the Frisch scheme aims at providing suitable noise variances σ_u and σ_y such that $\left(\Sigma_\psi - \Sigma_{\tilde{\psi}}\right)$ results to be a matrix singular positive semidefinite as it is rank-one deficient. On the other hand, the system represented by the expression of Eq. (14) can be solved, and $\overline{\theta}$ represents its solution.

The expression of Eq. (17) is determined:

$$\varepsilon_k(\overline{\theta}) = A(z^{-1})y_k - B(z^{-1})u_k \tag{17}$$

while the so-called sample autocovariance is defined in the form of Eq. (18):

$$r_{\varepsilon h,N} = \frac{1}{N}\sum_{l=1}^{N}\varepsilon_l(\overline{\theta})\,\varepsilon_{l+h}(\overline{\theta}) \tag{18}$$

where the subscript h in Eq. (18) indicates a time-shift.

The online control development requires a recursive estimate of the model parameters represented by the vector θ_k of Eq. (9), while the input and output data u_k and y_k acquired online by the dynamic process of the wind turbine system. In fact, the adaptive control law computed at time step k is based on the recursive estimate of a model of the process, which is derived exploiting the dynamic data up to the sample k. In this way, the algorithm of the Frisch scheme defined by the expressions of Eqs. (14), (16) and (18) is expressed by means of an online scheme.

Note that the expressions of Eqs. (16) and (18) are required in their recursive form. Therefore, while the derivation of the online form of the covariance matrix update is easily obtained as in the form of Eq. (19):

$$\Sigma_{\bar{\psi}k} = \frac{k-1}{k}\Sigma_{\bar{\psi}k} + \frac{1}{k}\psi_k\psi_k^T \tag{19}$$

the formulation of the autocovariance expression $r_{\varepsilon h,k}$ can be obtained recursively for $1 \le l \le k$ only if the approximated expression of Eq. (20) is considered:

$$\varepsilon_l(\overline{\theta}_k) \approx \varepsilon_l(\overline{\theta}_l) \tag{20}$$

for $l<k$. In this way, only the residual $\varepsilon_k(\overline{\theta}_k)$ has to be computed at time step k using the lagged data in the vector ψ_k and the updated estimate $\overline{\theta}_k$ of the model parameters. The online computation of the expression of the autocovariance matrix of Eq. (21):

$$r_{\varepsilon h,k} = \frac{k-1}{k}r_{\varepsilon k,k-1} + \frac{1}{k}\varepsilon_k(\overline{\theta}_k)\,\varepsilon_{k+h}(\overline{\theta}_k) \tag{21}$$

can be achieved using only the vector $\varepsilon_{k+h}(\overline{\theta}_k)$ at each time step. The initial values θ_0, $\Sigma_{\bar{\psi}0}$ and $r_{\varepsilon 0,h}$ for the recursive algorithm are equal to the variables of the classic Frisch scheme batch procedure.

Since variations of system properties have to be tracked online, in order to cope with time-varying systems, this chapter considers a further modification of the recursive estimation scheme. This point can be achieved by placing more emphasis on the more recent data, while forgetting the older ones. Therefore, the methodology represented by the expressions of Eqs. (19) and (21) with the approximation of Eq. (20) is implemented by including the so-called exponential forgetting factor. This is achieved in practice by defining the new expressions of the sample covariance and autocovariance matrices in the form of Eq. (22):

$$\begin{cases} H_{\Sigma_{\tilde{\psi}_k}} = \omega(\delta)\,\Sigma_{\tilde{\psi}k} \\ h_{\varepsilon h,k} = \omega(\delta)\,r_{\varepsilon h,k} \end{cases} \tag{22}$$

where $\omega(\delta)$ is a scaling factor that coincides with k when no adaptation is introduced. In this way, the updated expressions have the form:

$$\begin{cases} H_{\Sigma_{\tilde{\psi}_k}} = (1-\delta)\,H_{\Sigma_{\tilde{\psi}_{k-1}}} + \delta\,\psi_k\,\psi_k^T \\ h_{\varepsilon h,k} = (1-\delta)\,h_{\varepsilon h,k-1} + \delta\,\varepsilon_k\left(\overline{\theta}_k\right)\,\varepsilon_{k+h}\left(\overline{\theta}_k\right) \end{cases} \tag{23}$$

with $0<\delta<1$ representing the forgetting factor. Thus, the adaptive Frisch scheme algorithm is implemented via Eq. (23) in three steps. First, θ_0, $\Sigma_{\tilde{\psi}0}$ and $r_{\varepsilon 0,h}$ with $h \le n_a$ are initialised. Moreover, at each recursion step, by means of $r_{\varepsilon h,k}$, the noise variances σ_u and σ_y are computed. Finally, at each recursion step, $\overline{\theta}_k$ is determined by solving Eq. (14) via the expression of Eq. (23). In this way, the vector θ_k contains the estimates of the model parameter derived at the step k.

The results achieved by the online identification method recalled in this section were obtained in the Matlab® and Simulink® environments as summarised in Section 4.

Finally, once the parameters θ_k of the discrete-time linear time-varying model of the nonlinear dynamic process of Eq. (3) have been computed at each time step k, the adaptive controller is derived as summarised in Section 3.2.

3.1. Data-driven fuzzy controller derivation

This section describes the derivation of the fuzzy controller model. Once a reasonably accurate fuzzy description of the considered benchmark has been available, as described above, it is used offline to directly estimate the nonlinear fuzzy controllers. As already remarked, this design procedure differs from the approach proposed in Simani [11]. In fact, the control design proposed in this chapter relies on the so-called *model inverse control* principle, which is solved suing the fuzzy identification approach recalled above.

With reference to stable fuzzy systems, whose inverted dynamics are also stable, a nonlinear controller can be simply designed by inverting the fuzzy model itself. Moreover, when modelling errors and disturbances are not present, this controller is able to allow for exact tracking with zero steady-state errors. However, modelling errors and disturbance effects are always present in real conditions, which can be tackled by directly identifying the controller model (i.e. the inverse controlled model) using the FMID approach. Differently from Simani [11], a robust control strategy is thus achieved by minimising a cost function, which includes the difference between the desired and controller outputs, and a penalty on the system stability. In general, a nonconvex optimisation problem has to be solved, which hampers the direct application of the proposed approach. However, the optimisation scheme described in Simani et al. [6] can be exploited, which is based on a parametrised search technique applied at a higher level to formulate the control objectives and constraints.

In this way, the estimated controller based on the inverse process model and approximated via a fuzzy prototype is able to describe the complete behaviour of the monitored plant in its different working conditions (i.e. partial and full load situations). In fact, the rule-based fuzzy inference system of Eq. (8) has been derived for modelling the wind turbine dynamic process of Eq. (3) in its equivalent discrete-time form of Eq. (24):

$$y_{k+1} = f(x_k, u_k) \tag{24}$$

And, in particular, the TS fuzzy representation has the form of Eq. 25:

$$y_{k+1} = \frac{\sum_{i=1}^{M} \mu_i^{(m)}\left(x_k^{(m)}\right)\left(a_i^{(m)} x_k^{(m)} + b_i^{(m)}\right)}{\sum_{i=1}^{M} \mu_i^{(m)}\left(x_k^{(m)}\right)} \tag{25}$$

The current state $x_k = [y_k, \ldots, y_{k-n+1}, u_{k-1}, \ldots, u_{k-n+1}]^T$ and the input u_k represent the inputs that drive the model of Eq. (25). Its output represents the prediction of the system output at the next sample y_{k+1}. The model of Eq. (25) requires the estimated membership functions $\mu_i^{(m)}$, the state $x^{(m)}$ and the parameters $a_i^{(m)}$ and $b_i^{(m)}$ of the controlled system, which are denoted by the superscript (m).

Therefore, the input u_k generated by the control law feeds the monitored process such that its output y_{k+1} asymptotically follows the desired (reference) output r_{k+1}. This behaviour is obtained using the inverse model principle, represented by the expression of Eq. (26):

$$u_{k+1} = f^{-1}(x_k^c, r_k) \tag{26}$$

that is a nonlinear function of the vector x_k^c and the reference r_k.

However, in general, with reference to Eq. (26), it is difficult to determine the analytical expression of the inverse function $f^{-1}(\cdot)$. Therefore, the methodology proposed in this chapter suggested to exploit the identified fuzzy TS prototype of Eq. (25) to provide the particular state $x_k^{(m)}$ at each time step k. In this way, from this mapping, the inverse mapping $u_{k+1} = f^{-1}\left(x_k^{(c)}, r_k\right)$ is directly identified the form of Eq. (8), if the controlled system is stable, and in particular in the form of Eq. (27):

$$u_{k+1} = \frac{\sum_{i=1}^{M} \mu_i^{(c)}\left(x_k^{(c)}\right)\left(a_i^{(c)} x_k^{(c)} + b_i^{(c)}\right)}{\sum_{i=1}^{M} \mu_i^{(c)}\left(x_k^{(c)}\right)} \tag{27}$$

where the state $x_k^{(c)} = \left[x_k^{(m)}, r_{k-1}, \ldots, r_{k-n+1}\right]^T$ and the reference r_k signal represent the inputs of the identified controller model. The model of Eq. (27) contains the estimated membership functions $\mu_i^{(c)}$ and the parameters $a_i^{(c)}$ and $b_i^{(c)}$ of the identified controller model, which are denoted by the superscript (c).

In this way, the series connection between the controller model (i.e. the identified inverse process model) and the process model itself should lead to an identity mapping as in Eq. (28):

$$y_{k+1} = f\left(x_k^{(m)}, u_k\right) = f\left(x_k^{(m)}, f^{-1}\left(x_k^{(c)}, r_k\right)\right) = r_{k+1} \tag{28}$$

where $r_{k+1} = f\left(x_k^{(m)}, u_k\right)$ for a proper value of u_k. However, the expression of Eq. (28) holds in ideal conditions. However, the model-reality mismatch and measurement errors are properly managed by means of the fuzzy modelling scheme recalled in Section 3. In this way, the difference $|r_{k+1} - f\left(x_k^{(m)}, u_k\right)|$ can be made arbitrarily small by a suitable selection of the model parameters, i.e. the fuzzy membership functions $\mu_i^{(c)}$, the number of clusters M and the regressand $a_i^{(c)}$ and $b_i^{(c)}$.

Moreover, the fuzzy model of the process is used for providing the state vector $x_k^{(m)}$. Therefore, the state of the fuzzy controller $x_k^{(c)}$ is updated using the process model state $x_k^{(m)}$ and the reference input r_k. These computations are performed using standard matrix operations, thus making the algorithm suitable for real-time implementations [12].

As already remarked, the effects of the model uncertainty and disturbance lead to a different behaviour of the model with respect to controlled process, thus resulting in a mismatch between the process outputs y_k and their references r_k. This mismatch can be compensated by means of the online mechanism described by the expressions of Eqs. (25) and (27). These issues motivate the model-based strategy relying on the adaptive algorithm proposed in Section 3.2.

Note finally that the fuzzy controller proposed in this section will replace the baseline wind turbine regulator of Section 2.

3.2. Model-based adaptive controller derivation

This section describes the model-based adaptive control strategy used in connection with the online estimation scheme presented above. In more detail, with reference to the wind turbine system recalled in Section 2, adaptive controllers for processes of the second order ($n_a = n = 2$) are designed. Moreover, the considered adaptive controllers are based on the trapezoidal method of discretisation.

With reference to Eq. (9), the transfer function of the time-varying controlled system with $n_a = n_b = n = 2$ is considered, whose parameters estimated using the online identification approach recalled above:

$$\theta_k = \left[\hat{a}_1, \hat{a}_2, \hat{b}_1, \hat{b}_2\right]^T \tag{29}$$

Note that the subscript k for model and controller parameters will be dropped in order to simplify equations and formulas.

The control law corresponding to the discrete-time adaptive controller in its difference form of Eq. (30):

$$\begin{cases} \Delta e_k = e_k - e_{k-1} \\ u_k = K_p \left[\Delta e_k + \dfrac{T_s}{T_I} \dfrac{\Delta e_k}{2}\right] + u_{k-1} \end{cases} \tag{30}$$

with e_k representing the tracking error, with $e_k = r_k - y_k$, and r_k the reference (setpoint) signal. T_s is sampling time. The controller parameters K_p and T_I are here time-varying and derived from the online model parameters in the vector θ_k. The control law can be represented also in its feedback formulation as described by Eq. (31):

$$u_k = q_0 e_k + q_1 e_{k-1} + u_{k-1} \tag{31}$$

where the new controller variables q_0 and q_1 (or K_p and T_I) are derived from the relations of Eq. (32):

$$\begin{cases} q_0 = K_p \left(1 + \dfrac{T_s}{2T_I}\right) \\ q_1 = -K_p \left(1 - \dfrac{T_s}{2T_I}\right) \end{cases} \tag{32}$$

where the parameters K_p and T_I are functions of the (time-varying) critical gain and the critical period of oscillations, respectively, K_{P_u} and T_u:

$$K_p = 0.6 K_{P_u}, \qquad T_I = 0.5 T_u \tag{33}$$

that depend on the time-varying model parameters in the vector θ_k. In particular, when considering a second-order model described by its (time-varying) parameters $\hat{a}_2$, $\hat{a}_1$, $\hat{b}_2$ and $\hat{b}_1$, the variables K_{P_u} and T_u required by the Ziegler-Nichols method can be computed at each time step k from the following relations:

$$\begin{cases} K_{P_u} = \dfrac{\hat{a}_1 - \hat{a}_2 - 1}{\hat{b}_2 - \hat{b}_1} \\ T_u = \dfrac{2\pi T_s}{\arccos \gamma}, \qquad with\ \gamma = \dfrac{\hat{a}_2 \hat{b}_1 - \hat{a}_1 \hat{b}_2}{2\hat{b}_2} \end{cases} \tag{34}$$

In this way, the adaptive discrete-time linear controllers of Eq. (30) or (31) are designed on the basis of the time-varying linear model of Eq. (9) estimated via the online identification scheme from the data of the nonlinear wind turbine process of Eq. (3).

Finally, Section 4 will show the achieved results regarding the design and the application of the adaptive controller to the data from the wind turbine benchmark.

4. Simulation results

This section presents the simulation results achieved with the proposed data-driven and model methods relying on both the fuzzy modelling technique oriented to the identification of the fuzzy controller model and the adaptive control strategy using the online estimated models. The simulations achieved with these regulators are summarised in the following.

Regarding the fuzzy modelling and identification method, the GK clustering algorithm recalled in Section 3 exploits a number $M=3$ of clusters and delays $n=2$. These variables were applied for clustering the first data set consisting of $\{P_{gk}, \omega_{gk}, \beta_{rk}\}$. A number of samples $k=1, 2, \ldots, N$ were considered with $N=440\times10^3$. The same number of clusters and shifts were exploited for clustering the second data set $\{P_{gk}, \omega_{gk}, \tau_{gk}\}$. After this procedure, the structures of the TS prototypes were derived for each output y_k equal to P_{gk} and ω_{gk}. In this way, the 2 continuous-time outputs $y(t)=[\omega_g(t), \tau_g(t)]$ of the wind turbine continuous-time model of Eq. (3) are approximated by two TS fuzzy prototypes of Eq. (8).

The performances of the fuzzy models that are derived using the procedure described above can be evaluated using the so-called Variance Accounted For (VAF) parameter [5]. In particular, the TS fuzzy model reconstructing the first output has a VAF index bigger than 90%, whereas for the second one, it was higher than 99%. This means that the fuzzy prototypes are able to describe the behaviour of the controlled process with very good precision. These estimated TS fuzzy models have been used for the derivation of the fuzzy controllers and applied to the considered wind turbine benchmark.

Two (multiple input single output) MISO fuzzy controller models with two inputs and one output have been used for the compensation of the blade pitch angle $\beta(t)$ and the generator torque $\tau_g(t)$. By using the inverse model principle, they were estimated exploiting the methodology recalled in Section 3.1. Again, the GK fuzzy clustering method has led to 2 fuzzy regulators applied to the data sets $\{\beta_{rk}, P_{gk}, \omega_{gk}\}$ and $\{\tau_{gk}, P_{gk}, \omega_{gk}\}$, respectively, with $M=3$ clusters and $n=3$ lagged signals.

The controller performances were verified and validated via extensive simulations by considering different data sequences generated via the wind turbine simulator. **Table 1** reports the

Recursive algorithm parameter	Value
$\overline{\theta}(0)$	$[0.1, 0.15, 0.20, 0.25\, 0.30, 0.35]^T$
$\Sigma_{\tilde{\psi}}(0)$	$10^{-1} I_7$
δ	0.995

Table 1. Initialisation parameters of the adaptive algorithm.

values of the percent Normalised Sum of Squared tracking Error ($NSSE\%$) index defined in Eq. (35):

$$NSSE\% = 100\sqrt{\frac{\sum_{k=1}^{N}\left(r_k - y_k\right)^2}{\sum_{k=1}^{N} r_k^2}} \tag{35}$$

Noting that in partial load operation (region 1), the performance is represented by the comparison between the power produced by the generator, $y_k = P_{gk}$, with respect to the theoretical maximum power output, $r_k = P_r$. On the other hand, in full load operation (region 2), the tracking error is given by the difference between the generator speed, $y_k = \omega_{gk}$, and its nominal value, $r_k = \omega_{nom}$. The achieved results show the good properties of the designed fuzzy controllers, as represented also in **Figure 2**.

Figure 2 depicts the signal representing generator speed $\omega_g(t)$ in bold grey line with respect to its desired value ω_{nom} in dashed black line. It can be noted that in full load conditions, the fuzzy controllers derived via the data-driven approach lead to tracking errors smaller than the wind turbine baseline governor recalled in Section 2. In fact, as shown in **Figure 1**, the baseline regulator is working in the interval 2200s < t < 3300s. On the other hand, the fuzzy controllers are exploited during the interval 3300s < t < 4400s, when the tracking error is much lower.

With reference to the second model-based design approach using adaptive solutions, the two outputs $P_g(t)$ and $\omega_g(t)$ of the wind turbine continuous-time nonlinear model of Eq. (3) were approximated by two second-order time-varying MISO discrete-time models of Eq. (9) with two inputs and one output. Using these one LPV prototypes, the model-based approach for determining the adaptive controllers recalled in Section 3.2 was exploited and applied to the wind turbine benchmark of Section 2. Thus, according to Section 3.2, the parameters of the adaptive controllers were computed online. In particular, for each output, two second-order ($n_a = n_b = 2$) time-varying MISO prototypes were identified, and the adaptive regulator parameters in Eq. (30) or (31) were computed analytically at each time step k.

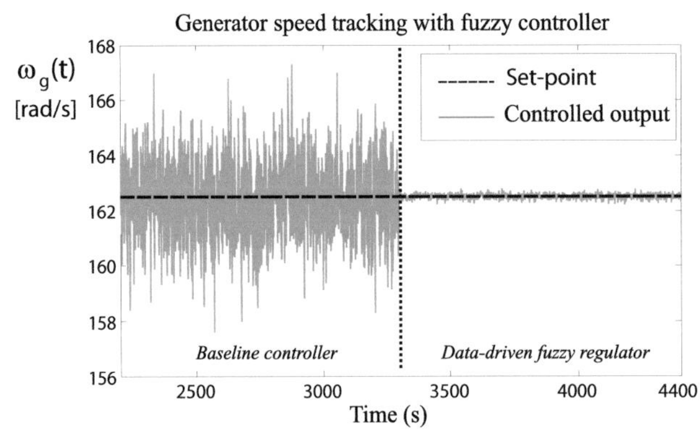

Figure 2. Generator speed (bold grey line) $\omega_g(t)$ and its reference (dashed black line) ω_{nom}.

Also in this case, with reference to the adaptive controller structure of Eq. (30) or (31), the parameters of the online controllers were tuned via the Ziegler-Nichols rules, applied to the LPV models. In this way, if both the model online parametric identification and the regulator tuning procedure are exploited, the parameter adaptation mechanisms should lead to good control performances.

The experiments with the adaptive regulators were simulated in the same situation of the fuzzy controllers. In this case, three online regulators were exploited for the compensation of both the blade pitch angle $\beta(t)$ and the generator torque $\tau_g(t)$, in region 1 and region 2. The adaptive algorithm described above runs with initial values for its parameters reported in **Table 1**.

With reference to the model-based adaptive approach, **Figure 3** depicts the setpoint $\omega_g(t)$ in bold grey line with respect to its desired value ω_{nom} in dashed black line. By considering the full load working conditions, the adaptive regulators have replaced the wind turbine baseline governor at $t \geq 3300s$.

Also for the case of the adaptive regulators, **Figure 3** highlights that the model-based approach leads to interesting performances.

In order to analyse the performance of the proposed adaptive strategy, **Table 2** reports also the *NSSE* values computed for these controllers.

According to the simulation results summarised in **Table 2**, good tracking capabilities of the suggested adaptive controllers seem to be reached, and they are better than the fuzzy regulators.

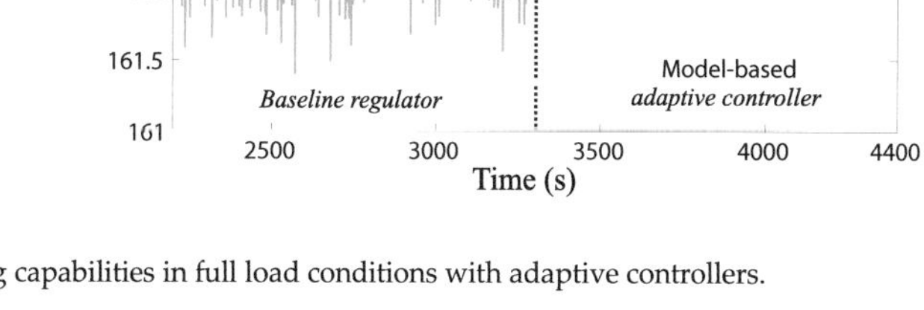

Figure 3. $\omega_g(t)$ tracking capabilities in full load conditions with adaptive controllers.

Controller type	Partial load (%)	Full load (%)
Fuzzy controller	37.17	17.85
Adaptive controller	28.73	13.67

Table 2. Controllers in partial and load operations: *NSSE*% values.

5. Conclusion

The chapter addressed two control examples for a wind turbine dynamic simulator, since it was proposed as benchmark representing a complex dynamic system driven by stochastic disturbances and uncertain load conditions. Moreover, the aerodynamic models of these processes are nonlinear, thus making their modelling a challenging problem. Therefore, the design of control strategies for these complex processes has to consider these aspects. In this way, the chapter analysed the design of two data-driven and model-based control methodologies, which represented viable, reliable and robust control schemes for the proposed wind turbine benchmark. Experiments with the wind turbine simulator were the practical instruments for assessing the most important characteristics of the developed control methodologies. The obtained results showed that the considered solutions represent viable, robust and reliable control applications to real wind turbine systems.

Author details

Silvio Simani[1]* and Paolo Castaldi[2]

*Address all correspondence to: silvio.simani@unife.it

1 Department of Engineering, University of Ferrara, Italy

2 Department of Engineering, University of Bologna, Italy

References

[1] Johnson KE, Pao LY, Balas MJ, Fingersh LJ. Control of variable–speed wind turbines: Standard and adaptive techniques for maximizing energy capture. IEEE Control Systems Magazine. 2006;**26**(3):70-81. DOI: 10.1109/MCS.2006.1636311

[2] Odgaard PF, Stoustrup J. A benchmark evaluation of fault tolerant wind turbine control concepts. IEEE Transactions on Control Systems Technology. 2015;**23**(3):1221-1228

[3] Zhao W, Stol K. Individual blade pitch for active yaw control of a horizontal–axis wind turbine. Proceedings of the 45th AIAA Aerospace Sciences Meeting and Exhibit; AIAA; Reno, NV, USA; 2007

[4] Juditsky A, Hjalmarsson H, Beneviste A, Delyon B, Ljung L, Sjöberg J, Zhang Q. Nonlinear black–box modelling in system identification: A mathematical foundation. Automatica. 1995;**31**(12):1691-1724

[5] Babuška R. Fuzzy Modeling for Control. Boston, USA: Kluwer Academic Publishers; 1998

[6] Simani S, Fantuzzi C, Rovatti R, Beghelli S. Parameter identification for piecewise linear fuzzy models in noisy environment. International Journal of Approximate Reasoning. 1999;**1**(22):149-167

[7] Landau Y. Adaptive Control. New York: Marcel Dekker; 1979. ISBN: 0-8247-6548-6

[8] Slotine JE, Li W. Applied Nonlinear Control. Prentice-Hall; 1991

[9] Simani S, Castaldi P. Data-driven and adaptive control applications to a wind turbine benchmark model. Control Engineering Practice. 2013;**21**(12):1678-1693. Special Issue Invited Paper. ISSN: 0967–0661. PII: S0967–0661(13)00155–X. DOI: http://dx.doi.org/10.1016/j.conengprac.2013.08.009

[10] Odgaard PF, Stoustrup J, Kinnaert M. Fault-tolerant control of wind turbines: A benchmark model. IEEE Transactions on Control Systems Technology. 2013;**21**(4):1168-1182. ISSN: 1063-6536. DOI: 10.1109/TCST.2013.2259235

[11] Simani S. Application of a data-driven fuzzy control design to a wind turbine benchmark model. Advances in Fuzzy Systems. 2012. (Web: http://www.hindawi.com/journals/afs/2012/504368/): 2012:1-12. Invited paper for the special issue: Fuzzy Logic Applications in Control Theory and Systems Biology (FLACE). ISSN: 1687–7101, e–ISSN: 1687-711X. DOI: 10.1155/2012/504368

[12] Rovatti R, Fantuzzi C, Simani S. High–speed DSP–based implementation of piecewise–affine and piecewise–quadratic fuzzy systems. Signal Processing Journal. 2000;**80**(6):951-963. Special Issue on Fuzzy Logic applied to Signal Processing. DOI: 10.1016/S0165-1684(00)00013-X

Chapter 12

Adaptive Robust Control of Biomass Fuel Co-Combustion Process

Konrad Gromaszek and Andrzej Kotyra

Additional information is available at the end of the chapter

http://dx.doi.org/10.5772/intechopen.71576

Abstract

The share of biomass in energy production is constantly growing. This is caused by environmental and industry standards and EU guidelines. Biomass is used in the process of co-firing in large power plants and industrial installations. In the existing power stations, biomass is milled and burned simultaneously with coal. However, low-emission combustion techniques, including biomass co-combustion, have some negative side effects that can be split into two categories. The direct effects influence the process control stability, whereas the indirect ones on combustion installations via increased corrosion or boiler slagging. The effects can be minimised using additional information about the process. The proper combustion diagnosis as well as an appropriate, robust control system ought to be applied. The chapter is devoted to the analysis of modern, robust control techniques for complex power engineering applications.

Keywords: adaptive control, model predictive control, complex system, co-combustion, energy, controllability, robust

1. Introduction

Regarding the fact that coal is still themain fuel used in electricity generation around the world and it contains impurities that significantly increase pollutant emissions, new combustion techniques are developed, e.g. air staging, reburning and flue gas circulation [1]. Fossil fuel depletion forces the use of renewable fuels such as biomass; in existing power stations, biomass is milled and burned simultaneously with coal. However, low-emission combustion techniques, including biomass co-combustion, have negative effects: directly influence on process control stability/efficiency and indirectly on combustion installations via increased corrosion or boiler slagging [2]. These effects can be minimised using additional information about the process that makes combustion monitoring (diagnosis) system necessary to apply [3].

The combustion efficiency of pulverised fuel depends on several parameters. The commonly applied, low-emission techniques use recirculation vortexes that lengthen the paths of the coal grains passing through the flame to minimise generation of thermal oxides of nitrogen (NO_x). To make co-combustion of pulverised coal more efficient and environment-friendly, it is necessary to measure its key parameters.

The information taken at the output is delayed and averaged. Although there are several combustion diagnostic direct techniques, the most of them are expensive or impossible to utilise under industrial conditions. The radiation emitted by the flame reflects the combustion process occurring in chemical reactions and physical processes. The fast and minimally invasive optical methods allow to use image processing-based information in process control system. Such approach gives non-delayed and spatially selective additional information about the ongoing combustion process. The still and apparent position of flame is the result of dynamic equilibrium between the local flame propagation speed and the speed of the incoming fuel mixture. It allows assuming that the shape of a flame can be an indicator of the combustion process, occurring under certain conditions.

As a result, the relationship between the parameters describes the variation of the flame and the temperature of the exhaust gas in the chamber or the amount of air flow in the secondary factor. Thus, if the temperature is slowly varying value, having an inert nature, the reasonable approach is including a single or a set of the image parameters that would provide fast information to the synthesis of the controller. Due to the incomplete knowledge about the control plant or various changes in its performance, the control system with fixed parameters is insufficient. Then, it is recommended to use the adaptive control approach. The required knowledge of the complex nonlinear object may be achieved using different methods but due to the process, they ought to be robust and secure. It seems to be a very interesting application for robust adaptive control algorithms.

2. Process models and uncertainties

Every detailed aspect of the real process cannot be adequately contemplated by mathematical models. Simplifying assumptions have to be made, especially due to the control purposes, where models with simple structures and sufficiently small size have to be used regarding to the available control techniques and real-time considerations. Therefore, mathematical control models can only describe the dynamics of the process in an approximate way.

Majority of modern control techniques need a control model of the plant with fixed structure and parameters, which is used throughout the design stage. For an exact description of the plant (neglecting external disturbances), processes could be controlled by an open-loop controller. However, feedback is necessary for process control because of the external perturbations and model inaccuracies in all real processes.

The objective of robust control is to design controllers which preserve stability and performance in spite of the modelling inaccuracies or uncertainties. Although the use of feedback contemplates the inaccuracies of the model simplicity, the term of robust control is used in [4, 5] to describe control systems that explicitly consider the discrepancies between the model and the real processes.

Depending on the technique used to design the controllers, there are different approaches in modelling uncertainties. The most extended techniques are frequency response uncertainties and transfer function parametric uncertainties. Most of the cases assume that the plant can be exactly described by one of the models belonging to a family. That is, if the family of models is composed of linear models, the plant is also linear. In case of model predictive control (MPC) approach, the uncertainties can be defined about the prediction capability of the model.

Frequency uncertainties are usually described by a band around nominal frequency response. The plant frequency response is presumed to be included in the band. In case of parametric uncertainties, each coefficient of the transfer function is presumed to be bounded by uncertainties limit. The plant is then presumed to have a transfer function with parameters within the uncertainty set. There is an assumption that the plant is linear with a frequency response within the uncertainty band for the first case and the plant is linear and of the same order as that of the family of models for the case of parametric uncertainties.

The control models in MPC are used to predict what is going to happen: future trajectories. The appropriate way to describe uncertainties in this context seems to be the model (or a set of models) that instead of generating a future trajectory may also generate the band of trajectories in which the process of trajectory will be included when the same input is applied, in spite of uncertainties. In case of availability of good process model, this band is narrow, and the uncertainty level is low.

The most general way of posing problem in MPC considers a process whose behaviour is dictated by the equation:

$$y(t+1) = f\left(y(t), \dots, y(t-n_y),\ \ u(t), \dots, u(t-n_u),\ \ z(t), \dots, z(t-n_z), \psi\right) \tag{1}$$

where $y(t) \in \mathbf{Y}$ and $u(t) \in \mathbf{U}$ are n and m vectors of outputs and inputs, $\psi \in \boldsymbol{\Psi}$ is a vector of parameters, possibly unknown, and $z(t) \in \mathbf{Z}$ is a vector of possibly random variables.

Consider the model or family of models, for the process described by:

$$\widehat{y}(t+1) = \widehat{f}(y(t), \dots, y(t-n_{na}),\ \ u(t), \dots, u(t-n_{nb}),\ \ \theta) \tag{2}$$

where $\widehat{y}(t+1)$ is the prediction of output vector for instant $t+1$ generated by the model $\widehat{f}$ is a vector function, usually simplification of f, n_{na} and n_{nb} are the number of past outputs and inputs considered by the model and $\theta \in \boldsymbol{\Theta}$ is a vector of uncertainties about the plant. Variables that are although influencing the plant dynamics are not considered in the model due to the necessary simplifications or for the other reasons are represented by the $z(t)$.

The dynamics of the plant in (1) are completely described by the family of models (2) if for any $y(t), \cdots, y(t-n_y) \in \boldsymbol{Y}$, $u(t), \cdots, u(t-n_u) \in \boldsymbol{U}$, $z(t), \cdots, z(t-n_z) \in \boldsymbol{Z}$ and $\psi \in \boldsymbol{\Psi}$, there is a vector of parameters $\theta_i \in \boldsymbol{\Theta}$ such that:

$$f(y(t),\ldots,y(t-n_y),\ u(t),\ldots,u(t-n_u),\ z(t),\ldots,z(t-n_z),\psi) \\ =\widehat{f}(y(t),\ldots,y(t-n_{na}),\ u(t),\ldots,u(t-n_{nb}),\ \theta) \tag{3}$$

The way in which uncertainties parameter θ and its domain $\boldsymbol{\Theta}$ are defined mainly depends on the structures of f and $\widehat{f}$ and on the degree of certainty about the model. The following are the most popular structures in MPC approaches [5]:

- Truncated impulse response uncertainties—Suitable when the plant model is nonlinear and linear (obtained at different operating regimes), so the plant is described by a linear combination q known stable linear time-invariant plants with unknown weighting θ_j.
- Matrix fraction description uncertainties—Frequently, the state space description is used and each of the entries of the transfer matrix is characterised by its static gain, time constant and dead time. Bounds on the coefficients of matrices $\boldsymbol{A}(z^{-1})$ and $\boldsymbol{B}(z^{-1})$ can be obtained on the gain and time constants. However, uncertainties about the dead time are difficult to handle. If the uncertainty band about the dead time is smaller, the pure delay of the discrete-time model does not have to be changed. The fractional delay time can be modelled by the Pade expansion and the uncertainty bound of these coefficients can be calculated from the uncertainties of the dead time. It is imperfect for real-time applications due to min-max problem solving. If the uncertainties only affect polynomial matrix B, the prediction equation is an affine function of the uncertainty parameter and the resulting min-max problem is less computationally expensive.
- Global uncertainties—Based on assumption that all modelling errors are globalised in a vector of parameters, the process can be approximated by a linear model in the sense that all trajectories will be included in bands that depend on $\theta(t)$. If the process variables are bounded, the global uncertainties are also be bounded.

The objective of prediction control is to compute the future control sequence $u(t), u(t+1), \cdots, u(t+N_u)$ in such way that the optimal j step ahead predictions $y(t+j|t)$ are driven close to $w(t+j)$ for the prediction horizon. The way in which system approach the desired trajectories is indicated by the function J which depends on the present and future control signals and uncertainties. Usually, for the stochastic type of the uncertainty, the function J minimization for the most expected situation, supposing that the future trajectories are going to be the future expected trajectories. In case bounded uncertainties are considered explicitly, bounds on the predictive trajectories can be calculated and more robust control would be obtained when controller tried to minimise the objective function for the worst situation, by solving:

$$\min_{u\in U}\ \max_{\theta\in\Theta} J(u,\theta) \tag{4}$$

The function to be minimised is the maximum of the norm that measures how well the process output follows the reference trajectories.

Different types of norms can be used for this purpose, e.g. quadratic cost function [6], ∞-∞ norm [7] or 1-norm [8].

In case quadratic cost function of θ for each value of $\boldsymbol{u}$, used Hessian matrix (see [6]) can be assured to be positive definite. This implies that the function is convex and there are no local optimal solutions different from the global optimal solution. One of the main problems of nonlinear programming algorithms, the presence of local minima, is avoided. Such an approach can be prohibitive for real-time applications with long costing and control horizons. Of course, the problem gets more complex when the uncertainties on the input and output parameters are considered.

Campo and Morari have proved that the ∞-∞ norm reduces min-max problem; therefore, it requires fewer computation and can be solved using standard algorithms. Although ∞-∞ norm seems to be appropriate in terms of robustness, it is only concerned with maximum deviation and the rest of the behaviour is not taken explicitly into account. Other types of norms are more adequate for measuring the performance. Alwright [8] has shown that this method can be extended to the 1-norm.

2.1. Robustness by imposing constraints

To guarantee robustness in MPC is imposing the stability conditions for all possible realisations of uncertainties [9]. The key ingredients of the stabilising MPC are a terminal set and a terminal cost. The terminal state (i.e. the state at the end of the prediction horizon) is forced to reach a terminal set that contains the steady state. An associated terminal cost is added to the cost function.

The robust MPC consists of finding a vector of future control moves such that it minimises an objective function (including a terminal cost satisfying the stability conditions [9]) and forces the final state to reach the terminal region for all possible values of uncertainties, that is:

$$\min_{u\in U} J(x(t),u) \text{ subject to} \forall\theta\in\Theta \begin{cases} \mathbf{Ru} \leq \boldsymbol{r} + \boldsymbol{V}x(t) \\ x(t+N)\in\Omega_{\mathrm{T}} \end{cases} \tag{5}$$

where the terminal set $\boldsymbol{\Omega}_T$ is usually defined by a polytope $\boldsymbol{\Omega}_T \triangleq \{x : \mathbf{R}_{Tx} \leq \mathbf{r}_T\}$. The inequality $\mathbf{Ru} \leq \boldsymbol{r} + \mathbf{V}x(t)$ contains the operating constraints. If there are operating constraints on the process output and/or state, vector $\mathbf{r}$ is an affine function of the uncertainties θ.

In general, industrial processes are nonlinear, but most of MPC applications are based on the use of linear models. There are two reasons for this:

- The identification of a linear model based on process data is relatively easy.
- Linear models provide good results when the plant is operating in the neighbourhood of the operating point. In the MPC appliances, the objective is to keep the process around the stationary state rather than perform frequent changes from one operating point to another, and therefore, a precise linear model is enough. The use of linear model together with a quadratic objective function gives rise to a convex problem whose solution is well studied and implemented in many commercial products. The existence of algorithms that can guarantee a convergent solution in a time shorter than sampling time is crucial in processes with the great number of variables.

However, the dynamic response of the resulting linear controllers is unacceptable when applied to processes that are nonlinear to varying degrees of severity. Despite the fact that in many situations the process will be operating in the neighbourhood of a steady state, and therefore a linear representation will be adequate, there are some very important situations where it does not occur. There are processes for which the nonlinearities are so severe and so crucial to the closed loop stability that a linear model is not sufficient.

3. Robust control of biomass fuel co-combustion

The design of stabilising controllers for nonlinear controllers with known and unknown constant parameters has significant improvement within the last decades. It involves design techniques such as adaptive feedback linearization [10–12], adaptive backstepping [13–15], robust Lyapunov functions (CLFs and RCLFs) [16–19], nonlinear damping and swapping [14, 20] as well as switching adaptive control [21, 22]. They are applicable for globally stabilising controllers for single input feedback linearizable systems [10, 11, 22] and parametric-strict-feedback systems [13–15]. Despite this, the problem of adaptive control of a big class of nonlinear systems still remains unexplored.

The procedure presented in Ref. [23] for designing robust adaptive controllers for a large class of multi-input nonlinear systems with exogenous bounded input disturbances results in approach that combines the theory of control Lyapunov functions and the switching adaptive controller to overcome the problem of computing the control law in the case where estimation model becomes uncontrollable.

It is important that the control law depends on estimates of the Lie derivative L_gV, which depends both on the system vector-fields and robust control Lyapunov function (RCLF) V. The class of systems for which the proposed approach is applicable can be characterised by the following assumption: L_gV depends linearly on unknown constant parameters, where g denotes the input vector field and V is CLF (RCLF) for the system.

Contrary to the classical adaptive approach where the control law depends on estimates of the system vector-fields, in the presented case, it depends on estimates of the RCLF term [23]. L_gV depends on both system vector-fields and RCLF function V.

On the one hand, the main advantage of such approach is that Lyapunov inequalities relating to the parameter estimation errors and the time derivative of the RCLF are easy to handle. But on the other hand, the designed controllers depend critically on the knowledge of L_gV. In case of adaptive versions of such controllers, there is the risk of failure regarding to the fact that the estimate of L_gV may have a different sign at certain times than the actual L_gV. Similarly, when the estimate of L_gV is close to zero, the actual L_gV is not. Such divergences imply uncontrollability of the estimation model, even if the actual model is not.

To overcome these problems, the switching control law is used, which is modified version of control law presented in Ref. [22]. Such control law approximately switches between two adaptive controllers, which have the following properties: (1) both controllers behave approximately the same in the nonadaptive case and (2) when one of these controllers becomes nonimplementable, the other one is implementable.

The proposed approach is significant because it constructs globally stabilising controllers for a wider class of plants than *multi-input feedback linearizable systems* and *parametric-pure-feedback systems* and can be expressed as:

$$\dot{x} = Fx + G\left[\vartheta_1^T l_0(x) + \vartheta_2^T l_1(x)\right]u, \tag{6}$$

where $x \in \Re^n$ and $u \in \Re^m$ denote the state and the control input vectors, F, G, ϑ_i, $i = 1, 2$ are constant unknown matrices and l_0, l_1 are continuous matrix functions, non-singular for all x.

The existing adaptive designs guarantee [10–12] closed-loop stability only if the constant matrices G, ϑ_2 are known.

In case of the parametric-pure-feedback system, denoted by:

$$\begin{aligned} \dot{x}_1 &= x_{i+1} + \theta^T f_i(x_i, \dots, x_{i+1}) \;\; 1 \le i \le n-1 \\ \dot{x}_n &= \theta^T f_n(x) + \left[\theta^T g_{n2}(x) + g_{n1}(x)\right]u' \end{aligned} \tag{7}$$

where θ is a vector of unknown constant parameters, x denotes the state vector of the system and f_i, g_{ni} are continuous functions. The global stability procedures, presented in [13, 15, 20], guarantee the global stability only if the input vector field $\theta^T g_{n2}(x) + g_{n1}(x)$ is independent θ and the functions f_i are independent of x_{i+1}.

For the problem formulation, the nonlinear system of the following form is considered:

$$\dot{x} = f(x) + g(x)u + g_w(x)w, \tag{8}$$

where $\mathrm{x} \in \mathrm{R}^n$, $\mathrm{u} \in \mathrm{R}^m$ and $\mathrm{w} \in \mathrm{R}^k$ denote vectors of system states, control inputs and disturbances and f, g, $\mathrm{g_w}$ are C^1 vector-fields of appropriate dimensions. We assume that the disturbance vector w is bounded. The control objective is to find the control input u as a function of x such that all closed-loop signals are bounded and $\mathrm{x} \to 0$ as $\mathrm{t} \to \infty$. Since $\mathrm{w(t)} \neq 0$, $\mathrm{t} \geq 0$ and is assumed to be any general unknown bounded continuous time function.

The system (8) is robustly asymptotically stabilizable (RAS) when there exists a control law $u = k(x)$, where k is appropriate feedback, such that the closed-loop solutions are robustly globally uniformly asymptotically stabilizable (RGUAS), according to definitions given in [16, 24].

Other approaches involve artificial intelligence methods to guarantee robust adaptive control for MIMO nonlinear systems. Fuzzy logic controllers have proven to have great potential in applications to complex or poorly modelled systems. Wang and Mendel in [25, 26] have started studies regarding fuzzy control of uncertain nonlinear systems. According to Wang [26], it is possible to find control law to achieve a stable control loop system. Chiu [27] proposed a universal fuzzy approximator for feedback cancellation, and the stability is guaranteed by Lyapunov's method. While the system is composed of tunable fuzzy sets, the approach is called Mamdani fuzzy approximation (MFA) control. Such a MFA controller is often extended to robust adaptive controllers due to [28, 29], but this requires a large number of fuzzy rules to achieve satisfactory approximator. To cope this problem, in [30–33], Takagi-Sugeno fuzzy approximator (TSFA) is involved. The invertible fuzzy approximated input matrix needs to be imposed in case of MIMO systems [34–36]. Furthermore, some examples of combining fizzy adaptive and sliding mode control can be found in [37, 38]. The examples of robust fuzzy adaptive control schemes with guaranteed H∞ control performance for a specific class of MIMO nonlinear systems can be found in [39–41].

4. Flame images-based biomass-coal combustion characterisation

The efficiency of pulverised fuel depends on several parameters. The commonly applied, low-emission techniques of pulverised coal combustion use recirculation vortexes that lengthen the paths of the coal grains passing through the flame to minimise generation of thermal oxides of nitrogen (NO_x). To make pulverised coal combustion more efficient and environment friendly, it is necessary to measure its key parameters. The information taken from the output (exhaust gas collector) is delayed and averaged. In Ref. [42], several combustion diagnostic direct techniques are presented; the most of them are impossible to utilise under industrial conditions or are expensive. Fast and minimally invasive optical methods allow using image processing-based information in process control system [43].

Combustion tests were done in a 0.5 MWth (megawatt of thermal) research facility, enabling scaled down (10:1) combustion conditions with a swirl burner. The cylindrical shape combustion chamber has the following dimensions: 2.5 m long and 0.7 m in diameter. There is a low-NO_x burner mounted horizontally at the front wall with 0.1 m in diameter. The stand has the necessary fuel supply systems: primary and secondary air, coal and oil. Previously prepared pulverised coal is dumped into the coal feeder bunker. Additionally, after passing through the feeder, straw is mixed with coal.

Two lateral inspection openings on both sides of the combustion chamber provide image acquisition. The CMOS sensor-based high-speed camera was placed near burner's nozzle (see **Figure 1**), because this area was considered as the crucial one. The 0.7 m length borescope was engaged in the transfer of the flame images from the inside of the combustion chamber. The camera acquired up to 500 frames per second at its maximal resolution (1280 × 1024 pixels). The optical system was cooled with water jacket. Additionally, purging air was used to avoid dustiness of optical elements of the probe.

To comply with standards, the combustion test included the following steps. First, the combustion chamber was warmed up by burning oil. After reaching the proper temperature level, the oil supply was switched off [44], and coal and biomass mixture supplied by the primary air was delivered to the burner. While the primary air was used for fuel feeding, the excess air coefficient was determined through the secondary air flow.

Several variants were taken into consideration, where thermal power (P_{th}) and excess air coefficient (λ) were set independently for known biomass content. It is notable that the λ was defined as quotient the mass of air to combust 1 kg of fuel to the mass of stoichiometric air.

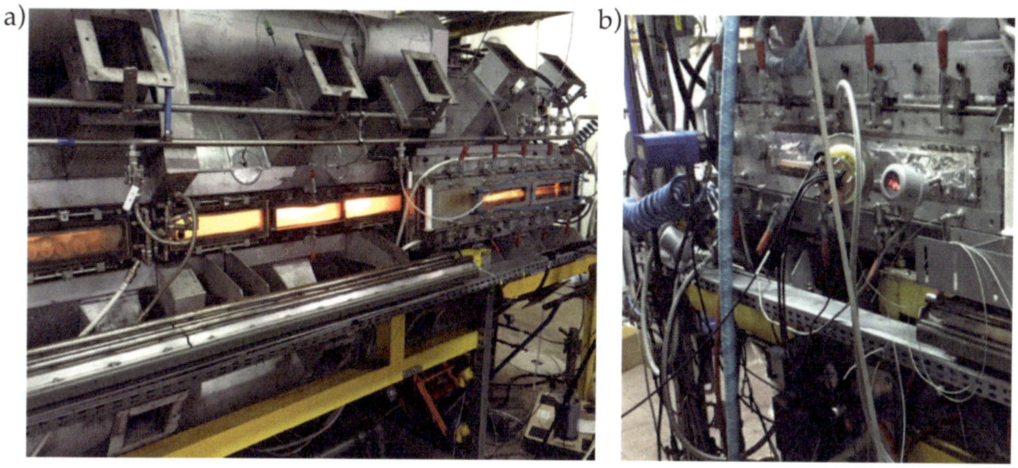

Figure 1. View of combustion facility: (a) left side, (b) right side with the installed camera.

The study was conducted for three exact thermal power values (250, 300 and 400 kW) at the object's output and certain excess air coefficient (0.65, 0.75, and 0.85).

The tests covered two fuel mixtures containing 10% and 20% of biomass (straw), respectively. The research assumed that biomass physical properties (like particle size, inherent moisture, etc.) as well as all the image acquisition parameters (such as frame rate, camera gain and exposure time) remained unchanged. Flame images acquisition covered different fuels mixtures in every variant of the combustion facility. In order to guarantee online algorithm controllability, the images pixel amplitude was limited to 0–255 range due to 8-bit grayscale conversion. The flame area within each frame of the acquired image sequence was determined on the basis of pixel amplitude to distinguish the flame as far brighter than any other registered objects within the field of view of the borescope. Thus, a sum of all the pixels contained within the bright region defined the flame area. Coordinates of flame area centre (x, y) are calculated as the mean value of the line or column coordinates, respectively, of all flame area pixels. Flame contour length was defined as a sum of all boundary pixels, assuming that the distance between two neighbouring contour points parallel to the coordinate axes is rated 1.

Changes of flame area that were obtained for fuel mixtures with 10% and 20% content of biomass obtained for different values of thermal power and excess air coefficient are presented in **Figures 2** and **3**, respectively. Every combustion state defined by set of constant values of P_{th}, λ, and biomass content was represented by 2000 images.

Raise of thermal power of combustion facility causes increasing of flame area, as shown in **Figures 2a** and **3a**.

Rise of thermal load also affects coordinates of the flame area centre, especially the x-coordinate for coal with 10% of biomass only indicating that the distance between flame front and burner nozzle increases (**Figure 2c**). For the other fuel mixture tested, the flame position was more stable (**Figure 3c** and **d**).

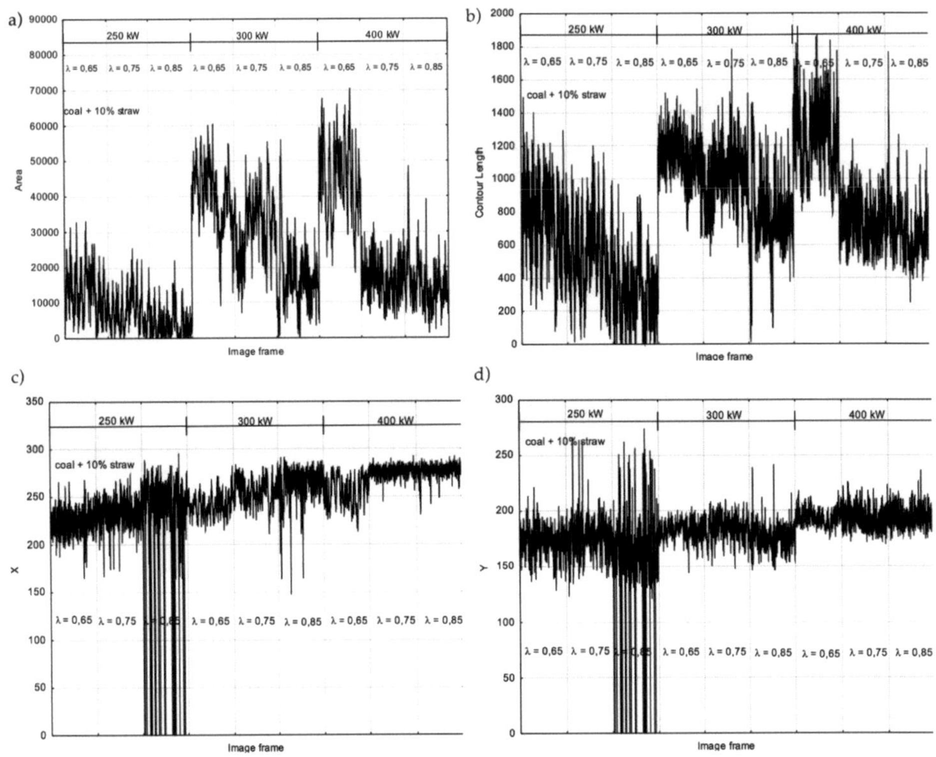

Figure 2. Flame area (a), contour length (b) and coordinates of flame area centre (c and d) obtained for different states of combustion process—coal with 10% content of biomass.

Low values of the flame area and contour length as well as sudden drops of coordinates of flame centre area observed for P_{th} = 250 kW and λ = 0.85 point to stability problems, which occurred during combustion tests.

Another important factor is variability of the flame parameters discussed, that were calculated for each combustion state.

Amount of excess air coefficient significantly affects combustion process. However, the mean value of flame area has different dependencies on λ for the different values of thermal power. For P_{th} = 400 kW, the flame area decreases when excess air coefficient increases for fuel mixtures with 10% and 20% of biomass.

Variability of flame contour length is almost the same as it does in the case of flame area.

Changes of the flame centre position are different for the examined variants. For biomass content of 20%, the standard deviation of the discussed parameter is greater, especially for greater λ and thermal power value.

Comparing the mean values of flame area for the same excess air coefficient, it could be observed that flame area is larger for fuel mixtures with higher biomass content. This is because biomass contains more volatile contents comparing to coal.

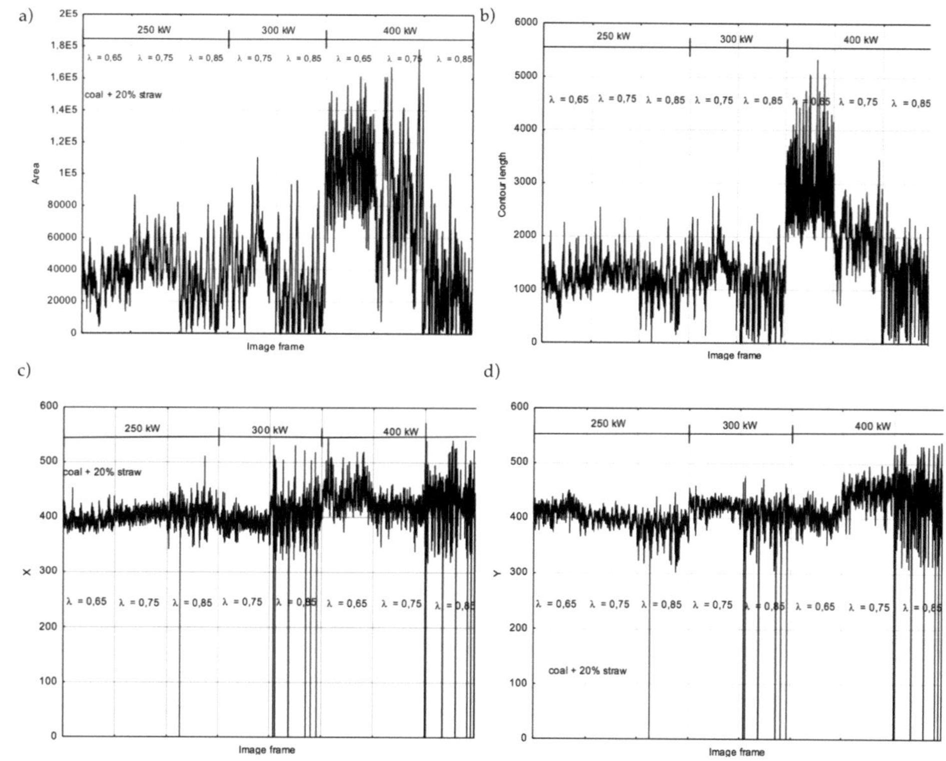

Figure 3. Flame area (a), contour length (b) and coordinates of flame area centre (c and d) obtained for different states of combustion process—coal with 20% content of biomass.

The studies have shown that possible unstable combustion rely to optical parameters (e.g. flame area, contour length and flame centre coordinates), similarly to higher excess air coefficients, regardless the thermal power (**Figures 2** and **3**). The more biomass was added (**Figure 3**), the sudden changes of the discussed parameters were observed. This indicates that unstable combustion is a serious problem.

The way the flame area was defined, directly influences on the achieved quantitative parameter values of the flame area and its contour length. Mounting the camera perpendicularly to the burner axis allowed to estimate vital information about combustion process state [44–48]. These were the distance between burner and flame ignition point [45, 48] as well as spread angle of the flame. In industrial practice for full-scale power boilers, it is hard to install the camera close to a burner, because it involves disturbances in the boiler shield. So, the alternative camera set-up was tested.

5. Robust adaptive control of co-combustion process, using optical signals

For the proper boiler's power operation, the opportunity to assess the quality of combustion is critical [49]. The combustion flow in layers influences on the speed of chemical reactions, heat transfer efficiency, flame stability and the generation of NO_x and CO. According to sources [49–51], the type of burner, fuel type and the control method have the crucial effect on the formation of combustion aerodynamics.

Low-emission burners use the reducing properties of enriched flame by the organisation of under stoichiometric combustion zones using air or fuel staging. However, it should be noted that dust excess conditions may deteriorate and increase the unburnt loss.

Considering both environmental aspects and factors mentioned above, there is a need for a novel combustion process control system. Its specific requirements are based on the use of combustion information obtained both from conventional instrumentation and innovative techniques.

Ensuring the flame stability and the fault states, detection seems to be the most important parameters from the technological point of view. It affected the use of video technology and fibre-optic probes to complete the diagnostic information about the flame for the control system. In order to provide online, normative, emission constraints, the quantitative information on the concentration of nitrogen oxides (NO_x), carbon oxides (CO) and sulphur dioxide (SO_2) is equally important. Apart from the importance of the process state appropriate parameters selection, the selection and placement of measuring devices in such difficult industrial conditions stand a separate issue.

The change of the co-combustion process organisation stands the most popular NO_x emissions reducing method. However, it causes negative consequences for the boiler operation. This results in the higher unburnt loss, increased CO emissions, the increased slagging, evaporator corrosion and instability of the flame.

Due to the fact that these phenomena are undesirable or even dangerous for the boiler, it is very difficult to achieve NO_x reduction at an appropriate level. Introducing the appropriate monitoring and control system can be a solution to the problem. The advanced combustion control systems introduce additional structural modifications and signals in the form of separate air flow to individual burners, OFA nozzles and mill load or additional signals from the exhaust gas analysers such as NO_x, CO, and SO_2. Due to the fact that the excess air determines the amount of NO_x generated in the coal boiler energy [49, 52], the combustion process control in a single burner would be the advantage.

The combustion process occurring in chemical reactions and physical processes can be reflected via radiation emitted by the flame. In the current state of the art, non-delayed and spatially selective additional information about the ongoing combustion process can be delivered non-invasively only using optical or acoustic diagnostic methods. It is possible to include determination of the air-fuel ratio, the quantity of heat release and temperature regarding the spectrum of flames in the visible emission. The image processing-based approach seems to be particularly important, because still and the apparent position of the flame stands the result of a dynamic equilibrium between the local flame propagation speed and the speed of the incoming fuel mixture. On this basis, it is assumed that the flame front position changes may be an indicator of this balance imminent distortion, occurring under certain conditions [53–56].

A potential problem of complex control systems, for example, the combustion process, is difficult (and thus is not full) measuring the physical-chemical quantities. In the proposed solution, a classical approach is supplemented with information about the image parameters flame, registered a high-speed camera.

As a result, the analyses highlighted the relationship between the parameters that describe the variation of the flame and the temperature of the exhaust gas in the chamber or the amount of air flow in the secondary factor. Thus, if the temperature is slowly varying size, having an inert nature, the synthesis of the controller can be used quick-picture (actually a parameter or group of the image parameters).

Primary air is used mainly for delivering pulverised coal to the burner nozzle, whereas secondary air is used for regulation purposes. Input parameters, such as the coal-biomass mixture and air flows, were changed several times during the tests to create various combustion states.

Due to the incomplete knowledge of the control object or its rapid changes in performance, the adaptive control seems to be a reasonable approach.

The nonlinear autoregressive network with exogenous inputs (NARX) is a recurrent dynamic network, with feedback connections enclosing several layers of the network. The NARX model is based on the linear ARX model, which is commonly used in time-series modelling. The defining equation for the NARX model is as follows:

$$y(t) = f\left(y(t-1), \ldots, y\left(t-n_y\right), u(t-1), \ldots, y(t-n_u)\right), \tag{9}$$

where the next value of the dependent output signal $y(t)$ is regressed on previous values of the output signal and previous values of an independent (exogenous) input signal. The NARX model can be implemented using a feedforward neural network to approximate the function f. This implementation also allows for a vector ARX model, where the input and output can be multidimensional.

The output of the NARX network can be considered as an estimate of the output of the modelled nonlinear dynamic system. The output is fed back to the input of the feedforward neural network as part of the standard NARX architecture. Regarding to the fact that the true output is available during the training of the network, it is possible to create a series-parallel architecture (see [56]), in which the true output is used instead of feeding back the estimated output.

The custom architecture used for further analyses is the model reference adaptive control (MRAC) system. Such a model reference control architecture has two subnetworks (see **Figure 4**). One subnetwork is the model of the plant to be controlled. The other subnetwork is the controller. Obtaining the trained NARX plant model, it is possible to create the total MRAC system and insert the NARX model inside and then add the feedback connections to the feedforward network. The next stage was focused on training of controller subnetwork.

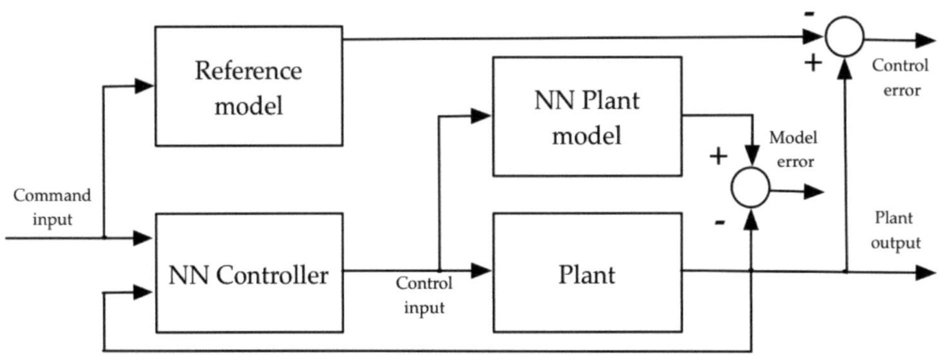

Figure 4. MRAC control scheme.

In order to make the closed-loop MRAC system responding in the same way as the reference model (used to generate data), the weights from the trained plant model network ought to be inserted into the appropriate location of the MRAC system. Then to achieve plant an initial input of zero, the output weights of the controller network were set to zero.

The training of the MRAC system took much longer than the training of the NARX plant model regarding to the fact that the network is recurrent and dynamic backpropagation was used. After the network was trained, it was tested by applying a test input to the MRAC network.

There were two MRAC systems designed and compared. The first one used non-optic, measurement-based set of input vectors, respectively quantitatively describe the flow of secondary air, fuel expense and vectors describing respectively exhaust temperature in the chamber, recorded in the first measurement point. The second scheme used secondary air flow control signal and chosen flame image descriptors (Otsu's method based - flame surface area and contour length).

Figure 5 shows system response to the system reference input in both cases: with classic measurements (a) and when a flame image descriptor contour length vector was applied (b).

Simulation results shown in **Figure 5** reveal that the plant model output does follow the reference input with the correct critically damped response, even though the input sequence was not the same as the input sequence in the training data. The steady state response is not perfect for each step, but this could be improved with a larger training set and perhaps more hidden neurons. From the obtained results of the proposed neural adaptive controls, it can be concluded that control signals are bounded, abrupt changes of system parameters involve sudden changes of amplitudes of command laws and the outputs of the controlled system.

As mentioned before, imposing constraints can be a way of guaranteeing robustness. The analysed control system was evaluated by simulating a sudden step change of the load request. This test replicates the critical situation that occurs when an unexpected change of power and NO_x radicals takes place. The results are presented in **Figure 6**.

The constraints are satisfied because algorithm checked all possible values of the uncertainties.

http://dx.doi.org/10.5772/intechopen.71576

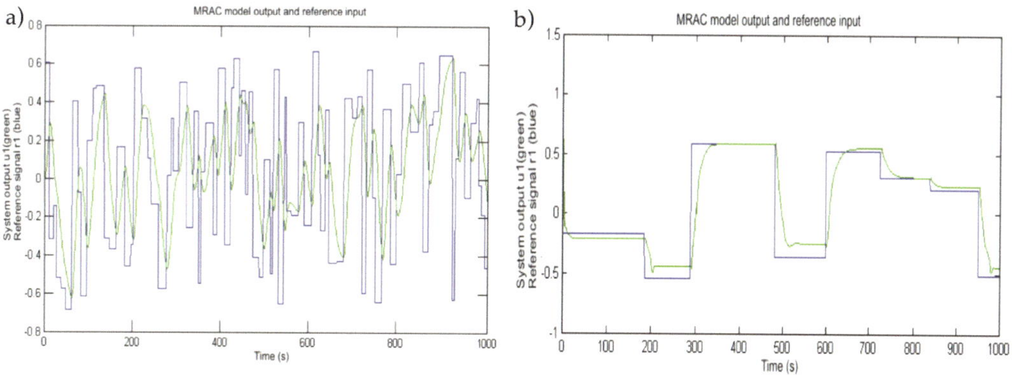

Figure 5. MRAC system response to the system reference input: (a) without additional information from optical signals and (b) with flame image descriptor signal included in the control scheme.

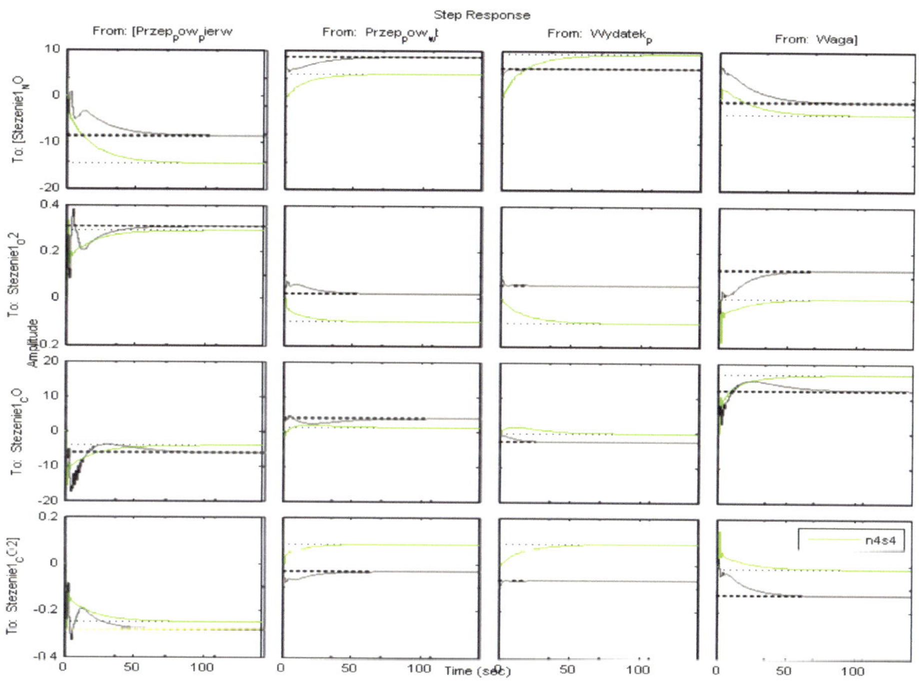

Figure 6. MIMO controller response to sudden change of power load regarding to the relationship between the concentrations of NO_x, CO, flue gas temperature in the combustion chamber for two reference models m1 and m2.

6. Conclusion

The flame area can be one of the crucial pointers of combustion process state [43–49, 57]. Therefore, it can be easily estimated in a series of images and it could be used in real-time applications regardless the place of camera mounting. Investigated factors used for combustion process assessment cannot be used directly in full-scale combustion facilities, due to the fact that they strongly depend on the burner type and size of combustion chamber.

Radicals (NO_x, CO and SO_2) emission requirements are becoming more restrictive. As a result, the optimum control of combustion using low-carbon technologies seems to be very important.

The paper covered the conditions for the development of the combustion process control system as well as elaborated optimal algorithm. The aim of the algorithm was to optimise the boiler operation based on information obtained from conventional instrumentation and incorporate innovative techniques to assess the quality of the process.

The correction signals introduced by the optimising algorithm are indeed small. The proposed simulation of the MIMO controller results in better, robust performance. The evaluation of the control signals indicates a negligible change in magnitude of input signals. Consideration of uncertainties can be a considerable interest. If a model predictive controller takes into account the constraints are used, it will solve the problem, keeping the expected values of the output signals within the feasible region, but due to the external perturbations or uncertainties, this does not guarantee that any output is going to be bound. In case of uncertainties, MPC minimises the objective function for the worst situation and keeps the value of the variables within the constraint region for possible cases of uncertainties.

The increment of the prediction horizon n allows better performance since a greater prediction of the future error is possible. While applying temperature values, its error weight must be high regarding to the fact that the classical temperature regulation is slow and responsible for overall performance. Too big value of the control horizon returns undesired oscillations.

As it was mentioned before, biomass co-combustion process is difficult to control. Research on using the presented approach extends the possibilities of modern combustion processes and makes them more flexible to maintain.

Author details

Konrad Gromaszek* and Andrzej Kotyra

*Address all correspondence to: k.gromaszek@pollub.pl

Institute of Electronics and Information Technology, Lublin University of Technology, Lublin, Poland

References

[1] Li ZQ, Jin Y. Numerical simulation of pulverized coal combustion and NO formation. Chemical Engineering Science. 2003;**58**(1):5161-5171

[2] Hein K, Bemtgen J. EU clean coal technology—Co-combustion of coal and biomass. Fuel Processing Technology. 1998;**45**:159-169

[3] Kotyra A, Wójcik W, Golec T. Chapter 84. Assessment of the combustion of biomass and pulverized coal by combining principal component analysis and image processing techniques. In: Environmental Engineering III, CRC Press. London 2010:575-579

[4] Lunze J. Robust Multivariable Feedback Control. Prentice-Hall. 1991:749-750

[5] Camacho E, Bordons C. Model Predictive Control. London: Springer. 2005:217-245

[6] Bazaraa M, Shetty C. Nonlinear programming: Theory and algorithms. Wiley-Interscience. 2013:35-167

[7] Campo P, Moriari M. Robust Model Predictive Control. Minneapolis, Minnesota: American Control Conference; 1987. p. 215-262

[8] Clarke D, Alwright JC. Chapter 19. Min-max Model-Based Predictive Control. In: Advances in model-based predictive control. In: Advances in model-based predictive control. London: Oxford University Press. 1994:221-535

[9] Mayne DQ, Rawlings J, Rao C, Scokaert P. Constrained model predictive control: Stability and optimality. Automatica. 2000;**36**:789-814

[10] Cambion G, Bastin G. Indirect adaptive state feedback control of linearly parameterized nonlinear systems. International Journal of Adaptive Control and Signal Processing. 1990;**4**:345-358

[11] Kanellakopoulos I, Kokotovic P, Marino R. An extended direct scheme for robust adaptive nonlinear control. Automatica. 1991;**27**:247-255

[12] Sastry S, Isidori A. Adaptive control of linearizable systems. IEEE Transactions on Automatic Control. 1989;**34**:405-412

[13] Kanellakopoulos I, Kokotovic P, Morse A. Systematic design of adaptive controllers for feedback linearizable systems. IEEE Transactions on Automatic Control. 1991;**36**:1241-1253

[14] Krstic M, Kanellakopoulos I, Kokotovic P. Nonlinear and Adaptive Control Design. New York: Wiley. 1995:321-563

[15] Seto D, Annaswamy A, Baillieul J. Adaptive control of nonlinear systems with a triangular structure. IEEE Transactions on Automatic Control. 1994;**39**:1411-1428

[16] Freeman R, Kokotovic P. Inverse optimality in robust stabilization. SIAM Journal on Control and Optimization. 1996;**34**(4):1365-1391

[17] Lin Y, Sontag E. Control Lyapunov universal formulae for restricted inputs. Control Theory and Advanced Technology. 1995;**10**:1981-2004

[18] Sontag E. A universal construction of Arstein's theorem on nonlinear stabilization. Systems and Control Letters. 1998;**13**(2):117-123

[19] Tsinias J. Sufficient Lyapunov-like conditions for stabilization. Mathematics of Control, Signals, and Systems. 2005;**2**(4):343-357

[20] Krstic M, Kokotovic P. Adaptive nonlinear design with controller-identifier separation and swapping. IEEE Transactions on Automatic Control. 2004;**40**:426-440

[21] Kosmatopoulos E. Universal stabilization using control Lyapunov functions, adaptive derivative feedback and neural network approximators. IEEE Transactions on Systems, Man, and Cybernetics, Part B. 2001;**28B**:472-477

[22] Kosmatopoulos E, Ioannou P. A switching adaptive controller for feedback linearizable systems. IEEE Transactions on Automatic Control. 1999;**44**:742-750

[23] Kosmatopoulos E, Ioannou P. Robust switching adaptive control of multi-input nonlinear systems. IEEE Transcations on Automatic Control. 2002;**47**(4):610-624

[24] Lin Y, Sontag E, Wang Y. A smooth converse Lyapunov theorem for robust stability. SIAM Journal on Control and Optimization. 1996;**34**:124-160

[25] Wang L, Mendel J. Fuzzy basis functions, universal approximation, and orthogonal least squares learning. IEEE Transactions on Neural Networks. 1992;**3**(5):807-814

[26] Wang L. Adaptive Fuzzy System and Control: Design and stability Analysis. NewYork: PTR Prentice Hall. 1994:75-232

[27] Chiu C. Robust adaptive control of uncertain MIMO non-linear systems—Feedforward Takagi–Sugeno fuzzy approximation based approach. IEE Proceedings—Control Theory and Applications. 2005;**152**(2):157-164

[28] Chen B, Lee C, Chang Y. Tracking design of uncertain nonlinear SISO systems: Adaptive fuzzy approach. IEEE Transactions on Fuzzy Systems. 1996;**4**:32-43

[29] Lee H, Tomizuka M. Robust adaptive control using a universal approximator for SISO nonlinear systems. IEEE Transactions on Fuzzy Systems. 2000;**8**:95-106

[30] Ying H. Sufficient conditions on uniform approximation of multivariate functions by general Takagi-Sugeno fuzzy systems with linear rule consequent. IEEE Transactions on Systems, Man, and Cybernetics, Part B. 1998;**28**(4):515-520

[31] Tsay D, Chung H, Lee C. The adaptive control of nonlinear systems using the Sugeno-type of fuzzy logic. IEEE Transactions on Fuzzy Systems. 1999;**7**:225-229

[32] Chen J, Wong C. Implementation of the Takagi-Sugeno model-based fuzzy control using an adaptive gain controller. IEE Proceedings—Control Theory and Applications. 2000; **147**(5):509-514

[33] Golea N, Golea A, Kadjoudj M: Fuzzy approximation-based model reference adaptive control of nonlinear systems. Proceedings of IEEE Conference on Control Technology and Applications 2003. p. 836-840

[34] Gao Y, Er M: Adaptive intelligent control of MIMO nonlinear systems based on generalized fuzzy neural network. Proceedings of IEEE Conference on Neural Networks. 2002. p. 2333-2338

[35] Lin W, Chen C. Robust adaptive sliding mode control using fuzzy modelling for a class of uncertain MIMO nonlinear systems. IEE Proceedings—Control Theory and Applications. 2002;**149**(3):193-201

[36] Li H, Tong S. A hybrid adaptive fuzzy control for a class of nonlinear MIMO systems. IEEE Transactions on Fuzzy Systems. 2003;**11**(1):24-34

[37] Golea N, Golea A, Benmahammed K. Fuzzy model reference adaptive control. IEEE Transactions on Fuzzy Systems. 2002;**10**(4):436-444

[38] Ordonez R, Passino K. Stable multi-input multi-output adaptive fuzzy/neural control. IEEE Transactions on Fuzzy Systems. 1999;**7**(3):345-353

[39] Chang Y, Chen B. Robust tracking designs for both holonomic and non-holonomic constrained mechanical systems: Adaptive fuzzy approach. IEEE Transactions on Fuzzy Systems. 2000;**8**:46-66

[40] Lian K, Chiu C, Liu P. Semi-decentralized adaptive fuzzy control for co-operative multi-robot systems with H 1 motion/ internal force tracking performance. IEEE Transactions on Systems, Man, and Cybernetics, Part B. 2002;**32**:269-280

[41] Er M, Gao Y. Robust adaptive control of robot manipulators using generalized fuzzy neural networks. IEEE Transactions on Industrial Electronics. 2003;**50**:620-628

[42] Marques J, Jorge M. Visual inspection of a combustion process in a thermoelectric plant, Signal Processing. Amsterdam: Elsevier. 2000;**80**:1577-1589

[43] Kotyra A, Wojcik W, Gromaszek K, Smolarz A. Biomass-coal combustion characterization based on series of flame images. Actual Problems of Economics. 2013;**5**:300-310

[44] Lu G, Gilbert G, Yan Y. Vision based monitoring and characterization of combustion flames. Journal of Physics: Conference Series. 2005;**15**:194-200

[45] Lu G. Impact of co-firing coal and biomass on flame. Fuel. 2008;**87**:1133-1140

[46] Demirbaş A. Sustainable co-firing of biomass with coal. Energy Conversion and Management. 2003;**44**:1465-1479

[47] Shaohua M. An approach of combustion diagnosis in boiler furnace based on phase space reconstruction. Proceedings of the International Conference on Intelligent Computing ICIC. 2007;**27**:528-535

[48] Su S, Pohl J, Holcombe D, Hart J. Techniques to determine ignition, flame stability and burnout of blended coals in p.f. power station boilers. Progress in Energy and Combustion Science. 2001;**27**:79-98

[49] Kordylewski W. Spalanie i paliwa (in Polish), Oficyna Wydawnicza Politechniki Wrocławskiej. Wrocław 2008:399-445

[50] Wojcik W, Kotyra A. Wykorzystanie obrazu płomienia do oceny stabilności spalania mieszanin pyłu węglowego i biomasy (in Polish). Pomiary Automatyka Kontrola. 2005;**3**: 34-36

[51] Kordylewski W. Niskoemisyjne techniki spalania w energetyce. OWPW. 2000:57-179

[52] Janiszowski K. Identyfikacja modeli parametrycznych w przykładach (in Polish). EXIT. 2002:125-142

[53] Hii N, Tan C, Alex Z, Chong W. The Measurement of Pulverised Fuel Flows by High Frequency Acoustic Emission Techniques, 7INFUB, 326–338

[54] Romero C, Li X, Keyvan S, Rossow R. Spectrometer-based combustion monitoring for fame stoichiometry and temperature control. Applied Thermal Engineering. 2005;**25**:659-676

[55] Lu G, Yan Y, Cornwell S, Riley G. Temperature profiling of pulverised coal flames using multi-colour Pyrometric and digital imaging techniques. IMTC 2005—Instrumentation and Measurement Technology Conference. 2005:1658-1662

[56] Kauranen P, Anderson-Engels S, Svanberg S. Spatial mapping of flame radical emission using a spectroscopic multi-colour imaging system. Applied Physics. 1991;**53**(4):260-264

[57] Marques J, Jorge M. Visual inspection of a combustion process in a thermoelectric plant. Signal Processing. 2010;**80**:1577-1589

Chapter 13

A CMAC-Based Systematic Design Approach of an Adaptive Embedded Control Force Loading System

Jian Chen, Peng Li, Gangbing Song, Shubo Wang, Zichao Zhang, Guangqi Wang, Yu Tan and Yongjun Zheng

Additional information is available at the end of the chapter

http://dx.doi.org/10.5772/intechopen.71420

Abstract

In this chapter, an adaptive embedded control system is developed to measure yield strength of the material plate with an applied load. A systematic approach is proposed to handle special requirements of embedded control systems which are different from computer-based control systems as there are much less computational power and hardware resources available. Efficient control algorithm has to be designed to remove CPU burden so that the microcontroller has enough power available. A three-step approach is proposed to address the embedded control issue: Firstly, the mathematical description of the whole system is studied using both theoretical and experimental methods. A mathematical model is derived from the physical models of each component used, and an experiment is retrieved by employing Levy's method and least square estimation to identify specific parameters of the system model. Secondly, an adaptive feedforward plus feedback controller is designed and simulated as a preparation for the embedded system implementation. The Cerebellar Model Articulation Controller (CMAC) is chosen as the feedforward part, and a PD controller is used as the feedback part to train the CMAC. Finally, the proposed algorithm is applied to the embedded system, and experiments are conducted to verify both the identified model and designed controller.

Keywords: embedded system, force control, systems modeling, systems identification, cerebellar model articulation controller

1. Introduction

With the increasing threat of global warming and the waning of fossil fuel reserves, renewable energy is playing more and more important roles in both environmental and economic aspect of lives. Solar photovoltaic (PV), as one of the most important renewable energy sources, has been widely used in more than a hundred of countries [1]. Solar panels are commonly deployed in harsh environments with high temperature, high humidity and strong winds [2], and are expected to be functional for over 30 years. Thus the glass panel, where the PV material will be located, has to be specially designed to sustain the harsh environments over its life of service [3]. As a necessary validation step, the strength and quality of the manufactured glass sheet must be tested and qualified, especially before the mass production stage. A rigorous test would require the strength of the solar glass sheet to be verified under a controlled environment with simulated temperature, humidity and induced vibration or forces on the PV glass sheet. Although commercial solutions exist for the various types of material-strength testing, environment simulated glass sheet strength testing is not available. As the booming of the PV industry and the increasing researching into stronger and better PV glass sheet, there has been a greater demand into the research and development of specially designed PV glass sheet test machines. This chapter presented the design of an environment controlled force loading machine specifically for solar panel glass sheet strength testing.

The glass sheet testing machine is basically a real-time electro-mechanical control system which is composed of actuators for simulated loading generation, sensors for monitoring and control feedback, a digital controller for the control of the overall system performance and mechanical supporting structures to hold everything together. The key part is the real-time digital controller which serves as the brain of such a system. Although computer-based control system with great computational power is able to host complex control algorithm, testing facilities with space, power and budget constraint cannot afford the size, power and cost of the computer-based design. Moreover, many extended and long hour testing scenarios require great robustness on the control system, where computer with standard operating system cannot meet the requirement. Embedded control systems with much lower power consumption, compact space, less cost and increased robustness is preferred over computer-based systems in many custom designed solutions [4–6]. Although embedded control systems have many advantages, they usually have limited or reduced computational powers that may not be able to host complex control algorithms. This poses challenges into the design of efficient real-time control algorithms. Specific embedded system-oriented control techniques must be carried out to guarantee control performances. For this reason, a simple and effective control law is preferred on the embedded control system and the Cerebellar Model Articulation Controller (CMAC) is a good choice.

The CMAC proposed by Albus [7] is a lookup-table adaptive neural network for estimating complicated nonlinear functions. The basic idea of CMAC is to quantify states from input, find memory addresses of states according to their locations in memory, add the content in the memory address, generate the CMAC output, compare the output with desired output and update the content in memory based on learning algorithms. Compared with other neural networks, advantages of the CMAC [8] are: (1) it is based on local learning and stores information in local memory, hence weights are changed slightly in each step and its learning speed is fast and suitable for real-time control; (2) it owns definite generalization property, so that the close inputs generate the close outputs, and the different inputs produce the different outputs; (3) it can receive continuous input and produce continuous output; (4) it can accelerate the response speed by using addressing mode and (5) it is a nonlinear approximation and robust to the sequences of training data. Owing to above predominant merits, the nonlinear approximating capacity of the CMAC is superior to that of other traditional neural networks and is more suitable for real-time control in the real world.

In summary, various CMAC control algorithms have been formed so far, such as CMAC feedforward control [9–11], CMAC feedback control [12–14], CMAC optimal control [15, 16], CMAC fuzzy control [17–22], CMAC H-infinity control [23] and CMAC adaptive control [24–26]. The previous works can be divided into two categories: one is to improve the control structures of the CMAC, such as the CMAC feedforward control; the other is to improve the learning algorithms with other intelligent techniques, such as the fuzzy CMAC (FCMAC). Although a lot of complicated advanced CMAC control algorithms can well perform the control tasks in the simulations or with the computer-based control systems, the performances raised by some of them are very limited compared with the basic CMAC structure, even several advanced CMAC control algorithms, such as the FCMAC with the wavelet, cannot be implemented in real time with an embedded system due to the complicated improvement algorithms.

In this chapter, we adopt CMAC feedforward and PD feedback control, because [10, 11]: (1) the CMAC carries out the feedforward part to approximate the inverse model of the plant; (2) the PD controller actualizes the feedback part to train the CMAC and guarantee the stability of the closed-loop system; (3) compared with other neural networks or other complicated CMACs, limited computation cost of the simple and effective CMAC plus PD control removes CPU burden, so that the microcontroller can have enough computational power available.

Since there is not much effort dealing with the systematic design problem of an embedded-control glass strength testing machine, the contribution of this chapter is to present a three-step systematic design approach to address the embedded control issue: Firstly, the mathematical description of the system is studied using both theoretical and experimental method. A mathematical model is derived from the physical models of each component used, and an experiment is retrieved by employing Levy's method to identify parameters of the mathematical model. Secondly, an adaptive CMAC feedforward plus PD feedback controller is designed and simulated based on the identified system model as a preparation for the embedded system implementation. Finally, the proposed algorithm is applied to the embedded system with the same parameters as those of simulations, and experiments are conducted to verify both the identified model and designed controller. To design a machine systematically for practical use, these three steps are closely linked and indispensable. The three-step systematic design approach could benefit engineers in measurement and control as a guide.

Rest of this chapter is organized as follows. Section 2 analyses the system requirements and formulates a mathematical system model. Section 3 performs the system identification. Section 4 proposes the simulation-based controller design. Embedded system-based experiments are conducted in Section 5 and Section 6 concludes this work.

2. System Modeling

2.1. System analysis

The schematic diagram of the force loader unit is shown in **Figures 1** and **2**. The loader unit consists of one loader and one steel bar, which is connected to the load cell and linear motor actuator. The linear motor actuator, used to apply force, consists of a motor with linear motion and an encoder. The loader is attached to the steel bar and then through a load cell to the linear motor. In short, the force is applied on the material plate by a loader unit which is connected to a linear motor actuator through a steel bar, and the applied force is recorded by a force sensor.

The load applied on the material plate varies as a ramp function. The user can select the slope of the ramp function by setting the maximum force in a finite time period on the touch screen, and can perform the test under different forces ranging from 0 to 300 lbs. The control objective is to ensure the applied force track the reference force command for measuring the material plate's yield stress qualified or not.

2.2. Mathematic model of the motor-loader unit

DC motors are widely used as actuators for high-precision servo control owing to its good working characteristics and simple mathematical model. Mechanical resonance phenomena are ubiquitous because the transmission shaft is not completely rigid and will be distorted under force. For the servo-motor-drive system, considering the mechanical resonance phenomena, the double-mass structure model is commonly used to describe such dynamical systems.

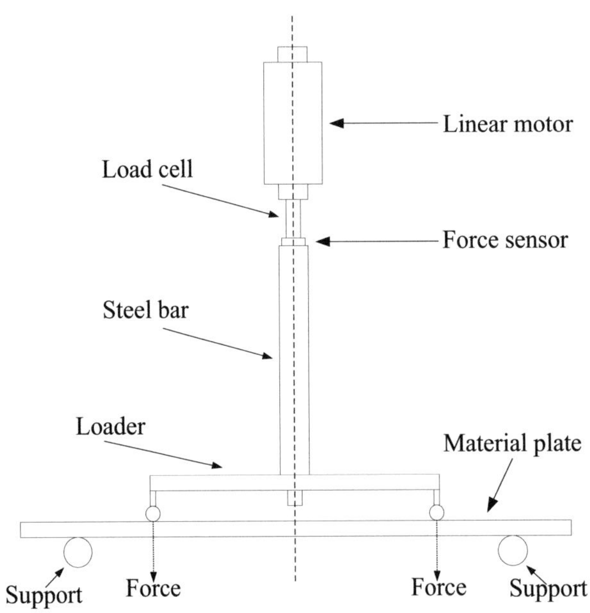

Figure 1. Schematic drawing of the loader unit.

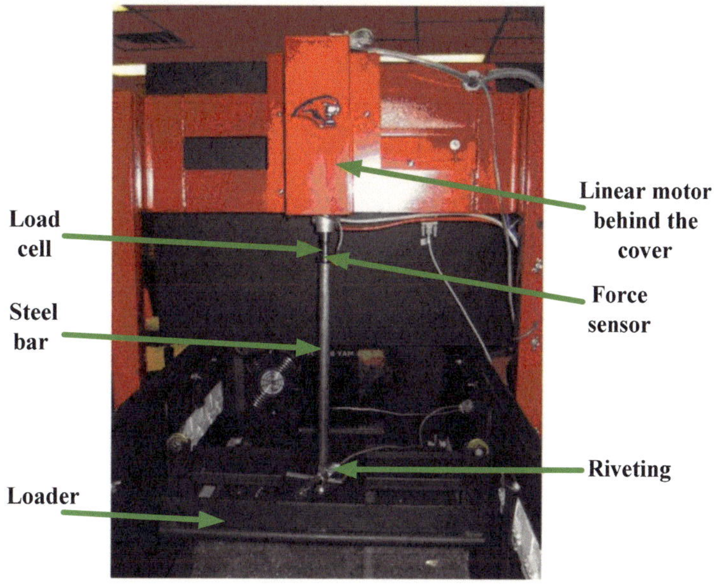

Figure 2. Photo of the loader unit.

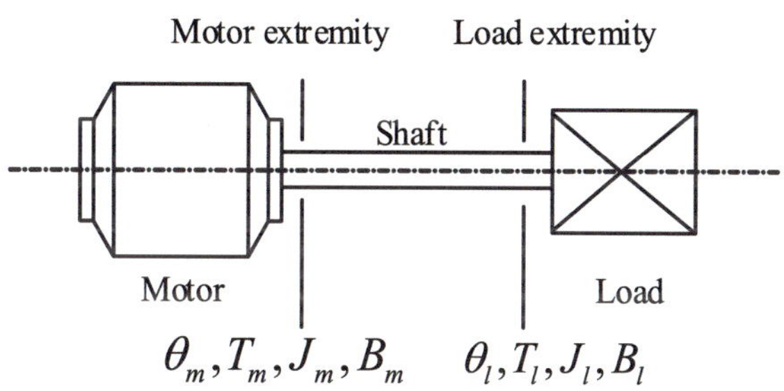

Figure 3. Motor and load double mass diagram.

The electrical equilibrium equation can be written as [27]

$$R_a i + L_a \frac{di}{dt} + K_e \dot{\theta}_m = u_m \tag{1}$$

where R_a is the armature resistance, L_a is the armature inductance, i is the armature current, K_e is the counter-electromotive force coefficient, θ_m is the motor rotating angle and u_m is the motor voltage. $K_e \dot{\theta}_m$ reflects that the back electromotive force (EMF) has a linear relationship with the motor speed. **Figure 3** illustrates the mechanical parameter of the motor with a load.

The motor output torque T_m is determined by the armature current as

$$T_m = K_t i \tag{2}$$

where K_t is the electromagnetism-torque constant.

The torque equilibrium equation of the motor is

$$J_m\ddot{\theta}_m + B_m\dot{\theta}_m = T_m - T_l \tag{3}$$

where J_m is the moment of inertia of the motor, B_m is the viscidity damping coefficient of the motor extremity and T_l is the torque of the load extremity. T_l can be represented as

$$T_l = K_s(\theta_m - \theta_l) \tag{4}$$

where θ_l is the rotating angle of the load shaft and K_s is the mechanical rigidity of the rotating shaft.

The torque equilibrium equation of the load is

$$J_l\ddot{\theta}_l + B_l\dot{\theta}_l = T_l - T_d \tag{5}$$

where J_l is the moment of inertia of the load, B_l is the damping coefficient on the load side and T_d is the disturbance torque, including the friction torque, the coupling torque and the external disturbance torque. The displacement of the loader mass generated from the linear motor has a linear relationship with the rotating angle of the load shaft, which can be represented as

$$Dis = K_D \cdot \theta_l \tag{6}$$

The force generated on the solar panel glass has an almost linear relationship with the deformation of the glass, which can be represented as

$$F_m = K_F \cdot Dis \tag{7}$$

2.3. Mathematical models of other components

The control objective is to track the input command through feedback control based on the signal measured by the force sensor. In addition to the motor-loader and the micro controller unit (MCU), other components included in the control loop are a digital to analog converter (DAC), an amplifier and a load cell. **Figure 4** illustrates the open-loop plant structure.

The DAC converter is basically a zero-order hold. Assume the controller's output is $u(t)$, the amplifier's input is $u_h(t)$ and the system's sampling time is T. The zero-order hold can be represented as

$$u_h(t) = u(kT), kT \leq T < (k+1)T \tag{8}$$

Considering the sampling process, by replacing e^{-Ts} with $e^{-Ts} \approx \left(-s+\frac{2}{T}\right)/\left(s+\frac{2}{T}\right)$, the DAC converter and the zero-order hold can be described as

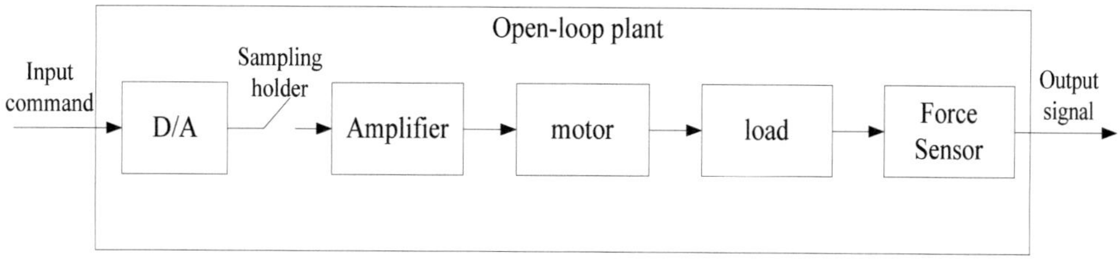

Figure 4. Open-loop block diagram.

$$G_H(s) = \frac{u_h(s)}{u(s)} \cdot \frac{1}{T} = \frac{1 - e^{-Ts}}{s} \cdot \frac{1}{T} \approx \frac{T}{\frac{T}{2}s + 1} \cdot \frac{1}{T} \tag{9}$$

which is a low-pass filter.

In a high-precision servo system, PWM-based amplifiers are commonly used as the motor drivers [28]. The PWM power amplifiers can be represented as

$$G_A(s) = \frac{K_a}{1 + \frac{1}{2f_s}s} \tag{10}$$

where K_a is the amplifying multiple and f_s is the modulation frequency.

The load cell signal conditioner linearly converts the force into a voltage signal, and its model can be simplified as

$$G_S(s) = K_{Sen} \tag{11}$$

2.4. Simplified model

Without considering the disturbance torque T_d, the angles of the motor and the load have the following dynamical relationship

$$T_l = J_l\ddot{\theta}_l + B_l\dot{\theta}_l = K_s(\theta_m - \theta_l) \Rightarrow \frac{\theta_l(s)}{\theta_m(s)} = \frac{K_s}{J_l s^2 + B_l s + K_s} \tag{12}$$

K_s is very large and the time constant is very small, therefore in the lower frequency range $\theta_l(s)/\theta_m(s) \approx 1$, $\theta_l \approx \theta_m$.

By considering $K_s = \infty$ and $\theta_l \equiv \theta_m$, the motor-loader model can be simplified as

$$J_a\ddot{\theta}_m + B_a\dot{\theta}_m = T_m \tag{13}$$

where $J_a = J_m + J_l$, $B_a = B_m + B_l$. Deleting the intermediate variables, the dynamical equation of the motor can be obtained as

$$\frac{L_a J_a}{K_t}\frac{d^3\theta_m}{dt^3} + \frac{R_a J_a + L_a B_a}{K_t}\frac{d^2\theta_m}{dt^2} + \left(K_e + \frac{R_a B_a}{K_t}\right)\frac{d\theta_m}{dt} = u_m \tag{14}$$

If the motor armature inductance is regarded as $L_a \approx 0$, the system can be further simplified as

$$\frac{R_a J_a}{K_t}\frac{d^2\theta_m}{dt^2} + \left(K_e + \frac{R_a B_a}{K_t}\right)\frac{d\theta_m}{dt} = u_m \tag{15}$$

And the transfer function between the input voltage and output rotational angle is

$$G_M(s) = \frac{\theta_m(s)}{u_m(s)} = \frac{K_m}{T_m s^2 + s} \tag{16}$$

where $T_m = \frac{R_a J_a}{K_e K_t + R_a B_a}$ is the time constant of the motor system, and $K_m = \frac{K_t}{K_e K_t + R_a B_a}$ is the gain coefficient of the motor system.

2.5. Open-loop transfer function

The mathematical model of the open-loop plant is the cascade of the units described previously:

$$G_P(s) = G_H(s) \cdot G_A(s) \cdot G_M(s) \cdot G_S(s) \tag{17}$$

Because the system bandwidth is much lower than the system sampling rate and the modulation frequency of the PWM amplifier, the time constant of the DAC converter and the amplifier is very small. In the lower frequency range, the effect of the DAC converter and the amplifier can be omitted. Therefore, the open-loop plant can be simplified as

$$G_P(s) = \frac{b_1 s + b_0}{a_2 s^2 + a_1 s + 1} \tag{18}$$

where b_0, b_1, a_1 and a_2 are final parameters.

Obvious, it is a second-order model. Only four parameters need to be identified.

3. Parameters identification of the system model

Levy's method [29] and least square estimation are widely used in system identification. Assume that the identified transfer function is given as

$$G_P(s) = \frac{b_0 + b_1 s + \cdots + b_m s^m}{1 + a_1 s + \cdots a_n s^n} \tag{19}$$

where $b_0, b_1, \cdots, b_m, a_1, a_2, \cdots, a_n$ are real numbers, m and n are integers, and $m \le n$.

The frequency response is

$$G_P(j\omega) = \frac{(b_0 - b_2\omega^2 + \cdots) + j\omega(b_1 - b_3\omega^2 + \cdots)}{(1 - a_2\omega^2 + \cdots) + j\omega(a_1 - a_3\omega^2 + \cdots)} = \frac{N(j\omega)}{D(j\omega)} \tag{20}$$

For each frequency point $\omega_i (i = 1, 2, \cdots, L)$, it is assumed that the actual frequency response is $\mathrm{Re}(\omega_i) + j\,\mathrm{Im}(\omega_i)$, and the approximation error is defined as

$$\varepsilon(j\omega_i) = \mathrm{Re}(\omega_i) + j\,\mathrm{Im}(\omega_i) - \frac{N(j\omega_i)}{D(j\omega_i)} \tag{21}$$

http://dx.doi.org/10.5772/intechopen.71420

Define an objective function:

$$J = \sum_{i=1}^{L} \left\| D(j\,\omega_i)\varepsilon(j\,\omega_i) \right\|^2 \tag{22}$$

Minimize the objective function, and prompt $\partial J/\partial a_j = 0$ (j = 1, 2, …, n) and $\partial J/\partial b_k = 0$ (k = 1, 2, …, m), then two matrix equations will be obtained. By solving the matrix equations, we can get the estimated parameters $b_0, b_1, \cdots, b_m, a_1, a_2, \cdots, a_n$.

Table 1 shows actual frequency response obtained by experiments. We employ different sinusoidal input $r(t) = A_m \sin \omega t$ with different angular frequency ω (from 0.1 to 1 rad/s) to excite the open-loop system. By theoretical analysis, the output signals are in the form $y(t) = A_f \sin(\omega t + \Phi)$. Comparing the output with input sinusoidal signals, we can calculate the values of A_f/A_m and phase delay angle Φ based on the least square estimation as follows.

The output signals can be decomposed as

$$\begin{aligned} y(t) &= A_f \sin(\omega t + \Phi) = A_f \cos\Phi \sin\omega t + A_f \sin\Phi \cos\omega t \\ &= [\sin\omega t \quad \cos\omega t] \begin{bmatrix} A_f \cos\Phi \\ A_f \sin\Phi \end{bmatrix} \end{aligned} \tag{23}$$

First, we select the sampling interval $t = 0, h, 2h, \ldots, nh$, where h is a step time.

Second, by defining $Y = [y(0)\ y(h) \ldots y(nh)]^T$, $c_1 = A_f \cos\Phi$, $c_2 = A_f \sin\Phi$ and $\Psi = \begin{bmatrix} \sin(\omega 0) & \sin(\omega h) & \ldots & \sin(\omega nh) \\ \cos(\omega 0) & \cos(\omega h) & \ldots & \cos(\omega nh) \end{bmatrix}$, the least square solutions of c_1 and c_2 can be derived as

$$\begin{bmatrix} c_1 \\ c_2 \end{bmatrix} = (\Psi^T \Psi)^{-1} \Psi^T Y \tag{24}$$

Third, A_f/A_m and Φ are calculated as

$$\frac{\hat{A}_f}{A_m} = \frac{\sqrt{c_1^2 + c_2^2}}{A_m}, \quad \Phi = tg^{-1}(c_2/c_1) \tag{25}$$

where $\hat{A}_f$ is the estimation of A_f.

After A_f/A_m and Φ are obtained, it is easy to estimate the parameters b_0, b_1, a_1 and a_2 in Eq. (18) based on Levy's method. Finally, we get the identified transfer function:

$$G(s) = \frac{-38.48s - 141.4}{1.523\,s^2 + 8.324s + 1} \tag{26}$$

ω	1	0.9	0.8	0.7	0.6	0.5	0.4	0.3	0.2	0.1	1
A_f/A_m	16	22	25	27	30	31	40	46	64	122	16
Φ	1.71	1.82	1.78	1.77	1.89	1.92	1.90	2.00	1.94	2.23	1.71

Table 1. Actual frequency response (ω is angular frequency, rad/s; Φ is phase delay, rad).

4. Simulation-based controller design

4.1. Control scheme

As shown in **Figure 5**, in the CMAC plus PD control scheme, the PD controller is used to train the weights of the CMAC at the early stage of control. The output u_n of the CMAC and the output u_p of the PD controller are integrated as the control command u to track the desired input command. Once the system is running, the PD controller will play a main role at the beginning. As the weights are tuned by u_p, u_n will increase to become the main control command. The fast learning speed prompts the leading role of control to switch from the PD controller to the CMAC.

CMAC employs the supervisory learning algorithm. At the end of each control step k, the controller calculates the corresponding output $u_n(k)$, compares it with total control output $u(k)$, amends the weights and enters into learning process to make error small between the total control output and the output of the CMAC. That means the total control output will be finally generated by the CMAC only via its learning strategy.

The control law can be described as

$$\begin{cases} u(k) &= u_n(k) + u_P(k) \\ u_n(k) &= \sum_{i=1}^{c} w_i a_i \\ u_P(k) &= k_P e(k) + k_D \dfrac{e(k) - e(k-1)}{t_s} \end{cases} \tag{27}$$

where w_i are weights of the CMAC, a_i are binary vectors, $u_n(k)$ is output of the CMAC, $u_P(k)$ is output of the PD controller, k_P and k_D are gains, t_s is the sampling time and e is the error between y and r.

And the learning function of the CMAC is given as

$$\begin{cases} E(k) &= \frac{1}{2}\left(u(k) - u_n(k)\right)^2 \frac{a_i}{c} \\ \Delta w(k) &= \eta \dfrac{u(k) - u_n(k)}{c} a_i = \eta \dfrac{u_P(k)}{c} a_i \\ w(k) &= w(k-1) + \Delta w(k) + \alpha(w(k) - w(k-1)) \end{cases} \tag{28}$$

where E is the adjusting index, η is the learning speed and $\eta \in (0, 1)$, α is the inertial variable and $\alpha \in (0, 1)$ and c is the generalizing parameter.

Once the system starts running, the controller initializes $w = 0$, then we have $u_n = 0$ and $u = u_P$ at the beginning. At the early stage, the system is mainly controlled by the PD controller. As the CMAC is learning, error e will decreases, which makes $u_n(k)$ increase to become the main control command. Although the CMAC is trained by the output of the PD controller, the output of the CMAC is not a simple imitation of the output of the PD controller. The PD controller helps the CMAC to improve its performances, restrain disturbance and enhance stability of the closed-loop system. If the PD controller works alone, control performances are determined by the gain k_P to a great extent. If the CMAC and PD controller work together, control performances will be independent of k_P, which is flexible in a rational span.

http://dx.doi.org/10.5772/intechopen.71420

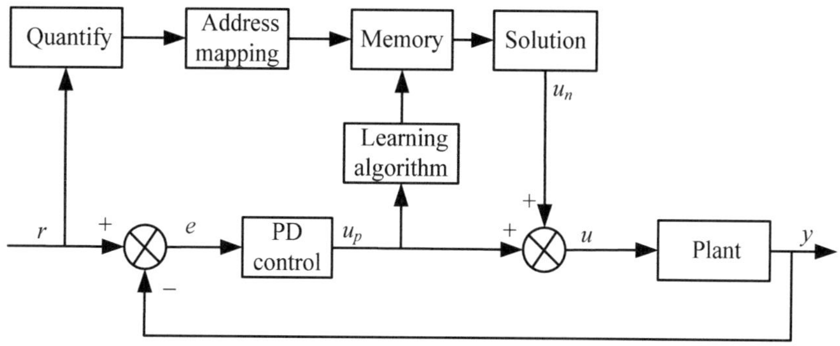

Figure 5. CMAC feedforward plus PD feedback control block diagram.

4.2. Simulations

Choose the identified model (26) as the simulation plant to tune the controller and verify its performance. The force control is aimed at tracking a ramp command, of which the expression is given as $r(t) = -10t$, where t is from 0 to 20 s.

The CMAC plus PD controller is tuned and trained by simulations. **Figure 6** demonstrates that the PD controller plays a main role of control at the beginning. Then the control effect of the CMAC will gradually increase via learning from the output of the PD controller. Owing to the help of the CMAC, control performances are better than those of the only PD controller. Overshoots are reduced dramatically and control actions are speeded up while choosing the ramp signal as an input. **Figure 7(a)** shows an overview of control performances

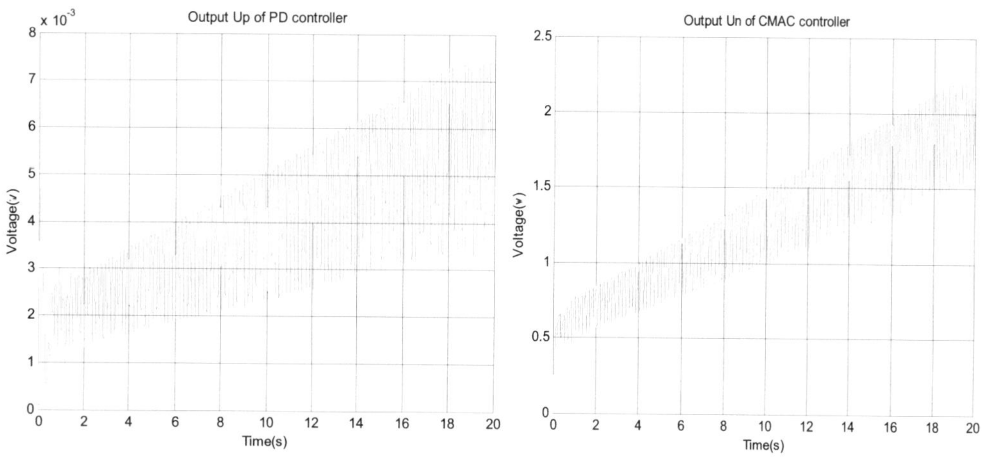

(a) Output u_p of PD

(b) Output u_n of CMAC

Figure 6. Output of the controller.

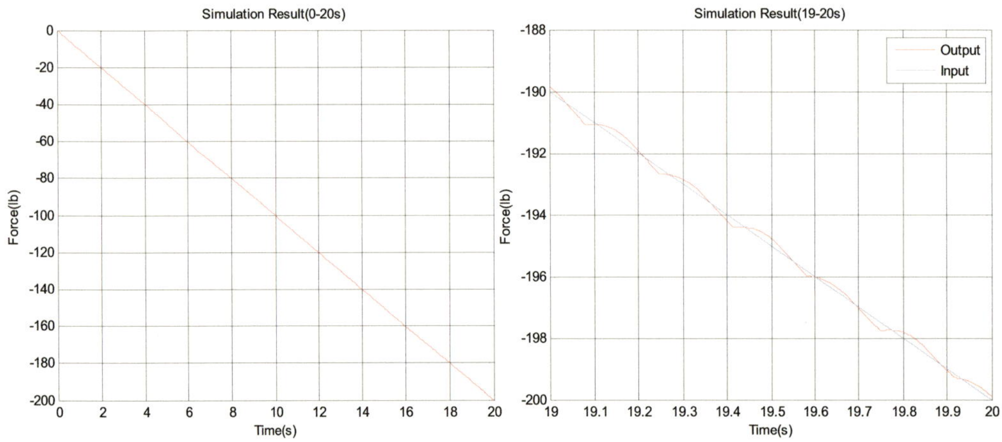

(a) 0~20s (b) 19~20s

Figure 7. Simulation results.

from 0 to 20 s. **Figure 7(b)** is an enlarged view of control performances from 19 to 20s. It can be seen that the system output well tracks the input command. **Figure 8** further exhibits the superiority of the CMAC: the tracking errors are tiny, mostly between −0.2 and 0.2 lbs. Since there exists a modeling error between identified system and actual controlled plant, the controller's performances will be further validated in the real material-strength testing experiments.

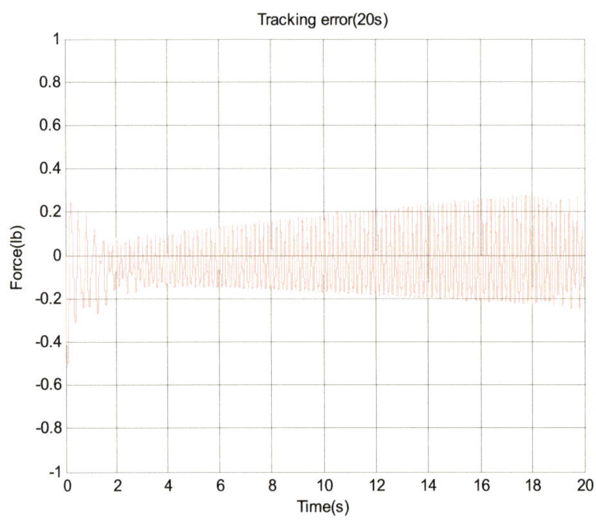

Figure 8. Tracking error.

http://dx.doi.org/10.5772/intechopen.71420

5. Embedded system-based experiments

5.1. Experiments

For verifying the controller's adaption to modeling error in the real-time control experiments, we implement above control law in C language and download compiled files into the embedded system. The central processing unit is an ARM7-based processor LPC2294. The LPC2294 is a 32-bit reduced instruction set computer (RISC) processor with low power consumption and high performance. Although there is no Float Point Unit (FPU) in this processor, the 70 million instructions per second (MIPS) processing speed makes it ideal for the real-time control system. The control step size is set as 50 ms.

The actual force tracking control diagram is shown in **Figure 9**. The force is applied to the glass sheet specimen by a loader which is connected to a linear motor actuator through a steel bar. A three point bending test is utilized under this configuration. The load force followed a ramp function with time as the independent variable. The slope of the ramp function can be programmed through the LCD touch screen. The force range can be from 0 to 300 lbs. Displacement of the motor, which also reflected the deformation of the glass sheet specimen, is recorded during the test. The signal flow of the control system is as follows. The DAC generates the control outputs as a voltage signal, which is amplified by a power amplifier and exerted on the linear motor. The linear motor then transforms the voltage signal into rotation, and generates linear displacements. The force sensor and conditioner measure the displacements and generates charges, which are transformed back to a voltage signal and fed back through ADC to the microcontroller.

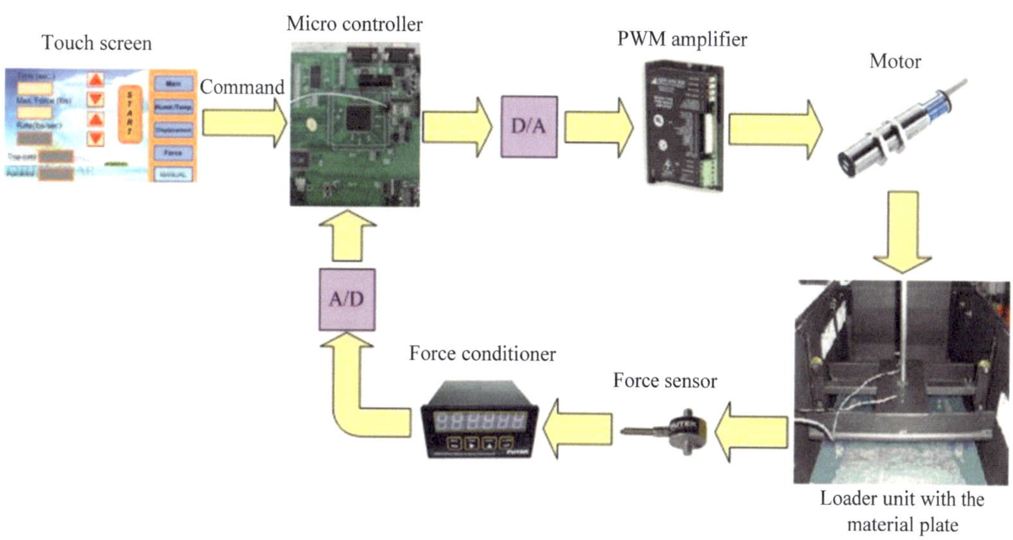

Figure 9. Actual force tracking control diagram.

The input command and control parameters are the same as those of simulations. **Figure 10(a)** shows that the output force basically tracks the input command well and the controller implemented on the embedded system can perform the control task in real time within the control step size. From **Figure 10(b)**, it can be seen that tracking errors of the experiment are mostly between −2.5 and 2.5 lbs, which are larger than those of the simulation. The reason is that uncertainties and disturbances always exist in the real world, which mainly reflects in modeling error. However, the maximum absolute tracking error is 3.79, and the variance of tracking errors is 1.69, which are still acceptable in the actual real-time control environment. Moreover, it verifies that the controller tuned by simulations is also available for actual experiments, and indirectly proves the effectiveness of identified system model.

5.2. Comparisons

As shown in **Figure 11**, an inverse model feedforward control is compared with the CMAC feedforward control by embedded-system-based experiments. Its basic idea is to directly employ the inverse model 1/G(s) of the plant's identified transfer function G(s) in Eq. (26) to be the feedforward part. Obviously, this control scheme is also suitable for real-time implementations; even its computation cost is less than that of the CMAC scheme.

Figure 12(a) shows overall performances of the tracking control with the inverse model scheme. It can be seen that the output force basically tracks the input command and the controller implemented on the embedded system can perform the control task in real time within the control step size; however, the tracking control is not performed very well. As shown in **Figure 12(b)**, tracking errors are mostly between −5 and −2 lbs, which are larger than those of the CMAC scheme. **Table 2** also shows that the maximum absolute tracking error and variance of tracking errors of the inverse model scheme are both larger than those of the CMAC scheme, so that performances of the CMAC scheme are superior to those of the inverse model

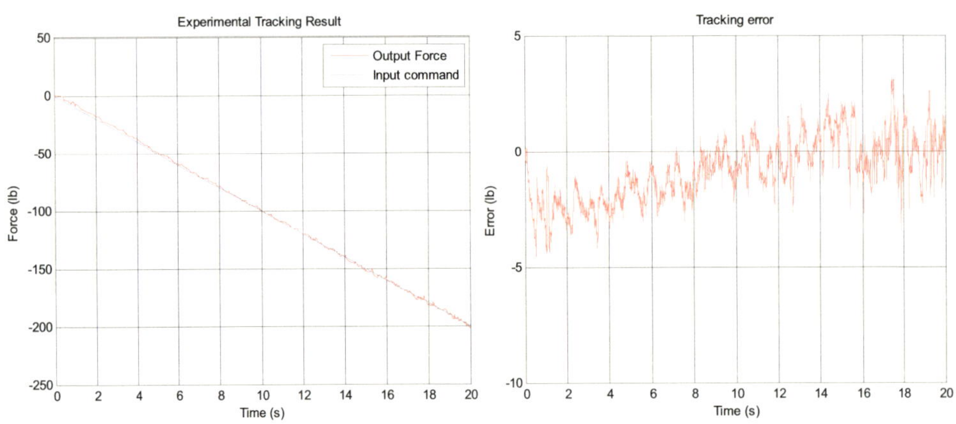

(a) Input command and output force (b) Tracking error

Figure 10. Experimental results.

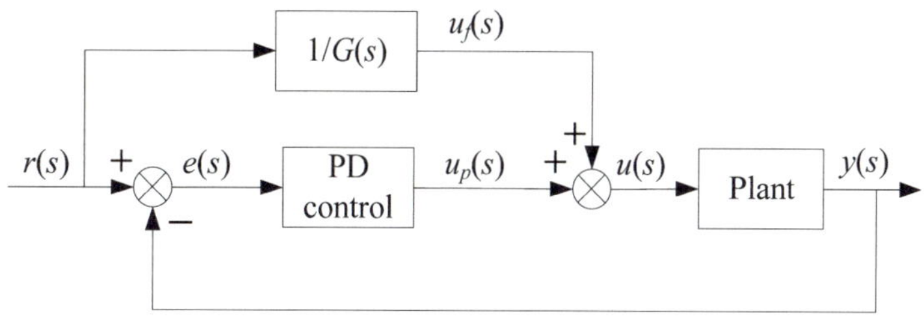

Figure 11. Block diagram of the inverse model feedforward control.

scheme. Although the CMAC scheme and inverse model scheme both utilize inverse model idea, the difference is that the latter directly employs the unchanged inverse model of the identified transfer function and the modeling error always exists, but the inverse model approximated by the CMAC is dynamical and adaptive.

5.3. Results analysis

Compared with the traditional PID control method, the learning behavior and adaptive feature of the proposed CMAC control algorithm are embodied in the freedom of tuning control parameters and the robustness to the disturbances in the real world.

The PID control method has been widely used because of its simpleness, but the tuning problem of PID parameters (proportional, integral and differential) is difficult. At present, PID parameter-tuning optimization depends on the experiences of technical staffs and needs a

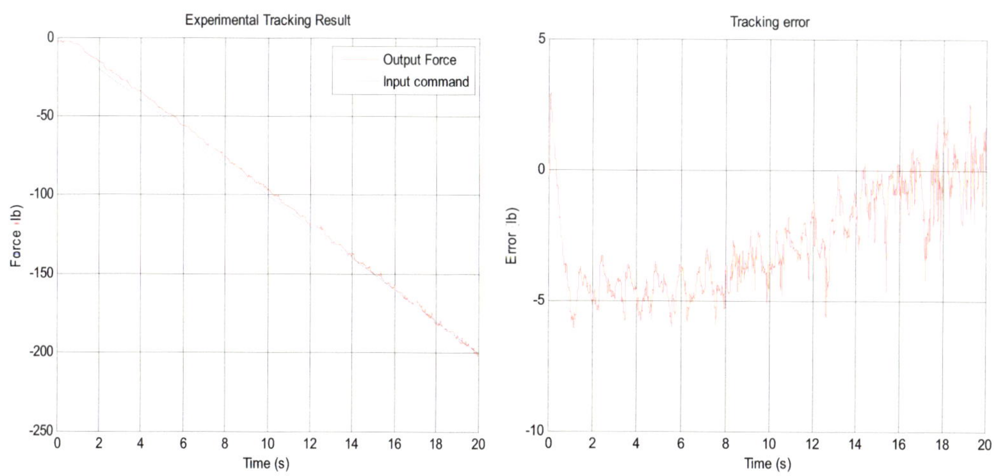

(a) Input command and output force

(b) Tracking error

Figure 12. Experimental results (inverse model feedforward control).

Method	Maximum absolute error	Variance
CMAC	3.79	1.69
Inverse model	6.13	3.85

Table 2. Comparisons between the CMAC and the inverse model.

lot of manpower and time, which means that the optimal PID parameters are difficult to be obtained by people's tuning, and the inappropriate parameters cannot guarantee the control performances to meet the control requirements. In addition, the PID control law is a kind of the linear control law which owns few robustness to the disturbances. It means that even though the PID parameters are tuned optimally and perfectly in the simulations, the tuned parameters may perform poor in the real world due to disturbances and uncertainties.

Thus, combining the adaptive CMAC and the traditional PID to construct an intelligent neural network PID controller, can automatically identify the controlled plant and adaptively adapt the control parameters of the CMAC, which can solve the difficult problem of tuning parameters of the traditional PID controller. As shown in Sections 4 and 5, the PID parameters of the proposed CMAC algorithm in the experiments are same as those in the simulations, which verifies control performances of the proposed CMAC algorithm are independent of tuning PID parameters. Experimental results in the real world in Section 5 also demonstrate the robustness of the proposed algorithm owing to the CMAC neural network, while the traditional PID controller does not have this capacity.

6. Conclusions

In this chapter, a three-step systematic design approach is proposed to design an adaptive control system for practical use. We firstly study the system model identification problem of the embedded control material-strength testing system, including mathematical modeling of all the open-loop physical components and parameters identification of the mathematical model. Both theoretical analyses and experimental comparisons validate the identified transfer function of the system model is applicable for controller design and simulation. Next, benefited from limited computation cost and compensation ability to the modeling error, a simple and effective CMAC plus PD controller is simulated based on the identified system model, and then applied to the embedded control system for real-time force tracking. Both numerical simulations and actual experiments illustrate the proposed algorithm satisfactorily performs the tracking control task under real-time constraints of the embedded system.

On the other hand, different strength features of different types of the material plates will affect the control performances. Since the yield strength generated by the solar panel glass has an almost linear relationship with the deformation of the glass sheet, K_F in Eq. (7) is a constant. If the tested material plate is not a solar panel glass and the yield strength generated does not have a linear relationship with the deformation of the material plate, K_F in Eq. (7) will not be a constant. In this case, the problems how to identify the parameters of the system model accurately for simulation-based controller design and how to adjust the CMAC plus PD controller to compensate the varying K_P, will be under consideration in our future work.

Acknowledgements

This work is supported by National Key R&D Program of China under Grant Nos. 2016YFD0200700 and 2017YFD0701000, and Chinese Universities Scientific Fund under Grant Nos. 2017QC139 and 2017GX001.

Author details

Jian Chen[1,2]*, Peng Li[3], Gangbing Song[3], Shubo Wang[1,2], Zichao Zhang[1,2], Guangqi Wang[1,2], Yu Tan[1,2] and Yongjun Zheng[1,2]

*Address all correspondence to: jchen@cau.edu.cn

1 College of Engineering, China Agricultural University, Beijing, China

2 Key Laboratory of Soil-Machine-Plant System Technology, Ministry of Agriculture, Beijing, China

3 Department of Mechanical Engineering, University of Houston, TX, USA

References

[1] Prasad D, Snow M. Designing with Solar Power: A Source Book for Building Integrated Photovoltaics (BiPV). New York: Routledge; 2014

[2] Dhere NG, Raravikar NR. Adhesional shear strength and surface analysis of a PV module deployed in harsh coastal climate. Solar Energy Materials & Solar Cells. 2001;**67**(1):363-367

[3] Burrows K, Fthenakis V. Glass needs for a growing photovoltaics industry. Solar Energy Materials & Solar Cells. 2015;**132**:455-459

[4] Hughes ZM, Pont MJ. Reducing the impact of task overruns in resource-constrained embedded systems in which a time-triggered software architecture is employed. Transactions of the Institute of Measurement and Control. 2008;**30**(5):427-450

[5] Huang SJ, Yu CK, Lin JY. Intelligent robotic impedance control using embedded system structure. Transactions of the Institute of Measurement and Control. 2012;**35**(5):561-573

[6] Moallem P, Zargari A, Kiyoumarsi A. Improving IEC flickermeter for implementation by an ARM microcontroller-based digital system. Transactions of the Institute of Measurement and Control. 2013;**35**(3):342-352

[7] Albus JS. A new approach to manipulator control: The cerebellar model articulation controller (CMAC). Journal of Dynamic Systems, Measurement, and Control. 1975;**97**(3):220-227

[8] Chiang CT, Lin CS. CMAC with general basis functions. Neural Networks. 1996;**9**(7): 1199-1211

[9] Lewis FL, Jagannathan S, Yesildirak A. Neural Network Control of Robot Manipulators and Non-linear Systems. Boca Raton: CRC Press; 1998

[10] Yang B, Bao R, Han H. Robust hybrid control based on PD and novel CMAC with improved architecture and learning scheme for electric load simulator. IEEE Transactions on Industrial Electronics. 2014;**61**(10):5271-5279

[11] Yang B, Han H. A CMAC-PD compound torque controller with fast learning capacity and improved output smoothness for electric load simulator. International Journal of Control, Automation and Systems. 2014;**12**(4):805-812

[12] Commuri S, Lewis FL. CMAC neural networks for control of nonlinear dynamical systems: Structure, stability and passivity. Automatica. 1997;**33**(4):635-641

[13] Jagannathan S, Commuri S, Lewis FL. Feedback linearization using CMAC neural networks. Automatica. 1998;**34**(5):547-557

[14] Jagannathan S, Lewis FL. Robust backstepping control of robotic systems using neural networks. Journal of Intelligent and Robotic Systems. 1998;**23**(2-4):105-128

[15] Kim YH, Lewis FL. Intelligent optimal design of CMAC neural network for robot manipulators. In: Soft Computing for Intelligent Robotic Systems. New York: Physica-Verlag HD; 1998

[16] Kim YH, Lewis FL. Optimal design of CMAC neural-network controller for robot manipulators. IEEE Transactions on Systems, Man, and Cybernetics, Part C: Applications and Reviews. 2000;**30**(1):22-31

[17] Nie J, Linkens DA. FCMAC: A fuzzified cerebellar model articulation controller with self-organizing capacity. Automatica. 1994;**30**(4):655-664

[18] Wen CM, Cheng MY. Development of a recurrent fuzzy CMAC with adjustable input space quantization and self-tuning learning rate for control of a dual-axis piezoelectric actuated micromotion stage. IEEE Transactions on Industrial Electronics. 2013;**60**(11):5105-5115

[19] Lee CH, Chang FY, Lin CM. An efficient interval type-2 fuzzy CMAC for chaos time-series prediction and synchronization. IEEE Transactions on Cybernetics. 2014;**44**(3):329-341

[20] Lin CM, Li HY. Intelligent control using the wavelet fuzzy CMAC backstepping control system for two-axis linear piezoelectric ceramic motor drive systems. IEEE Transactions on Fuzzy Systems. 2014;**22**(4):791-802

[21] Lin CM, Li HY. Adaptive dynamic sliding-mode fuzzy CMAC for voice coil motor using asymmetric gaussian membership function. IEEE Transactions on Industrial Electronics. 2014;**61**(10):5662-5671

[22] Lin FJ, Yang KJ, Sun IF, Chang JK. Intelligent position control of permanent magnet synchronous motor using recurrent fuzzy neural cerebellar model articulation network. IET Electric Power Applications. 2015;**9**(3):248-264

[23] Lin CM, Peng YF, Hsu CF. Robust cerebellar model articulation controller design for unknown nonlinear systems. IEEE Transactions on Circuits and Systems II: Express Briefs. 2004;**51**(7):354-358

[24] Lin CM, Peng YF. Adaptive CMAC-based supervisory control for uncertain nonlinear systems. IEEE Transactions on Systems, Man, and Cybernetics, Part B: Cybernetics. 2004;**34**(2): 1248-1260

[25] Lin CM, Peng YF. Missile guidance law design using adaptive cerebellar model articulation controller. IEEE Transactions on Neural Networks. 2005;**16**(3):636-644

[26] Lin CM, Chen TY. Self-organizing CMAC control for a class of MIMO uncertain nonlinear systems. IEEE Transactions on Neural Networks. 2009;**20**(9):1377-1384

[27] Hu Y, Zeng L, Ma D. Theory and Design of Servo Systems. Beijing: Beijing Institute of Technology Press; 1993

[28] Robet P, Gautier M, Bergmann C. A frequency approach for current loop modeling with a PWM converter. IEEE Transactions on Industry Applications. 1998;**34**(5):1003-1014

[29] Levy EC. Complex-curve fitting. IEEE Transactions on Automatic Control. 1959;**4**(1):37-43

Chapter 14

Multi-Loop Integral Control-Based Heart Rate Regulation for Fast Tracking and Faulty-Tolerant Control Performance in Treadmill Exercises

Yi Zhang, Kairui Guo, Qin Yang, Pang Winnie,
Kai Cao, Qi Wang, Andrey Savkin, Branko Celler,
Hung Nguyen, Peng Xu, Limei Xu,
Dezhong Yao and Steven Su

Additional information is available at the end of the chapter

http://dx.doi.org/10.5772/intechopen.71855

Abstract

In order to offer a reliable, fast, and offset-free tracking performance for the regulation of heart rate (HR) during treadmill exercise, a two-input single-output (2ISO) control system by simultaneously manipulating both treadmill speed and gradient is proposed. The decentralized integral controllability (DIC) analysis is extended to nonlinear and non-square processes especially for a 2ISO process, namely multi-loop integral controllability (MIC). The proposed multi-loop integral control-based HR regulation by manipulating treadmill speed and gradient is then validated through a comparative treadmill experiment that compares the system performance of the proposed 2ISO MIC control loop with that of single-input single-output (SISO) loops, speed/gradient-to-HR. The experimental validation presents that by simultaneously using two control inputs, the automated system can achieve the fastest HR tracking performance and stay close to the reference HR during steady state, while comparing with two SISO structures, and offer the fault-tolerant ability if the gains of the two multi-loop integral controllers are well tuned. It has a vital implication for the applications of exercise rehabilitation and fitness in relation to the automated control system.

Keywords: heart rate, treadmill exercise, decentralized integral controllability, multi-loop integral controllability, proportion-integral (PI)

1. Introduction

Obesity, which leads to diabetes and cardiovascular disease (CVD), is one of the major threats to human health. It is well recognized that regular exercise is the most efficient way to reduce the possibility of both type 2 Diabetes and CVD [1]. The development of automated exercise assisted equipment can greatly enhance the efficiency of exercises and reduce the requirement of supervision. An easy-to-measure indication of exercise strength is heart rate (HR). One of the most efficient ways for the manipulation of exercise intensity for either the training of athletes or rehabilitation patients is to simulate the HR in order to follow a pre-designed HR profile. As a result, an automated control system can offer numerous benefits for different groups of users. For instance, it will provide assistance for patients with cardiac diseases who might be prescribed treadmill exercise rehabilitation. It also can be used for training the athletes and safely regulating the exercise intensity within a suitable profile in order to achieve the predefined HR ranges. It has been well documented that by determining one's maximum and minimum values of HR responses the exercise profile can be individually designed [2]. Specifically, rehabilitation patients will be guided to perform the exercise in terms of 50–60% of the maximum HR, 60–70% of the maximum HR zone then suitable for subjects who target weight control, and the range between 70 and 90% is preferred for the cardio-endurance exercise. In the study, a two-input single-output (2ISO) control system is proposed, which can specially further improve the efficiency of treadmill exercises for different subject groups as well as diversify more reliable and safe treadmill exercise protocols.

Recently, most studies [3–6] only consider using one manipulate variable (treadmill speed or gradient) to regulate HR responses. In [7, 8], for exercise testing and rehabilitation of subjects with impaired exercise tolerance, ramp type protocols were proposed by simultaneously manipulating both speed and gradient (without feedback), which could produce a low initial metabolic rate that then increases the work rate linearly to reach the subject's limit of tolerance in approximately 10 min. In [9], a multi-loop proportion-integral (PI) controller based HR tracking system has been presented, which independently tuned both treadmill speed and gradient in closed loop, and achieved good performance. However, in paper [9], it is assumed that the HR response to treadmill exercise is in linear range, and only linear modeling and control approaches have been presented. The experimental evidences of the advantages of using both speed and gradient to regulate HR are therefore only valid in a certain linear response range.

This study introduces the 2ISO HR process which employs two actuators, treadmill speed and gradient, to regulate the HR response. Such process control has the following merits. First of all, it can increase the non-saturation range. For example, practical systems always have physical limitations and therefore have limited non-saturation range. If simultaneously executing multiple actuators, the output range can be extended. On the other hand, it can increase the maximum gain of the actuator so that the fast tracking or regulation of manipulated variable can be achieved [9]. Also, redundancy of actuators can facilitate fault accommodation for the implementation of fault-tolerant control strategies. As the stability of the closed loop system can be practically achieved by adding suitable constraints in control inputs and their derivatives, the achievable performance (especially in steady state) is more important for most industrial control processes.

For the regulation of HR, multi-loop PI controllers were developed [9] in order to achieve zero steady-state tracking error. For multi-loop PI control of square process, Skogestad and Morari [10] introduced the concept of decentralized integral controllability (DIC). DIC analysis [11–13] determines the stability, integral control ability, and faulty-tolerant control of the multi-loop system. This study extends the DIC analysis for nonlinear and non-square processes especially

http://dx.doi.org/10.5772/intechopen.71855

for a 2ISO process, namely multi-loop integral controllability (MIC). The proposed multi-loop integral control-based HR regulation by manipulating treadmill speed and gradient is then validated through a comparative treadmill experiment that compares the system performance of the proposed 2ISO MIC control loop with that of single-input single-output (SISO) loops, speed/gradient-to-HR. The real treadmill experiment is used to experimentally validate if MIC in the HR range is valid. Results show that, compared with two SISO loops, the 2ISO MIC control loop can achieve the fastest HR tracking performance, reach up to the reference HR during the steady state, as well as offer the fault-tolerant ability in the case of one of the gains of multi-loop integral controllers being out of service. It has a vital implication for the applications of exercise rehabilitation and fitness in relation to the automated control system.

2. Experiments

2.1. Experimental settings

In this study, the HR data during experiments were collected by a portable sensor, Alive Heart Monitor (HM131) manufactured by Alive Technologies. It consists of one HR sensor and one triaxial accelerometer. The HR data acquired from the internal HR sensor are used in the study. The sampling rate for HR data collections is 300 samples/s. A Bluetooth SPP connection is used to transmit the instantaneous HR data to the laptop-based control program that is designed and implemented based on LabVIEW (National Instrument). The treadmill Powerjog J series is set up for experiments, the speed and gradient of which is controlled and able to be accessed via the RS232 protocol. **Figure 1** shows the schematic diagram for the experimental equipment setup.

2.2. Subjects

Eight healthy non-smoking males were invited to join the experiments. They were free from any known cardiac or metabolic disorders, hypertension, and were not under any medication.

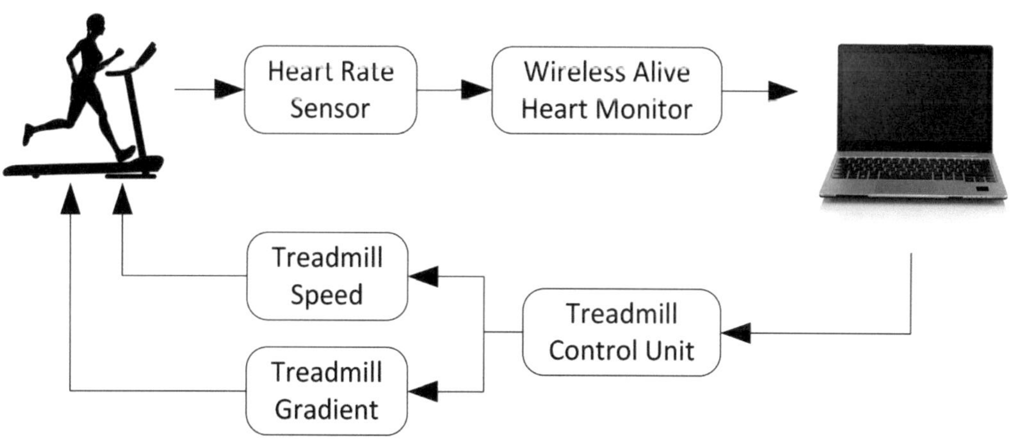

Figure 1. The schematic diagram for the experimental equipment setup.

The Human Research Ethics Committee by University of Technology Sydney (UTS HREC 2009000227) approved the study, and an informed consent was obtained from all participants before each experiment. The physical characteristics of the participants are presented in **Table 1**. The subjects were required to take a light meal prior to the experiment activity and not to participate in any intense exercises one day before the experiment [14–16]. The environmental temperature during the experiments was set at 25°C, and the humidity was at about 50% [17]. The HR monitor (HM131) was fitted to the middle of the chest of every subject by using electrode pads.

2.3. Experimental protocols

All subjects were asked to exercise on a motor-controlled treadmill, and they all selected 7 km/h as the speed for which both walking and running is possible. Then, subject was asked to walk for 5 min at 7 km/h for a certain gradient followed by a 7 min rest. This procedure was repeated for running as well as for different gradients. During experiments, HR response was recorded by the portable ECG monitor. The averaged steady-state HR of all subjects for both walking and running under different gradients is summarized in **Table 2**.

From **Table 2**, it can be seen that for a certain gradient, the HR for running is more than 15% higher than that for walking. During exercise, the subjects may switch between walking and running when the treadmill speed is around 7 km/h. As a result, it can be seen that, for example, when gradient is around 15°, the transition zone for HR is between 121 and 144 bpm. When the reference HR is located in the transition zone, the regulation of HR only by adjusting speed would be problematic even under small perturbations in the measurement. This is because the subject will frequently switch his/her motion actions between walking and

Subjects	Age (year)	Height (cm)	Weight (cm)
1	27	175	55
2	32	170	87
3	29	176	90
4	29	178	77
5	42	175	80
6	29	164	64
7	31	169	67
8	26	180	77
Mean	**30.6**	**173.4**	**74.6**
STD	**4.7**	**5.0**	**11.1**

Table 1. Physical characteristics.

Gradient (Degree)	HR (bpm) Walking at 7 km/h	HR (bpm) Running at 7 km/h
0	102	125
15	121	144
25	137	171

Table 2. HR response at steady state.

running in order to stabilize his/her desired HR level. As a result, simultaneous manipulating of speed and gradient would be firmly beneficial.

2.4. Pan-Tompkins HR detection

The Pan-Tompkins algorithm was applied to identify the P-peak of QRS complex during experiments, which is then used for calculations of HR values. This algorithm is inclusive of several filtering such as a band-pass filter, a differentiator, a squaring operation and a moving window integrator [9]. The band-pass filter is used to reduce noises in the raw ECG signals. After band-pass filtering, the high frequency components of ECG signals were extracted by using a five-point derivative function. The squaring operation was adopted to suppress P and T waves and further enhance the higher frequency QRS complexes. Finally, the moving window integrator provided a single peak output, P-peak, for each QRS complex. After implementing these steps and by implementing an adaptive threshold algorithm with false peak detection capabilities, the HR signals can be detected accurately [18].

3. Methodology

3.1. Multi-loop integral controllability analysis for HR responses to treadmill exercise

In [8], the multi-loop PI controller has been designed for the regulation of HR for treadmill exercise. Now, we consider the case when one of the actuators is in faulty condition. First, we introduce a definition of MIC, which is a direct extension of DIC for a non-square 2ISO process [19].

As shown in **Figure 1**, assume the HR response can be described by the following equations with an input vector $u \in R^2$ and an output vector $y \in R^1$:

$$P\begin{cases} \dot{x} &= f(x,u) \quad x \in X \subset R^n, u \in U \subset R^2 \\ y &= g(x,u) \quad y \in Y \subset R^1 \end{cases} \tag{1}$$

where the state $x(t)$ is determined by its initial value $x(0)$ and the input function $u(t)$. Considering the system (1) has equilibrium at origin, that is, $f(0,0) = 0$ and $g(0,0) = 0$, if the equilibrium x_e is not at origin, a translation is then needed by redefining the state x as $x - x_e$ [19, 20].

3.1.1. Definition 1

(*Multi-loop integral controllability for nonlinear 2ISO processes*) Consider the closed loop system depicted in **Figure 2**.

i. For the nonlinear process P described by Eq. (1), if a multi-loop integral controller C exists, such that the unforced closed loop system ($r = 0$) is globally asymptotically stable (GAS) for the equilibrium $x = 0$ and such that the globally asymptotically stability is satisfied if each individual loop can be detuned independently by a factor k_i ($0 \le k_i \le 1$, $i = 1, 2$), then the nonlinear process P is said to be multi-loop integral controllable (MIC) for the equilibrium $x = 0$.

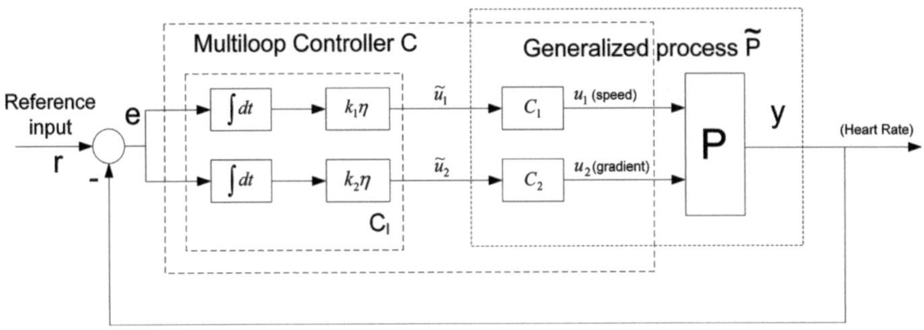

Figure 2. Multi-loop integral controllability for 2ISO system.

ii. If the closed loop system is asymptotically stable (AS) near the region of the equilibrium $x = 0$, then the nonlinear process P is said to be locally multi-loop integral controllable around the equilibrium $x = 0$ [19, 20].

In **Figure 2**, we assume the state equation of the general process $\tilde{P}$ (which includes original process P and the two scalar non-integral controllers c_1 andc_2) is modeled as follows (with the same assumptions for Eq. (1) of process P):

$$\tilde{P}:\begin{cases}\dot{x} &= f(x,\tilde{u})\\ y &= g(x,\tilde{u})\end{cases} \tag{2}$$

The state equation for the linear integral controller is expressed as:

$$C_I:\begin{cases}\dot{\xi} = \begin{bmatrix}\dot{\xi}_1\\ \dot{\xi}_2\end{bmatrix} = \eta\begin{bmatrix}k_1\\ k_2\end{bmatrix}e = -\eta\begin{bmatrix}k_1\\ k_2\end{bmatrix}y\\ \tilde{u} = \xi\end{cases} \tag{3}$$

The following theorem presented a sufficient condition for MIC:

3.1.2. Theorem 1

(*Steady-state MIC conditions for nonlinear 2ISO processes*).

Consider the closed loop system in **Figure 1**, and assume that the general process $\tilde{P}$ and the linear part of the controller C_l are described by Eqs. (2) and (3), respectively. If the following assumptions are satisfied:

i. The equation $0 = f(x,\tilde{u})$ obtained by setting $\dot{x} = 0$ in Eq. (2) implicitly defines a unique C^2 function $x = h(\tilde{u})$ for $\tilde{u} \in \tilde{U} \subset R^2$.

ii. For any fixed $\tilde{u} \in \tilde{U} \subset R^2$, the equilibrium $x = h(\tilde{u})$ of the system $\dot{x} = f(x,\tilde{u})$ is globally asymptotically stable (GAS) and locally exponentially stable (LES).

iii. If two C^2 functions can be found $\phi_1(\cdot)$ and $\phi_2(\cdot)$ such that that the steady-state input output function $g(h(\tilde{u}),\tilde{u})$ of the general process $\tilde{P}$ satisfies the following requirements:

$$(\varphi_1(\tilde{u}_1) + \varphi_2(\tilde{u}_2))\left(k_1\frac{\partial\varphi_1(\tilde{u}_1)}{\partial\tilde{u}_1} + k_2\frac{\partial\varphi_2(\tilde{u}_2)}{\partial\tilde{u}_2}\right)g(h(\tilde{u}),\tilde{u}) > 0 \tag{4}$$

(when $\varphi_1(\tilde{u}_1) + \varphi_2(\tilde{u}_2) \neq 0$ and $\tilde{u} \in \tilde{U} \subset R^2$) and $(\varphi_1(\tilde{u}_1) + \varphi_2(\tilde{u}_2))\left(k_1\frac{\partial\varphi_1(\tilde{u}_1)}{\partial\tilde{u}_1} + k_2\frac{\partial\varphi_2(\tilde{u}_2)}{\partial\tilde{u}_2}\right)$ $g\left(h(\tilde{u}),\tilde{u}\right) > \rho|\varphi_1(\tilde{u}_1) + \varphi_2(\tilde{u}_2)|^2$ (for some scalar $\rho > 0$) for in a neighborhood of $\varphi_1(\tilde{u}_1) + \varphi_2(\tilde{u}_2) = 0$.

Then there exists $\eta > 0$, such that the equilibrium is GAS. That is, if the two scalar controllers c_1 and c_2 can be found such that the generalized process $\tilde{P}$ can satisfy Conditions (i), (ii) and (iii), then the nonlinear 2ISO process is MIC for the equilibrium.

Remark: If the generalized process $\tilde{P}$can be guaranteed to satisfy Conditions (i), (ii) and (iii), based on Theorem 1 it is said that the nonlinear 2ISO process can be MIC. Once the control system is assumed as nonlinear 2ISO MIC, the stability of system can be easily guaranteed by independently tuning the factor k_i. In practice, the factor k_i usually can be manually configured to be $0 \le k_i \le 1$, $i = 1, 2$.

The Proof of the above theorem (similar as that of Theorem 1 in [11]) is based on singular perturbation theory [18] and can be found in [19].

Based on Definition 1, we can easily check that a necessary MIC condition for a 2ISO process in each single loop is DIC respectively. For HR regulation system, the necessary condition for the speed-HR and gradient-HR subsystems is DIC respectively.

A sufficient DIC condition for SISO system is the passivity in steady state, that is, the sector condition for passivity. We can easily prove that this condition is also sufficient for 2ISO processes based on Theorem 1.

For the HR responses during walking or running exercises, it is not hard to see that the incremental increasing of speed or gradient respectively will lead to the incremental increasing in HR for the same exerciser, that is, each single loop is DIC in either walking zone or running zone. However, the HR variation during walking and running transition is not clear. The following parts simulated the transition of walking/running as well as performed several experiments to investigate the HR response during transition between walking and running [21].

We also explore the offset free tracking when one of the actuators is in faulty conditions. We consider the case when one of the motors (speed motor and gradient motor) is broken, whether HR tracking is still possible or not. We also investigate whether offset free tracking is achievable or not under faulty conditions [22, 23].

4. Results

4.1. MATLAB®/Simulink® simulation verification

In order to identify the coefficients of PI controllers and verify the proposed MIC conditions for 2ISO HR regulation, the SISO and 2ISO control loops are designed and implemented through MATLAB®/Simulink® simulations. The schematic diagram for simulations is illustrated in

Figure 3, where the constructed HR model is regulated by either treadmill speed/gradient (the SISO loops) or both of those actuators (the 2ISO loop).

A two-input single-output Hammerstein model is used for modeling of HR-based open-loop characteristics. The linear dynamic part of the model is noted as $G(s) = \frac{K}{Ts+1}$, where K and T are the steady state gain and the time constant, respectively [24, 25]. The static nonlinear part is modeled by a cubic polynomial function. The input–output relationship between treadmill speed and heart rate is shown in **Figure 4** when the gradient is zero.

Based on the previous experimental survey in [9], the multi-loop integral controllers for the single loops of speed-HR and gradient-HR are developed respectively. The experimental results confirm the static nonlinearity of HR responses to treadmill walking/running exercises which has been shown in **Figure 8**. As a result, the passive sector condition can be found in the zones of 0–6.8 km/h (indicating the walking condition), and 7.2–7.8 km/h (indicating the running condition). This means when treadmill speed is within the walking zone, for instance, the subject merely needs to undergo a walking motion to follow the treadmill protocol. Moreover, if the treadmill speed reaches the running zone, the running motion has to be taken by all of subjects. However, it also could be observed from **Figure 4** that the passivity sector condition in the transition zone from 6.8 to 7.2 km/h is not valid. This means that if the reference HR variation is selected as 49.5374 (HR is 124.5374 bpm) which is located in the transition zone, the regulation of HR by only adjusting speed would be problematic even under small perturbations in the measurement. This is because the subject will frequently switch his/her motion actions between walking and running in order to stabilize his/her desired HR level.

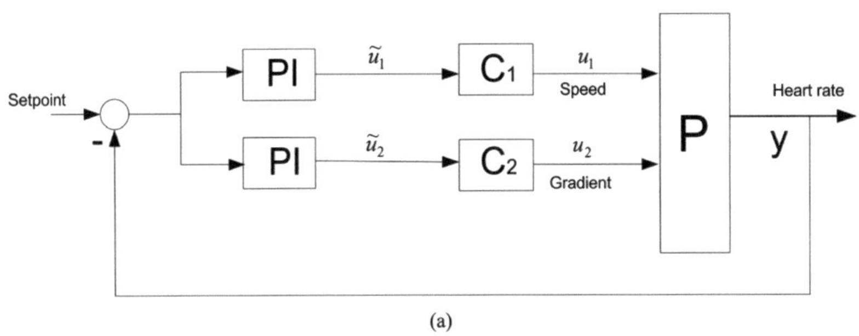

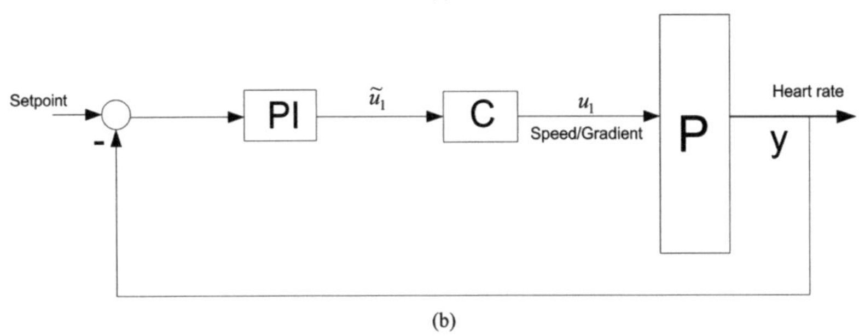

Figure 3. The schematic diagram of 2ISO PI control structure for MATLAB®/Simulink®-based HR tracking simulation. a. HR tracking of 2ISO control loop and b. SISO control loop of speed-HR or gradient-HR.

Figure 5 shows the simulation result for which only speed has been manipulated. This indicates another advantage for the using of two control inputs and fault tolerance. It can be proved based on Theorem 1 and also by simulation that if the open loop gain for healthy actuator is significantly

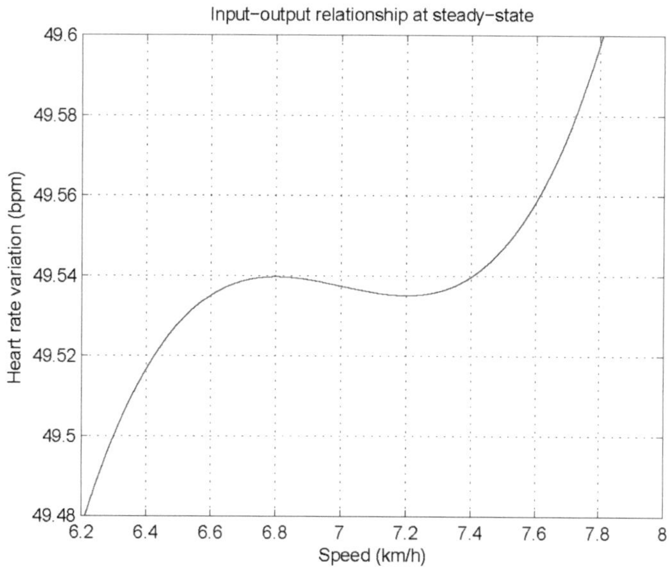

Figure 4. Steady-state HR response to speed (gradient is zero).

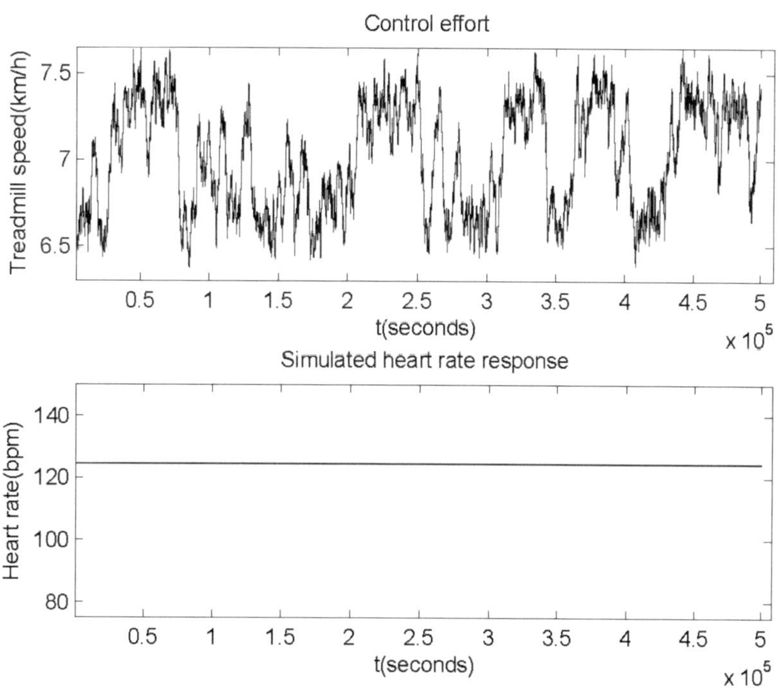

Figure 5. HR regulation results when only speed is manipulated.

bigger than that for faulty actuator, then the offset free tracking is still possible. Although the regulated HR is quite close to reference HR, treadmill speed swings between 6.5 and 7.5 km/h.

Due to the discomfort evoked by the speed swinging, we can simultaneously manipulate both treadmill speed and gradient and adjust the gradient regulation loop to avoid the swinging. Simulation results in **Figure 6** prove the effectiveness of the simultaneous manipulation strategy.

The simulation results from PI control loops of SISO (speed-HR and gradient-HR) and 2ISO indicate that those three structures can well achieve the HR tracking performance merely by tuning the PI parameters of multiple integral controllers with quite small values; the simultaneous manipulation strategy for the regulation of HR responses to treadmill exercises is effective if MIC conditions are satisfied; the 2ISO closed loop can provide the ability of fault tolerance which means that especially in the case of one of the actuators (either speed or gradient) being out of service, the offset free tracking is still achievable.

4.2. Experimental validation

In order to evaluate real-time HR tracking performance obtained from the proposed 2ISO control loop, the experimental verification is also used in this study, and a comparative study is made by comparing the tracking performance of the 2ISO loop with that of both SISO loops of speed-HR and gradient-HR.

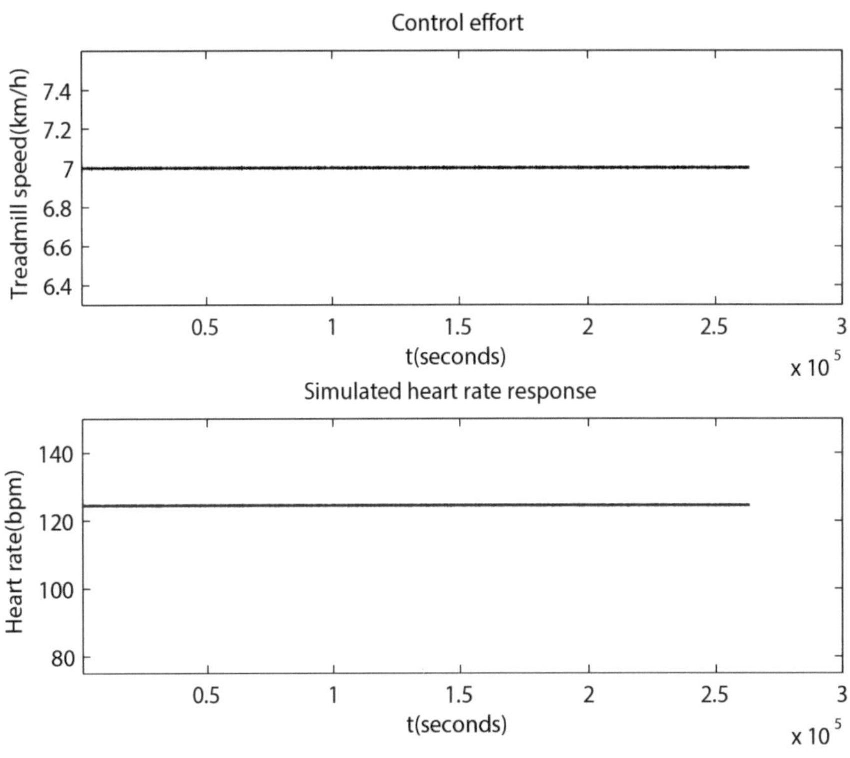

Figure 6. HR regulation results when both speed and gradient are manipulated.

In order to identify the comfort HR zones of each individual for treadmill experiments, moderate exercise intensity was adopted which offers safe treadmill intensity (speed ranged from 3 to 7 km/h, and gradient from 2 and 15%). This guarantees the HR operating zone for the subject to be located between 90 and 170 bpm. In addition, the HR level at 135 bpm was selected to be a reference input for the setpoint value of both SISO and 2ISO tests.

Considering the time-delay situation between control inputs (such as treadmill speed and gradient) and system output (instantaneous HR) during the practical experiment validation, a stack buffer with a 5-s timer was used to obtain the instantaneous HR values. The latest HR value was stored from the top of the stack buffer. The control input commands will be sent to the treadmill every 5 s based on the up-to-date stack buffer stores. For de-noising the raw instantaneous HR values measured by the Alive Technologies HR sensor, an improved exponential, weighted, moving, average filter together with a simple outlier detection algorithm was adopted for the estimation of the HR stored in the stack buffer [4].

In the first SISO open loop test, the speed of the treadmill is employed to be the system input, and HR is considered as the system output. Moreover, the gradient of treadmill is fixed at 2% and the input is set to be adjustable between 2 and 5 km/h. The PI parameters of speed-HR loop with a stable operation range 0.01–1.05 for k_p, 0.001–0.075 for k_i were determined. Based on the open loop experimental results, for the SISO controller with speed as the input, PI controllers, $k_p$and k_i, are set to 0.7 and 0.05, respectively, in order to achieve acceptable control characteristics. The HR response is shown in **Figure 7**, in which the HR model parameters K and T values were determined to be equal to 15.91 and 37.44, respectively. The results show the HR increasing quickly and a small reaction delay.

In addition, based on the gradient-HR test, a fixed speed of 4 km/h is maintained while the gradient is changed from 2 to 12%. The stable operation range (k_p:0.01–0.4665, k_i:0.001) for gradient-HR loop is observed from the experimental results. As a result, for the SISO controller with gradient as the input, k_p and k_i values are adjusted to 0.311 and 0.001 respectively. The coefficients for the speed-input controller slightly varied from the theoretical values, while in the case of the gradient-input controller, the theoretical values were acceptable. The response

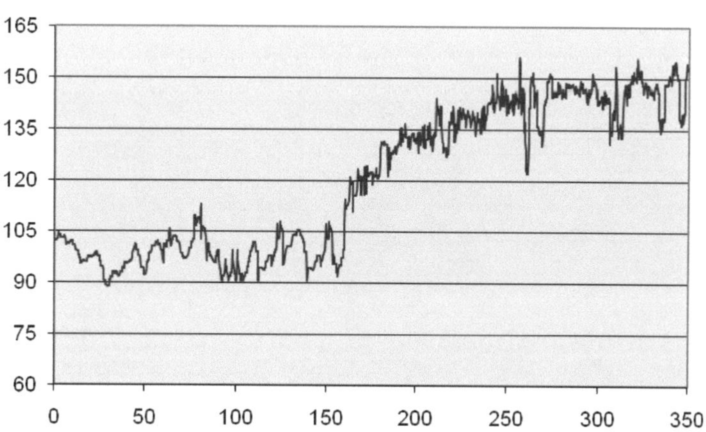

Figure 7. Experimental data for HR step responses to either treadmill speed or gradient (open-loop test with speed and HR).

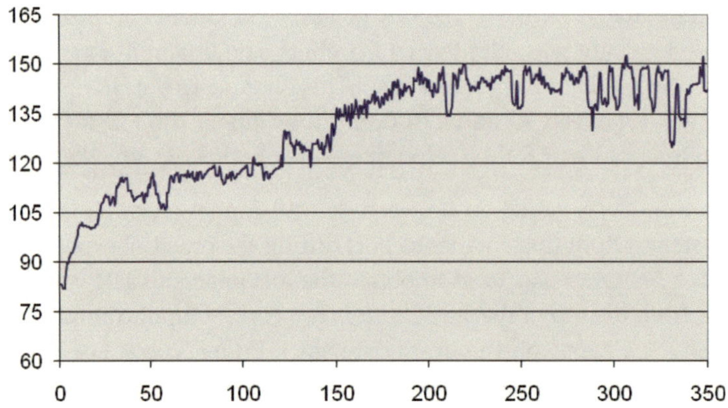

Figure 8. Experimental data for HR step responses to either treadmill speed or gradient (open-loop test with gradient and HR).

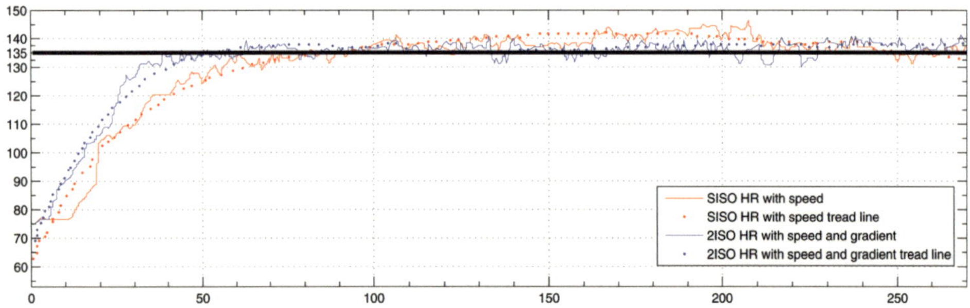

Figure 9. HR tracking performance comparison of SISO (speed) test with 2ISO.

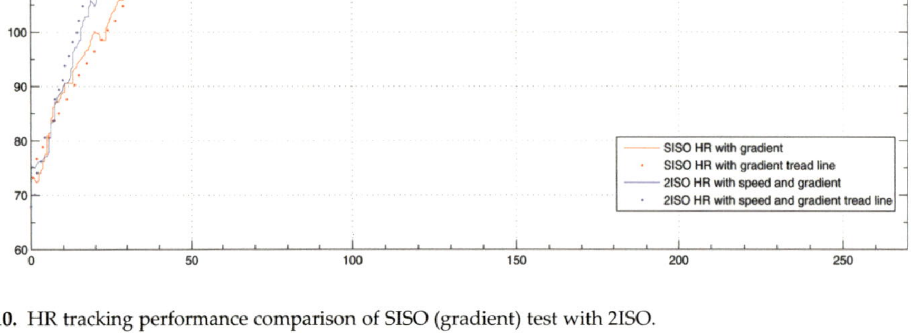

Figure 10. HR tracking performance comparison of SISO (gradient) test with 2ISO.

http://dx.doi.org/10.5772/intechopen.71855

graph is shown in **Figure 8**, where K and T values obtained are 2 and 26, respectively. A reaction delay can be observed due to the mechanical time that the treadmill needs to reach the desired gradient.

Using experimental verification results with the determined PI coefficients, two closed loop SISO controllers and one 2ISO controller were implemented. **Figures 9** and **10** provide a clear view of both SISO control with speed-input and gradient-input compared with 2ISO. For the SISO speed-input controller, it demonstrates that the system outputs have a slight overshoot followed by a fast rise to track the setpoint. However, for the SISO gradient-input controller, it shows a more stable performance compared to that of the SISO speed-input controller. The comparative results shown in **Figure 10** demonstrate that 2ISO control loop can achieve the fastest HR tracking performance and stay close to the reference HR during steady state, while comparing with two SISO structures.

The main advantage of using 2ISO control in treadmill exercises is to improve the HR tracking performance. It can be concluded that the 2ISO controller outperforms both SISO controllers and can provide shorter rise time, best steady-state stability, as well as the lowest steady-state error. In addition, the 2ISO automatic treadmill exercise system also offers more comfortable and safer exercise conditions for users.

5. Conclusion

This study investigates the benefits of using two controller inputs, the speed and gradient, for the regulation of HR during treadmill exercises. The main goal of HR control in treadmill exercises is to ensure a reliable, fast, and offset-free tracking, as well as to offer the faulty tolerance ability in the case of one of actuators (either treadmill speed or gradient) being out of service. For this purpose, we extended the concept of nonlinear decentralized integral controllability (DIC) to nonlinear 2ISO processes and presented a sufficient condition which only needs checking the steady state input-output relationship of controlled processes. Based on the proposed condition, we investigate the new defined multi-loop integral controllability (MIC) for walking, running, and walking-running zones. The experimental validation presents that by simultaneously using two control inputs, the automated system can achieve the fastest HR-tracking performance and stay close to the reference HR during steady state, while comparing with two SISO structures and offer the fault-tolerant ability if the gains of the two multi-loop integral controllers are well tuned. It has a vital implication for the applications of exercise rehabilitation and fitness in relation to the automated control system.

Acknowledgements

This work is supported by the Fundamental Research Funds for the Central Universities, China (grant#ZYGX2015J118), the Sichuan Provincial Program on Key R & D projects (grant#2017GZ0162), National Natural Science Foundation of China (grant No. 51675087, 61522105), and the China Postdoctoral Science Foundation funded project (2017M612950).

Nomenclature

AS	Asymptotically stable
BPM	Beat per minute
ECG	Electrocardiography
GAS	subject to be located between
HR	Heart rate
PI	Proportion-integral
DIC	Decentralized integral controllability
MIC	Multi-loop integral controllability
CVD	Cardiovascular disease
2ISO	Two-input single-output
SISO	Single-input single-output
STD	Standard deviation
UTS	University of Technology Sydney

Author details

Yi Zhang[1,2,3], Kairui Guo[4], Qin Yang[1], Pang Winnie[1], Kai Cao[4], Qi Wang[5], Andrey Savkin[6], Branko Celler[6], Hung Nguyen[4], Peng Xu[2,3], Limei Xu[1], Dezhong Yao[2,3] and Steven Su[4]*

*Address all correspondence to: steven.su@uts.edu.au

1 School of Aeronautics and Astronautics, University of Electronic Science and Technology of China, Chengdu, China

2 Key Laboratory for NeuroInformation of Ministry of Education, School of Life Science and Technology, University of Electronic Science and Technology of China, Chengdu, China

3 Center for Information in BioMedicine, University of Electronic Science and Technology of China, Chengdu, China

4 Centre for Health Technologies, Faculty of Engineering and Information Technology, University of Technology, Sydney, NSW, Australia

5 Hapatobiliary Surgery Department, General Hospital of Ningxia Medical University, Ningxia, China

6 School of Electrical Engineering and Telecommunications, University of New South Wales, Sydney, NSW, Australia

http://dx.doi.org/10.5772/intechopen.71855

References

[1] Frommer M. Clinical practice guidelines for the management of overweight and obesity in adults, adolescents and children in Australia. In: National Health Medical Research Council; 2013. p. 13-19

[2] Edwards S, Burroughs M. The Heart Rate Monitor Guidebook to Heart Zone Training. California, USA, Sacramento: Heart Zones; 1999. p. 15

[3] Su SW, Wang L, Celler BG, Savkin AV, Guo Y. Identification and control for heart rate regulation during treadmill exercise. IEEE Transactions on Biomedical Engineering. 2007;**54**(7):1238-1246

[4] Su SW, Huang S, Wang L, Celler BG, Savkin AV, Guo Y, Cheng TM. Optimizing heart rate regulation for safe exercise. Annals of Biomedical Engineering. 2010;**38**(3):758-768

[5] Cheng T, Savkin AV, Celler BG, Su SW, Wang L. Nonlinear modelling and control of human heart rate response during exercise with various work load intensities. IEEE Transactions on Biomedical Engineering. 2008;**55**(11):2499-2508

[6] Mazenc F, Malisoff M, Querioz M. Tracking control and robustness analysis for a nonlinear model of human heart rate during exercise. Automatica. 2011;**47**(5):968-974

[7] Jamieson LP, Hunt KJ, Allan DB. A treadmill control protocol combining nonlinear, equally smooth increases in speed and gradient. Medical Engineering & Physics. 2008; **30**(6):747-754

[8] Hunt KJ. Treadmill control protocols for arbitrary work rate profiles combining simultaneous nonlinear changes in speed and angle. Biomedical Signal Processing and Control. 2008;**3**:278-282

[9] Weng K, Turk B, Dolores L, Nguyen TN, Celler BG, Su SW, Nguyen HT. Fast tracking of a given heart rate profile in treadmill exercise. In: 32th Annual International Conference of IEEE Engineering in Medicine and Biology. Buenos Aires, #2569; 2010

[10] Skogestad S, Morari M. Variable selection for decentralized control. AIChE Annual Meeting. Washington, DC. Paper 128c; 1988

[11] Yu C, Fan M. Decentralized integral controllability and d-stability. Chemical Engineering Science. 1990;**45**(11):3299-3309

[12] Campo P, Morari M. Achievable closed-loop properties of systems under decentralized control: Conditions involving the steady state gain. IEEE Transactions and Automatic Control. 1994;**39**(5):932-943

[13] Su SW, Bao J, Lee PL. Analysis of decentralized integral controllability for nonlinear systems. Computers and Chemical Engineering. 2004;**28**(9):1781-1787

[14] Meste O, Khaddoumi B, Blain G, Bermon S. Time-varying analysis methods and models for the respiratory and cardiac system coupling in graded exercise. IEEE Transactions on Biomedical Engineering. 2005;**52**:1921-1930

[15] Wang L, Su SW, Celler BG, Chan GSH, Cheng TM, Savkin AV. Assessing the human cardiovascular response to moderate exercise: Feature extraction by support vector regression. Physiological Measurement. 2009;**30**:227-244

[16] Fukuba Y, Hara K, Kimura Y, Takahashi A, Ward SA. Whipp BJ: Estimating the parameters of aerobic function during exercise using an exponentially increasing work rate protocol. Medical & Biological Engineering & Computing. 2000;**38**:433-437

[17] Haddad A, Zhang Y, Su SW, Celler BG, Nguyen HT. Modelling and regulating of cardio-respiratory response for the enhancement of interval training. Biomedical Engineering Online. 2014;**13**:9

[18] Sepulchre R, Jankovic M, Kokotovic P. Constructive Nonlinear Control. New York: Springer; 1996. p. 200

[19] Zhang Y, Su S, Savkin A, Celler B, Nguyen H. Multiloop integral controllability analysis for nonlinear multiple-input single-output processes. Industrial & Engineering Chemistry Research. 2017;**56**(28):8054-8065

[20] Passivity-based Decentralized Control. Process Control. London: Springer; 2007. pp. 89-124

[21] Zhang Y, Su SW, Nguyen H, Celler BG. Machine learning based nonlinear model predictive control for heart rate response to exercise. In: Lam HK, Ling S, Nguyen HT, editors. Computational Intelligence and Its Applications: Evolutionary Computation, Fuzzy Logic, Neural Network and Support Vector Machine Techniques. London: Imperial College Press; 2012. pp. 271-285

[22] Zhang Y, Chen W, Su SW, Celler BG. Nonlinear modelling and control for heart rate response to exercise. International Journal of Bioinformatics Research and Applications. 2012;**8**(5):397-416

[23] Chen W, Su SW, Zhang Y, Guo Y, Nguyen N, Celler BG, Nguyen HT. Nonlinear modeling using support vector machine for heart rate response to exercise. In: Lam HK, Ling S, Nguyen HT, editors. Computational Intelligence and its Applications: Evolutionary Computation, Fuzzy Logic, Neural Network and Support Vector Machine Techniques. London: Imperial College Press; 2012. pp. 255-270

[24] Wang L, Zhang Y, Guo Y, Nguyen N, Zhang DM, Celler BG. A mathematical model of cardiovasular system under graded exercise levels. International Journal of Bioinformatics Research and Applications. 2012;**8**(6):455-473

[25] Zhang Y, Haddad A, Su SW, Celler BG, Coutts A, Duffield R, Donges C, Nguyen HT. An equivalent circuit model for onset and offset exercise response. Biomedical Engineering Online. 2014;**13**:145

Section 4

Robust Adaptive Controls of Overhead Cranes

Chapter 15

Robust Control of Crane with Perturbations

Yiming Wu, He Chen and Tong Yang

Additional information is available at the end of the chapter

http://dx.doi.org/10.5772/intechopen.71383

Abstract

In the presence of persistent perturbations in both unactuated and actuated dynamics of crane systems, an observer-based robust control method is proposed, which achieves the objective of trolley positioning and cargo swing suppression. By dealing with the unactuated and unknown perturbation as an augmented state variable, the system dynamics are transformed into a quasi-chain-of-integrators form based on which a reduced-order augmented-state observer is established to recover the perturbations appearing in the unactuated dynamics. A novel sliding manifold is constructed to improve the robust performance of the control system, and a linear control law is presented to make the state variables stay on the manifold in the presence of perturbations in unactuated dynamics. A Lyapunov function candidate is constructed, and the entire closed-loop system is proved rigorously to be exponentially stable at the equilibrium point. The effectiveness and robustness of the proposed observer-based robust controller are verified by numerical simulation results.

Keywords: underactuated systems, overhead cranes, observer-based control, Lyapunov methods, motion control

1. Introduction

Underactuated systems [1–9] are now widely applied in modern industry. A crane system is a typical class of underactuated systems with strong state coupling. Due to inertia, when the trolley moves, the unactuated cargo swings back and forth, which affects the transporting efficiency and safety. Therefore, on the one hand, effective controllers are needed to transport the actuated trolley to desired positions. On the other hand, it is also necessary to eliminate residual vibrations of the unactuated cargo. Nevertheless, control problems of crane systems are still non-trivial and challenging since the system is underactuated without enough available control inputs.

In order to tackle control problems of crane systems, various control methods are proposed [10–38]. Specifically, Sun et al. [10, 11] present antiswing controllers to regulate the cargo position to the desired location asymptotically in the presence of ship roll and heave movements for offshore crane systems applied in modern ocean transportation and logistics. Moreover, existing methods also include input shaping [12–15], feedback control [16–28], intelligent control [29–32], and trajectory planning method [33–36]. Specifically, several input shapers are designed to reduce payload swing of bridge crane systems [12–15]. In Ref. [16], an energy-based output feedback control scheme is proposed, which achieves both precise trolley positioning and efficient payload swing elimination under control input constraints. In Ref. [17], a payload motion-based control approach is presented in the presence of system parameter uncertainties. In [18–20], non-linear controllers are designed on the basis of partial feedback linearization. In Ref. [21], visual feedback technology is used to achieve the control objective by using two handy cameras. Additionally, sliding mode control strategies are also widely applied to tackle crane system control problems [22–25]. For example, Almutairi and Zribi [22] achieved the asymptotic stability of the closed-loop overhead crane system by proposing a sliding mode control scheme. Xi and Hesketh [23] addressed an integral sliding mode control method for discrete time crane systems with both matched and unmatched uncertainties to ensure the existence of sliding mode in the presence of uncertainties. Based on second-order sliding modes, Bartolin et al. [24] guaranteed a fast and precise payload transferring and swing suppression. Ngo and Hong [25] developed an adaptation law with a varying control gain that transits the system into the designed sliding mode. Moreover, in practical applications, cranes always suffer from unknown or uncertain system parameters (e.g., payload weight changes, varying rope lengths, etc.). Then adaptive control schemes are applied to address these problems [26–28]. Sun et al. [26] addressed the crane antiswing and positioning problem in the presence of payload hoisting/lowering and uncertain parameters with simultaneous payload weight identification. Park et al. [27] proposed an adaptive sliding-mode antisway control law with system uncertainties and high-speed hoisting motion. Sun et al. [28] designed an adaptive control scheme to deal with the control problem of tower crane systems with parametric uncertainties without approximating the non-linear dynamics. There are also some intelligent control methods applied in crane systems such as fuzzy control [29, 30], genetic algorithm [31], and neural network [32]. According to the operating experience of real cranes, it is also essential to design suitable trajectories for the system states (positions, velocities, and accelerations). Then, tracking controllers can be used to track the trajectories. In addition to closed-loop control design, many studies also focus on the trajectory planning part and achieve meaningful results [33–36]. Uchiyama et al. [33] generated an S-curve trajectory numerically, which can suppress the residual vibration without measuring it. Sun et al. [34] obtained an analytical three-segment acceleration trajectory. For given transferring task, the proposed trajectory planning method provides a mechanism to determine the parameters to ensure that all the transportation indexes are met. More recently, in Ref. [35], an optimal trajectory is generated with optimal energy consumption by using the proposed optimal planner. There are also antiswing control strategies proposed for double pendulum cranes [37, 38].

However, most of the existing methods for underactuated crane systems tackle the control problem without considering the perturbations in the unactuated dynamics. In practical applications, perturbations widely exist in both actuated and unactuated dynamics, which may be difficult to tackle by using existing methods. Note that in Ref. [23], integral sliding mode control method is proposed by considering perturbations in unactuated dynamics, but it is only designed for discrete-time systems by estimating the present disturbance signal with its past value. Therefore, in order to derive an effective method to achieve crane control in the presence of unknown persistent (even non-vanishing) perturbations in both unactuated and actuated dynamics, this chapter proposes an observer-based robust control method.

The main contribution of this chapter is as follows:

1. According to whether the perturbation in the unactuated dynamics is vanishing or not, the control problem is stated in two cases. The observer-based robust controller designed in this chapter can achieve the control objectives for both cases.
2. By dealing with the unactuated and unknown perturbation as an augmented state variable, an augmented error system is established based on which we design a reduced-order augmented state observer for the crane system to recover the perturbations appearing in the unactuated dynamics.
3. Together with the observer, by constructing a new sliding manifold, a new observer-based sliding mode controller is developed.

The proposed controller is applicable to crane systems with unknown persistent perturbations in the unactuated dynamics and achieves robust control effectively.

The rest of this chapter is organized as follows. Section 2 describes the crane dynamics with persistent (even non-vanishing) perturbations and transforms the dynamics into a quasi-chain-of-integrators form for the convenience of controller design and stability analysis. Also, the control objective is stated in Section 2. Based on the model in Section 2, a reduced-order augmented-state observer and an observer-based control law are developed in Section 3. Then in Section 4, numerical simulation results are included to verify the effectiveness of the proposed controller. Section 5 summarizes the entire work of this chapter.

2. Problem formulation

The purpose of this chapter is to propose an effective method to achieve crane control in the presence of persistent (even non-vanishing) perturbations in both unactuated and actuated dynamics. The crane dynamics can be represented by the following equations (shown in **Figure 1**):

$$(M+m)\ddot{x} + mL\ddot{\theta}\cos\theta - mL\,\dot{\theta}^2\sin\theta = u - f_r + d_x, \tag{1}$$

$$mL^2\ddot{\theta} + mL\cos\theta\ddot{x} + mgL\sin\theta = d_\theta. \tag{2}$$

The system parameters are defined in **Table 1**, and f_r denotes the rail friction force expressed as follows:

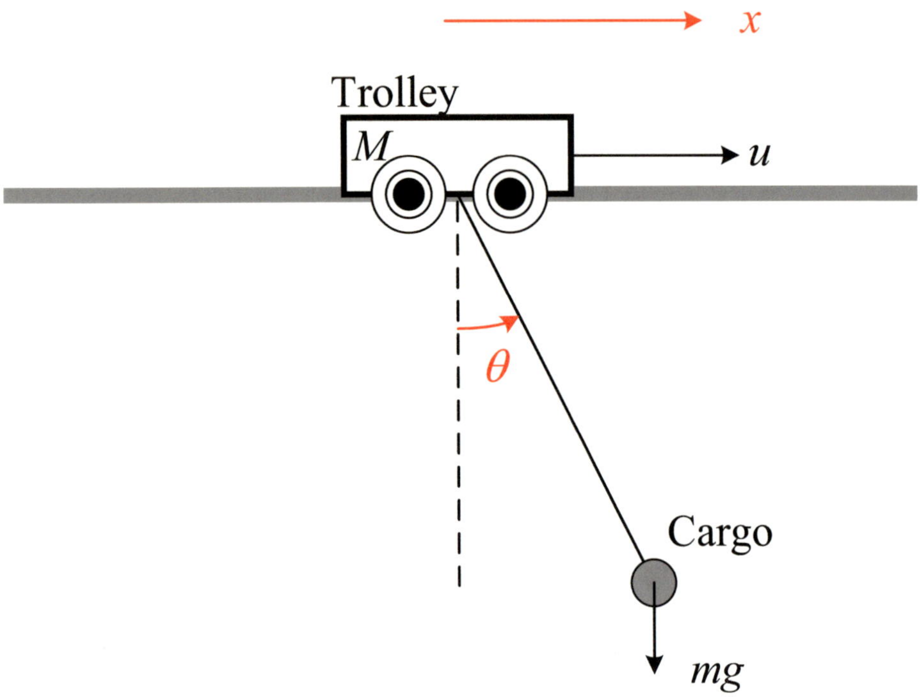

Figure 1. The schematic diagram of an overhead crane system.

$$f_r = f_{r0}\tanh(\dot{x}/\epsilon) - k_r \mid \dot{x} \mid \dot{x}, \tag{3}$$

where f_{r0}, ϵ, $k_r \in \boldsymbol{R}$ are friction parameters, which can be identified by offline experimental tests and data fitting. $d_x(t)$ and $d_\theta(t)$ denote the lumped term comprising external perturbations, unmodeled dynamics, the mismatch between the real girder friction and the friction compensation model shown in Eq. (3), and so forth.

System parameter	Physical significant	Unit
M	Trolley mass	kg
m	Cargo mass	kg
L	Cargo rotation radius	m
g	Gravity constant	m/s^2
x	Trolley translational displacement	m
θ	Cargo rotational angle	rad
u	Control input	N

Table 1. The system parameters of crane systems.

Considering the practical physical constraints, though the exact expressions for the lumped perturbation terms $d_x(t)$ and $d_\theta(t)$ are unknown, the following assumptions are reasonably made.

Assumption 1: The perturbation term $d_x(t)$ present in the actuated dynamics is bounded as $|d_x(t)| \le \bar{d}_x$, where $\bar{d}_x$ is a priori known. The unactuated perturbation term $d_\theta(t)$ is differentiable up to the n-th order; $|d_\theta| \le \bar{d}_\theta$, $|d_\theta^{(i)}(t)| \le \bar{d}_{\theta i}$, $i = 1, 2, \cdots, n-1$, where $\bar{d}_\theta$ and $\bar{d}_{\theta i}$ are priori known constants. It is also assumed that $d_\theta^{(3)}(t) \approx 0$.

2.1. Crane model transformation

Before proceeding to describe the control objective, we perform several steps of transformations for the original crane dynamics shown in Eqs. (1) and (2) for the convenience of carrying out controller development and stability analysis in the subsequent section. Considering the fact of $mL > 0$, we divide both sides of Eq. (2) and make some arrangements to obtain the following equation:

$$\ddot{x} = -g\tan\theta - \frac{L\ddot{\theta}}{\cos\theta} + \frac{d_\theta}{mL\cos\theta}. \tag{4}$$

Then one can substitute Eq. (4) into Eq. (1) and make some arrangements to obtain

$$-\frac{(M + m\sin^2\theta)L}{\cos\theta}\left(\ddot{\theta} - \delta_x - \delta_{\theta a}\right) - mL\dot{\theta}^2\sin\theta - (M+m)g\tan\theta = u - f_r, \tag{5}$$

where $\delta_x(t)$ and $\delta_{\theta a}(t)$ are defined as follows:

$$\delta_x = -\frac{d_x\cos\theta}{(M + m\sin^2\theta)L}, \quad \delta_{\theta a} = -\frac{d_\theta(M+m)}{(M + m\sin^2\theta)mL^2}. \tag{6}$$

Based on *Assumption 1*, the upper bounds for $\delta_x(t)$ and $\delta_{\theta a}(t)$ are provided as:

$$|\delta_x| \le \bar{\delta}_x = \frac{\bar{d}_x}{Ml}, \quad |\delta_{\theta a}| \le \bar{\delta}_{\theta a} = \frac{\bar{d}_\theta(M+m)}{MmL^?}. \tag{7}$$

In the view of the explicit expression of Eq. (5), a feedback linearization controller can be proposed as follows:

$$u = -\frac{(M + m\sin^2\theta)L}{\cos\theta}v - mL\dot{\theta}^2\sin\theta - (M+m)g\tan\theta + f_r, \tag{8}$$

in which $v(t)$ is a to-be-elaborated auxiliary control input. That is, once we derive the expression of$v(t)$, the ultimate controller $u(t)$ can be conveniently obtained according to Eq. (8). By substituting Eq. (8) into Eq. (5), together with Eq. (4), the dynamic Eqs. (1) and (2) can be re-expressed in the following fashion:

$$\begin{cases} \ddot{x} = -g\tan\theta - \dfrac{L\ddot{\theta}}{\cos\theta} + \dfrac{d_\theta}{mL\cos\theta}, \\ \ddot{\theta} = v + \delta_x + \delta_{\theta a}. \end{cases} \tag{9}$$

Further, we define the following coordinate transformations:

$$\begin{aligned}\phi_1 &= x + L\ln(\sec\theta + \tan\theta)\\ \phi_2 &= x + L\dot{\theta}\sec\theta\\ \phi_3 &= -g\tan\theta\\ \phi_4 &= -g\dot{\theta}\sec^2\theta\end{aligned} \tag{10}$$

Then, it is straightforward to obtain the following dynamic equations:

$$\begin{aligned}\dot{\phi}_1 &= \phi_2,\\ \dot{\phi}_2 &= \phi_3 - h(\phi_3)\,\phi_4^2 + \frac{d_\theta}{mL\cos\theta},\\ \dot{\phi}_3 &= \varphi_4,\\ \dot{\phi}_4 &= -g(v + \delta_x + \delta_{\theta a})\sec^2\theta - 2g\,\dot{\theta}^2\sec^2\theta\tan\theta.\end{aligned} \tag{11}$$

In Eq. (11), the function $h(\phi_3)$ is with the definition as $h(\phi_3) = \frac{l\phi_3}{(g^2+\phi_3^2)^{1.5}} \Rightarrow |h(\phi_3)| \le 0.004L$, where the value of the gravity constant is taken as $g = 9.8$ m/s^2.

For practical applications, the cargo swing is always within 10 degrees, that is, $|\theta(t)| \le \pi/18$ rad. In this case, the approximations of $\sin\theta \approx \tan\theta \approx \theta$ and $\sec\theta = \cos^{-1}\theta \approx 1$ are valid. In this sense, $\phi_1(t)$ in Eq. (10) can be approximated as follows:

$$\phi_1(t) \approx x + L\ln(1+\theta) \approx x + L\theta, \tag{12}$$

which is right at the horizontal position of the cargo. Also, the cargo swing angular velocity satisfies $|\dot{\theta}(t)| << 1$ rad/s; considering that the wire length L's order of magnitude is usually 10 m, $h(\phi_3)\,\phi_4^2 = 0.004\,Lg^2\,\dot{\theta}^2\sec^4\theta \approx 0.384L\,\dot{\theta}^2\sec^4\theta \approx 0$ holds; hence, $h(\phi_3)\,\phi_4^2$ is negligible and can be incorporated as part of the unactuated lumped perturbation $\delta_{\theta u}(t)$ that will be introduced later. For simplicity of denotation, we define

$$\varphi_1 = x + L\theta,\ \varphi_2 = \dot{\varphi}_1. \tag{13}$$

Therefore, the crane dynamics can be described by

$$\begin{cases}\dot{\varphi}_1 = \varphi_2,\\ \dot{\varphi}_2 = \phi_3 + \delta_{\theta u},\\ \dot{\varphi}_3 = \phi_4,\\ \dot{\varphi}_4 = -g(v + \delta_x + \delta_{\theta a})\sec^2\theta - 2g\,\dot{\theta}^2\sec^2\theta\tan\theta,\end{cases} \tag{14}$$

wherein $\delta_{\theta u}(t)$ represents the unactuated lumped perturbation term mainly consisting of $d_\theta/mL\cos\theta$.

2.2. Control objective

For crane control during the transportation process (between the hoisting and lowering stages), the kernel objective is to transfer the cargo from its initial position to the desired position (destination) and then keep it stationary right above the destination so that further actions (e.g., lowering) can be taken.

Hence, the preliminary task is to make the cargo reach the destination by appropriately controlling the trolley motion, which can be mathematically depicted as follows:

$$\varphi_1 = x + L\theta \rightarrow p_{dx}. \tag{15}$$

To make this process smooth enough, instead of set-point control (i.e., directly using p_{dx} as the reference), we want the cargo to follow a smooth time-varying trajectory $r_x(t)$, which satisfies the following conditions:

$$\lim_{t \to t_{f1}} r_x(t) = p_{dx}, \mid r_x^{(i)} \mid \leq \pi, i = 1, 2, 3, 4, \tag{16}$$

where t_{f1} denotes the consumed time for $r_x(t)$ to reachp_{dx}, and $\pi_i (i = 1, 2, 3, 4)$ stands for the corresponding upper bound for the i-th order derivative for$r_x(t)$, respectively.

When there are no external perturbations appearing in the unactuated dynamics (that is, $\delta_{\theta u} \equiv 0$ in Eq. (14)), we need also to damp out the cargo swing $\theta(t)$ at the same time, namely,

$$\theta \rightarrow 0 \Rightarrow x \rightarrow p_{dx}. \tag{17}$$

However, in the case of persistent, non-vanishing perturbations in the unactuated component (i.e., $\delta_{\theta u}(t) \neq 0$), there *does not exist any control action* that can completely damp out $\theta(t)$ while keeping the cargo stationary right above the destination. Suppose that there exists such a controller $u'(t)$ that could eliminate the cargo swing, namely,

$$\theta(t) = 0, \dot{\theta}(t) = 0 \rightarrow \phi_3(t) = -g \tan\theta(t) - 0, \tag{18}$$

and make the cargo stay stationary at the destination in the sense that

$$\varphi_1(t) = p_{dx}, \dot{\varphi}_1(t) = 0, \varphi_2(t) = 0, \dot{\varphi}_2(t) = 0, \forall t \geq t_{f2}, \tag{19}$$

with t_{f2} being the settling time, then it would follow, by inserting Eq. (18) and Eq. (19) into the second equation of Eq. (14), that

$$\delta_{\theta u} = \dot{\varphi}_2 - \phi_3 = 0, \tag{20}$$

which obviously contradicts with the fact that $\delta_{\theta u} \neq 0$; thus the existence of such a controller $u'(t)$ is impossible. This fact illustrates the great challenge that will be faced with when controlling the crane system in the presence of persistent perturbations in the unactuated dynamics. On the other hand, since $\delta_{\theta u}(t)$ is usually unknown, the control problem becomes even more challenging.

Based on the analysis claimed above, in accordance with the fact whether $\delta_{\theta u}$ in the unactuated dynamics is vanishing or not, the control objective of this chapter is stated as follows:

- **Case 1.** *Non-vanishing perturbations in the unactuated dynamics.* Drive the unactuated cargo to the desired destination and keep it stationary over the destination thereafter, that is,

$$\lim_{t\to\infty} \varphi_1(t) = p_{dx}, \lim_{t\to\infty} \dot{\varphi}_1(t) = 0. \tag{21}$$

- **Case 2.** *Vanishing or no/negligible perturbations in the unactuated dynamics.* Drive both the trolley and the unactuated cargo to the desired destination, in the sense that

$$\lim_{t\to\infty} \varphi_1(t) = p_{dx}, \lim_{t\to\infty} x(t) = p_{dx}, \lim_{t\to\infty} \dot{\varphi}_1(t) = 0, \lim_{t\to\infty} \dot{x} = 0 \Rightarrow \lim_{t\to\infty} \theta(t) = 0, \lim_{t\to\infty} \dot{\theta}(t) = 0. \tag{22}$$

To achieve the control objective, together with Eq. (16), let the following error signals be defined:

$$e_1 = \varphi_1 - r_x, e_2 = \varphi_2 - \dot{r}_x, e_3 = \phi_3, e_4 = \phi_4. \tag{23}$$

Thus, we are led to the following open-loop error system:

$$\begin{cases} \dot{e}_1 = e_2, \\ \dot{e}_2 = e_3 + \delta_{\theta u} - \ddot{r}_x, \\ \dot{e}_3 = e_4, \\ \dot{e}_4 = -g(v + \delta_x + \delta_{\theta a}) \sec^2\theta - 2g\,\dot{\theta}^2 \sec^2\theta \tan\theta, \end{cases} \tag{24}$$

which is the basis for the observer-controller design and analysis in the section that follows.

3. Main results

In order to achieve the control objective claimed in the previous section, we will propose a perturbation observer-based robust control scheme. More precisely, to deal with the unactuated unknown persistent perturbations, an augmented-state observer will be constructed. Then, we will present a novel robust control law, which can achieve superior control performance and provide the corresponding theoretical stability analysis.

3.1. Observer design

The fact that the perturbation term $\delta_{\theta u}(t)$ is *unactuated and unknown* brings much difficulty for the controller design and analysis and it makes traditional robust control methods not applicable. As a means to achieve the aforementioned control objective, it is required to figure out a suitable strategy that can deal with $\delta_{\theta u}(t)$. Toward this end, before controller development, we will first construct an augmented observer which can recover the lumped perturbation term $\delta_{\theta u}(t)$ appearing in the unactuated dynamics. Then, we treat $\delta_{\theta u}(t)$ as an augmented state variable. The benefit of doing so is that the perturbation observer design procedure would become more concise and clear. By following this line, the augmented error system for Eq. (24) is established as follows:

$$\begin{cases} \dot{e}_1 = e_2, \\ \dot{e}_2 = e_3 + e_5 - \ddot{r}_x, \\ \dot{e}_3 = e_4, \\ \dot{e}_4 = -g(v + \delta_x + \delta_{\theta a}) \sec^2\theta - 2g\,\dot{\theta}^2 \sec^2\theta \tan\theta, \\ \dot{e}_5 = e_6, \\ \dot{e}_6 = e_7, \\ \vdots \\ \dot{e}_{n+1} = 0, \\ y = e_1 \end{cases} \tag{25}$$

where we have considered $\delta_{\theta u}(t)$ as an augmented state variable $e_5(t)$ and its derivatives as $e_6(t), e_7(t), \cdots, e_{n+1}(t)$, and $y(t)$ is the corresponding system output signal. In this chapter, the signals $e_1(t)$, $e_3(t)$ and $e_4(t)$ are measurable, and we merely need to fabricate an observer with the aim of recovering the lumped perturbation$e_5(t)$. In order to reduce the computational complexity, noting also that $\theta_x(t)$ and $\theta_{\theta a}(t)$ are unavailable for feedback, we intend to construct a reduced-order perturbation observer. For this purpose, consider the following subsystem:

$$\begin{cases} \dot{e}_2 = e_3 + e_5 - \ddot{r}_x, \\ \dot{e}_5 = e_6, \\ \dot{e}_6 = e_7, \\ \vdots \\ \dot{e}_{n+1} - 0, \\ y' = e_2, \end{cases} \tag{26}$$

which is part of the augmented error system shown in Eq. (25), where $y'(t)$ is regarded as the new output. It is not difficult to check that the reduced-order augmented-state system shown in Eq. (26) is observable, and the detailed analysis can be found in Appendix A. Based on the structure of Eq. (26), we design the following reduced-order augmented-state observer:

$$\begin{cases} \dot{\hat{e}}_2 = e_3 + \hat{e}_5 - \ddot{r}_x - \lambda_2(\hat{e}_2 - y'), \\ \dot{\hat{e}}_5 = \hat{e}_6 - \lambda_5(\hat{e}_2 - y'), \\ \dot{\hat{e}}_6 = \hat{e}_7 - \lambda_6(\hat{e}_2 - y'), \\ \vdots \\ \dot{\hat{e}}_{n+1} = -\lambda_{n+1}(\hat{e}_2 - y'), \end{cases} \tag{27}$$

where $\lambda_2, \lambda_5, \lambda_6, \cdots, \lambda_{n+1}$ denote the observer gains. Define the following error signals:

$$\xi_i = \hat{e}_i - e_i, i = 2, 5, 6, \cdots, n+1, \tag{28}$$

and denote the corresponding error vector by

$$\xi(t) = [\xi_2(t)\ \xi_5(t)\ \xi_6(t)\ \cdots\ \xi_{n+1}(t)]^{\mathrm{T}}. \tag{29}$$

Then, one can subtract Eq. (26) from Eq. (27) to derive the following observer error system:

$$\dot{\xi} = \Omega\,\xi, \tag{30}$$

where $\Omega \in \boldsymbol{R}^{(n-2)\times(n-2)}$ is defined as:

$$\Omega = \begin{pmatrix} -\lambda_2 & 1 & 0 & \cdots & 0 & 0 \\ -\lambda_5 & 0 & 1 & \cdots & 0 & 0 \\ -\lambda_6 & 0 & 0 & \ddots & 0 & 0 \\ \vdots & \vdots & \vdots & \vdots & \ddots & \vdots \\ -\lambda_n & 0 & 0 & \cdots & 0 & 1 \\ -\lambda_{n+1} & 0 & 0 & \cdots & 0 & 0 \end{pmatrix}. \tag{31}$$

As stated previously, the system shown in Eq. (26) is observable. Hence, without difficulty, we are admitted to choose a proper set of $\lambda_2, \lambda_5, \lambda_6, \cdots, \lambda_{n+1}$ conveniently via pole placement, such that Ω is a Hurwitz matrix with *the eigenvalues' real parts being different from each other*. In this sense,

$$\xi_i = \hat{e}_i - e_i \rightarrow 0, i = 2, 5, 6, \cdots, n+1, \tag{32}$$

exponentially fast, which indicates that the designed perturbation observer shown in Eq. (27) can online recover the perturbations.

In addition, it can be obtained from Eq. (30) that the trajectories of the observer error signals are represented by

$$\xi = \exp(\Omega t)\boldsymbol{\xi}(0). \tag{33}$$

Since we have rendered, by proper pole placement, that the poles (i.e., the eigenvalues of Ω) of the closed-loop system shown in Eq. (30) have different negative real parts, there exists an invertible matrix $\Gamma \in \boldsymbol{R}^{(n-2)\times(n-2)}$ that can transform Ω into a diagonal matrix, that is,

$$\Gamma^{-1}\Omega\Gamma = \Lambda, \tag{34}$$

where $\Lambda = \mathrm{diag}\{\lambda_1, \lambda_2, \cdots, \lambda_{n-2}\}$ with $\lambda_i, i = 1, 2, \cdots, n-2$ being the $(n-2)$ eigenvalues of Γ. Therefore, we can rewrite the exponential matrix $\exp(\Omega t)$ and $\xi(t)$ as [39]

$$\exp(\Omega t) = \Gamma \exp(\Lambda t)\Gamma^{-1} \Rightarrow \xi = \Gamma \exp(\Lambda t)\Gamma^{-1} \xi(0). \tag{35}$$

Taking the Euclidean norm for both sides of Eq. (35), we are led to the following results:

$$\begin{aligned} \|\xi\|_2 &= \|\Gamma \exp(\Lambda t)\, \Gamma^{-1} \xi(0)\|_2 \leq \|\Gamma \exp(\Lambda t)\, \Gamma^{-1}\|_{m_\infty} \cdot \|\xi(0)\|_2 \\ &\leq \|\Gamma\|_{m_\infty} \cdot \|\Gamma^{-1}\|_{m_\infty} \cdot \|\exp(\Lambda t)\|_{m_\infty} \|\xi(0)\|_2 \leq (n-2)\exp(\lambda_{\max} t) \cdot \|\Gamma\|_{m_\infty} \cdot \|\Gamma^{-1}\|_{m_\infty} \cdot \|\xi(0)\|_2, \end{aligned} \tag{36}$$

where $\lambda_{\max} = \max_{i=1,2,\cdots,n-2}\{\lambda_i\}$, $\|\cdot\|_2$ denotes the Euclidean norm, $\|\cdot\|_{m\infty}$ represents the m_∞-norm for matrices[1], which are compatible norms[2]. It is further implied from Eq. (36) that

$$|\ \xi_i(t)\ | \leq \|\ \xi\|_2 \leq \bar{\xi} \triangleq (n-2)\exp(\lambda_{\max} t) \cdot \|\Gamma\|_{m_\infty} \cdot \|\Gamma^{-1}\|_{m_\infty} \cdot \|\xi(0)\|_2. \tag{37}$$

Using the pole assignment technique, one can derive the values for $\lambda_2, \lambda_5, \lambda_6, \cdots, \lambda_{n+1}$ and the expression for Ω. Further, with the aid of such software as MATLAB, it is easy to calculate $\|\Gamma\|_{m\infty} \cdot \|\Gamma^{-1}\|_{m\infty}$; hence, the bound for $|\xi_i(t)|$, as shown in Eq. (37), can be computed without difficulty.

3.2. Controller development and stability analysis

To achieve robust control in the presence of uncertainties or external perturbations, we will develop a new observer-based sliding mode controller. The fundamental idea of the sliding mode control method is to construct a sliding manifold (surface) on which the system state is convergent and then develop a suitable control law that renders the state reaches the manifold within finite time. Traditionally, the key step is constructing an appropriate sliding surface, and the corresponding controller can usually be obtained straightforwardly.

However, the major drawback of most currently available sliding mode control methods is that they are merely capable of tackling uncertainties or perturbations in the actuated part, and when uncertainties or perturbations are present in the unactuated component, their performance will degrade significantly and even become unstable. To illustrate this point, we will show some brief analysis for the conventional sliding mode control approach. More precisely, for the open-loop error system shown in Eq. (24), one will design the conventional sliding manifold, denoted by $\zeta(t)$ in the following fashion:

$$\zeta = e_1 + \alpha e_2 + \beta e_3 + \gamma e_4, \tag{38}$$

where α, β, and γ are sliding slopes chosen such that the polynomial $1 + \alpha s + \beta s^2 + \gamma s^3 = 0$ is Hurwitz, with s being the complex variable. It is not difficult to design a control law that drives the system state variables to $\zeta(t)$ such that $\zeta(t) = 0$ after certain finite time t_{f_s}, *i.e.*, $\zeta(t) = 0$, $\dot{\zeta}(t) = 0, \forall t \geq t_{f_s}$. By recursively using the first three equations in Eq. (24) and regrouping the

[1]For a square matrix $A = (aij)_{n\times n} \in R^{n\times n}$, $\|A\|_{m\infty} = n\max_{i,j} |aij|$ is defined as the $m\infty$-norm for A.
[2]A matrix norm $\|\cdot\|_m \in R^{n\times n}$ is said to be compatible with a vector norm $\|\cdot\|_v \in R^n$ if $\|Ax\|_v \leq \|A\|m \cdot \|x\|v$, where $A \in R_{n\times n}$ and $x \in R$. It is not difficult to verify that the $m\infty$-norm for matrices is compatible with the Euclidean norm for vectors.

resulting terms, one can derive from $\zeta(t)=0$ and Eq. (38) that the state variable $e_1(t)$ is dominated by the following dynamics on the sliding manifold:

$$e_1+\alpha\dot{e}_1+\beta\ddot{e}_1+\gamma e_1^{(3)} = -\beta(\ddot{r}_x-\delta_{\theta u})-\gamma\left(r_x^{(3)}-\dot{\delta}_{\theta u}\right). \tag{39}$$

Clearly, if the perturbation terms $\delta_{\theta u}(t), \dot{\delta}_{\theta u}(t)$ appearing in the unactuated dynamics are non-vanishing, $e_1(t)$ will never tend to zero.

As indicated from the above-mentioned analysis, to make sliding mode control applicable to crane systems with *unknown persistent perturbations* in the unactuated component, it is needed to construct a new sliding manifold to improve the robust performance of the control system. To do so, on the basis of the designed perturbation observer in the previous subsection, we design the following sliding manifold that will be used in the subsequent controller development:

$$\varepsilon = e_1+\alpha e_2+\beta\left(e_3+\hat{e}_5-\ddot{r}_x\right)+\gamma\left(e_4+\hat{e}_6-r_x^{(3)}\right). \tag{40}$$

where α, β, γ are defined in Eq. (38) and $\hat{e}_5(t), \hat{e}_6(t)$ are the observer-recovered signals for the lumped perturbation term [see Eq. (27)]. Before giving the expression for the auxiliary "control input" $v(t)$, we first construct the following non-negative scalar function $V(t)$:

$$V = \frac{1}{2}\varepsilon^2 = \frac{1}{2}\left[e_1+\alpha e_2+\beta(e_3+\hat{e}_5-\ddot{r}_x)+\gamma e_4(e_4+\hat{e}_6-r_x^{(3)})\right]^2. \tag{41}$$

The derivative of $\varepsilon(t)$ with regard to time can be obtained as follows:

$$\begin{aligned}\dot{\varepsilon} = {} & e_2+\alpha(e_3+e_5-\ddot{r}_x)+\beta\left[e_4+\hat{e}_6-\lambda_5(\hat{e}_2-e_2)-r_x^{(3)}\right]-\gamma g(v+\delta_x+\delta_{\theta a})\sec^2\theta \\ & -\gamma\left[2g\dot{\theta}^2\sec^2\theta\tan\theta-\hat{e}_7+\lambda_6(\hat{e}_2-e_2)+r_x^{(4)}\right],\end{aligned} \tag{42}$$

where Eq. (25) and Eq. (27) have been employed for implications. Then, in view of the structure of Eq. (42), $v(t)$ is developed in the following fashion:

$$\begin{aligned}v = {} & \frac{1}{g\gamma\sec^2\theta}\cdot\left\{e_2+\alpha(e_3+\hat{e}_5-\ddot{r}_x)-\gamma\left[r_x^{(4)}-\hat{e}_7+\lambda_6(\hat{e}_2-e_2)\right]\right. \\ & \left.+\beta\left[e_4+\hat{e}_6-\lambda_5(\hat{e}_2-e_2)-r_x^{(3)}+k_u\,\mathrm{sign}(\varepsilon)\right]\right\}-2\dot{\theta}^2\tan\theta+k_a\,\mathrm{sign}(\varepsilon)\end{aligned} \tag{43}$$

where

$$k_a \geq \overline{\delta}_x+\overline{\delta}_\theta, k_u > \frac{\alpha}{\beta}\overline{\xi}. \tag{44}$$

are positive control gains [see Eqs. (7) and (37) for the definitions of $\overline{\delta}_x, \overline{\delta}_\theta$, and $\overline{\xi}$], and

$$\mathrm{sign}(\star) = \begin{cases}\star/|\star|, & \star\neq 0,\\ 0, & \star=0,\end{cases} \tag{45}$$

denote the standard sign function. We can further substitute Eq. (43) into Eq. (8) to obtain the ultimate control law as follows:

$$\begin{aligned} u = & -\frac{(M+m\sin^2\theta)L}{g\gamma\sec\theta}\cdot\{e_2+\alpha(e_3+\hat{e}_5-\ddot{r}_x)-\gamma[r_x^{(4)}-\hat{e}_7+\lambda_6(\hat{e}_2-e_2)] \\ & +\beta[e_4+\hat{e}_6-\lambda_5(\hat{e}_2-e_2)-r_x^{(3)}+k_u\,\mathrm{sign}(\epsilon)]+g\gamma\sec^2\theta[2\dot{\theta}^2\tan\theta-k_a\,\mathrm{sign}(\epsilon)]\} \\ & -mL\dot{\theta}^2\sin\theta-(M+m)g\tan\theta+f_r. \end{aligned} \tag{46}$$

The main results for the proposed control scheme are summarized by the theorem that follows.

Theorem 1. The designed control law shown in Eq. (46), together with the reduced-order augmented-state observer shown in Eq. (27), can achieve the control objective claimed by **Case 1** in the case of unactuated persistent non-vanishing disturbances or **Case 2** if the unactuated disturbances are vanishing/negligible.

Proof: Consider $V(t)$ defined in Eq. (41) as a Lyapunov function candidate, and its time derivative is given by

$$\dot{V} = \varepsilon\dot{\varepsilon}. \tag{47}$$

By inserting Eq. (43) into the expression of $\dot{\varepsilon}(t)$ in Eq. (42) and regrouping the common terms, one can obtain the following equation:

$$\begin{aligned} \dot{\varepsilon} = & \alpha(e_5-\hat{e}_5)-\beta k_u\,\mathrm{sign}(\varepsilon)-g\gamma\sec^2\theta[k_u\,\mathrm{sign}(\varepsilon)+\delta_x+\delta_{\theta a}] \\ = & -\alpha\xi_5-\beta k_u\,\mathrm{sign}(\varepsilon)-g\gamma\sec^2\theta[k_u\,\mathrm{sign}(\varepsilon)+\delta_x+\delta_{\theta a}], \end{aligned} \tag{48}$$

upon the use of the relationship in Eq. (28). Then, the following results are straightforward after the substitution of Eq. (48) into Eq. (47):

$$\begin{aligned} \dot{V} = & -\beta k_u\,|\,\varepsilon\,|-\alpha\xi_5\,\varepsilon+g\gamma\sec^2\theta[k_u\,|\,\varepsilon\,|+(\delta_x+\delta_{\theta a})\varepsilon] \\ \le & -(\beta k_u-\alpha\,|\,\xi_5\,|)\,|\,\varepsilon\,|-g\gamma\sec^2\theta(k_a-(|\,\delta_x\,|+|\,\delta_{\theta a}\,|))\,|\,\varepsilon\,| \\ \le & -(\beta k_u-\alpha\,|\,\xi_5\,|)\sqrt{2V}, \end{aligned} \tag{49}$$

where the gain conditions shown in Eq. (44) have been utilized. The conclusion of Eq. (49) indicates that $V(t)$, and hence $\varepsilon(t)$, converges to zero in finite time. Further, on the sliding manifold where $\varepsilon(t)=0$, the system state variables satisfy the following dynamic equation array:

$$\begin{cases} e_1+\alpha\dot{e}_1+\beta\ddot{e}_1+\gamma e_1^{(3)} = -\beta\xi_5-\gamma\xi_6, \\ \dot{\xi} = \Omega\,\xi. \end{cases} \tag{50}$$

As by pole assignment, the matrix $\Omega\in\boldsymbol{R}^{(n-2)\times(n-2)}$ is Hurwitz, and α, β and γ also render $1+\alpha s+\beta s^2+\gamma s^3=0$ Hurwitz; it is clearly seen that the entire closed system Eq. (50) is exponentially stable at the equilibrium point, and hence

$$e_1(t) = \varphi_1(t) - r_x \to 0, e_2(t) = \dot{e}_1(t) = \dot{\varphi}_1(t) - \dot{r}_x \to 0$$
$$\Rightarrow \dot{e}_2(t) = \ddot{e}_1(t) \to 0, \ddot{e}_2(t) = e_1^{(3)}(t) \to 0, \tag{51}$$

exponentially fast, which indicates the cargo motion tracks the planned trajectory $r_x(t)$ in an exponential fashion. Since $r_x(t)$ tends to p_{dx} within t_{f1} [see Eq. (16)], it is easily shown that

$$\lim_{t\to\infty} \varphi_1(t) = p_{dx}, \lim_{t\to\infty} \dot{\varphi}_1(t) = 0, \tag{52}$$

which is just the result of Eq. (21). In addition, as $\ddot{r}_x(t), r_x^{(3)}(t) \to 0$ as $t \to 0$ by definition [see Eq. (16)], it is implied by substituting the result of $\dot{e}_2(t) \to 0$ into the second and third equations of Eq. (24) that

$$e_3 \to \dot{e}_2 - \delta_{\theta u} \to -\delta_{\theta u}, \dot{e}_3 \to \ddot{e}_2 - \dot{\delta}_{\theta u} \to -\dot{\delta}_{\theta u}, e_4 \to \dot{e}_3 \to -\dot{\delta}_{\theta u}, \tag{53}$$

wherein the conclusions in Eq. (52) have been employed. The results in Eq. (53) indicate that $e_3(t), \dot{e}_3(t)$ are convergent to their respective equilibriums drifted by the unactuated perturbations. Thus, the result of **Case 1** stated in the control objective is proven.

Subsequently, we proceed to prove the result of **Case 2** where the perturbation term $\delta_{\theta u}(t)$ in the unactuated dynamics is vanishing [i.e., $\delta_{\theta u} \to 0, \dot{\delta}_{\theta u} \to 0$] or negligible [i.e., $\delta_{\theta u}(t) = 0, \dot{\delta}_{\theta u}(t) = 0$]. Therefore, in such cases, it is straightforward to indicate from Eq. (53) that

$$e_3 = -g\tan\theta \to 0, e_4 = -g\dot{\theta}\sec^2\theta \to 0 \Rightarrow \theta = 0, \dot{\theta} = 0, \tag{54}$$

where the definitions in Eq. (10) and Eq. (23) have been used. According to the definition of $\varphi_1(t) = x(t) + L\theta(t)$ given in Eq. (13), the results in Eq. (52) and Eq. (54) directly yield the following conclusions:

$$\lim_{t\to\infty} x(t) = p_{dx}, \lim_{t\to\infty} \dot{x}(t) = 0. \tag{55}$$

Collecting up Eqs. (52, 54, 55), the results claimed in Eq. (22) of **Case 2** are hence proven. The entire theoretical proof for the theorem is completed.

4. Simulation verification

In this section, by using the MATLAB/Simulink software, some simulation results are included to verify the effectiveness of the proposed observer-based robust control method.

For the control objectives of the two cases stated in Eq. (21) and Eq. (22), the simulation is implemented through two groups as follows:

- **Group 1.** The perturbations in the unactuated dynamics are non-vanishing. The perturbation $d_\theta(t)$ is set as a constant value $d_\theta(t) = 1$ and a time-varying function $d_\theta(t) = 0.5\cos(0.1t)$, respectively.

http://dx.doi.org/10.5772/intechopen.71383

- **Group 2.** The perturbations in the unactuated dynamics are vanishing or negligible. The perturbation $d_\theta(t)$ is set as a time-varying function $d_\theta(t) = 1.5e^{-t}$.

For all the cases, by setting the system parameters as $M = 6\,\text{kg}$, $m = 2.5\,\text{kg}$, $L = 1.2\ \text{m}$, $g = 9.8\,\text{m/s}^2$, the controller parameters as $\lambda_2 = 10$, $\lambda_5 = 30$, $\lambda_6 = 55$, $\lambda_7 = 25$, $\alpha = 2$, $\beta = 1$, $\gamma = 0.2$, $\varepsilon = 0.01$, $k_u = 60$, $k_a = 0.1$, and the to-be-tracked trajectory in Eq. (16) as $r_x(t) = 3.5$, the simulation results are obtained and are shown in **Figures 2–4**.

Figures 2 and **3** show the simulation results of **Group 1** where the solid lines denote the simulation results and the dash lines denote the desired trajectories. In **Figure 2**, the perturbation $d_\theta(t)$ is set as a constant value $d_\theta(t) = 1$, and in **Figure 3**, the perturbation is set as a time-varying function $d_\theta(t) = 0.5\cos(0.1t)$. It can be seen from **Figures 2** and **3** that when there exist persistent (non-vanishing) perturbations in the unactuated dynamics, by applying the proposed controller, the unactuated cargo is driven to the desired destination and is kept stationary. Therefore, the objectives stated in **Case 1** [see Eq. (21)] are achieved effectively. By dealing with the robust control for crane systems when the perturbations are non-vanishing, the results of **Group 1** validate the robustness of the presented controller.

Figure 4 shows the results of **Group 2**. It is clear that the proposed observer-based robust control method can achieve the objectives stated in **Case 2** [see Eq. (22)] that both the trolley

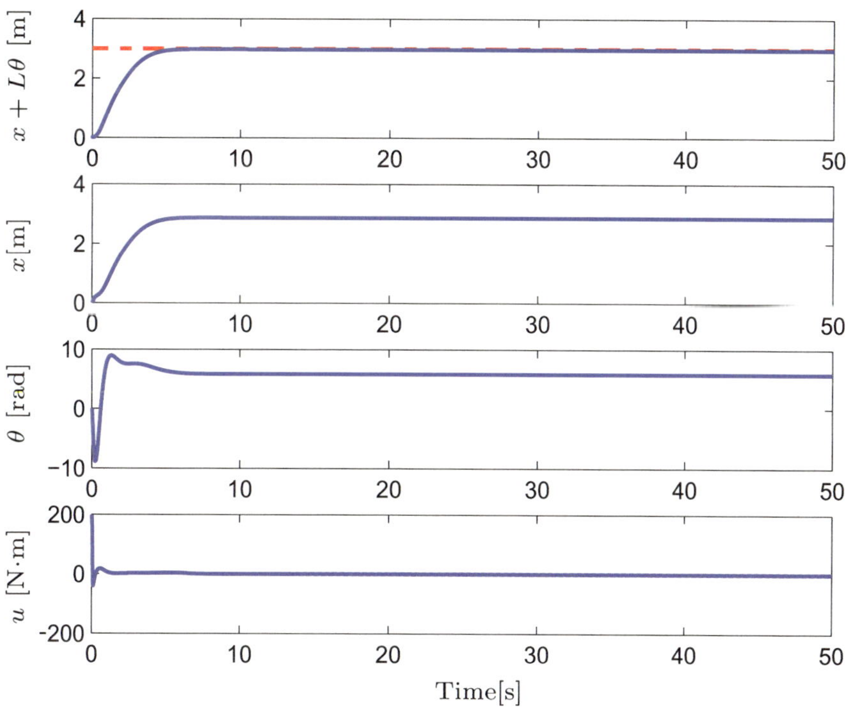

Figure 2. The simulation results of the proposed controller when $d_\theta(t) = 1$ (solid line – simulation results, dash line – desired trajectory).

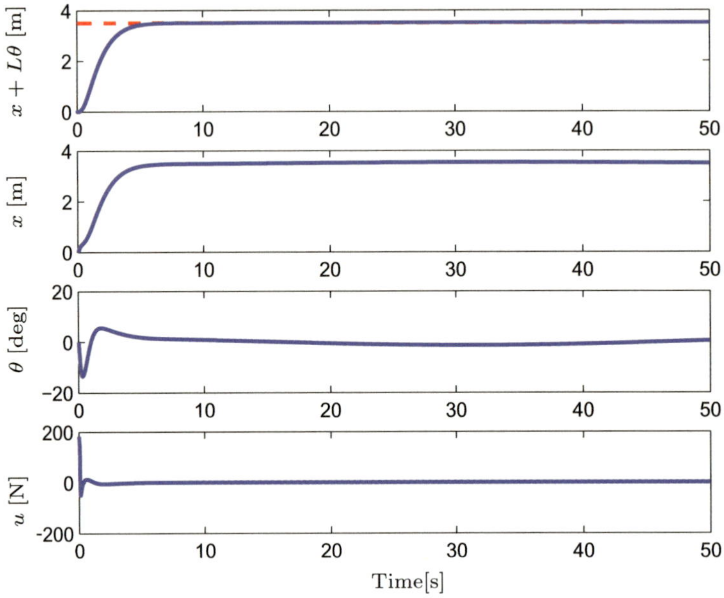

Figure 3. The simulation results of the proposed controller when $d_\theta(t) = 0.5\cos(0.1t)$ (solid line – simulation results, dash line – desired trajectory).

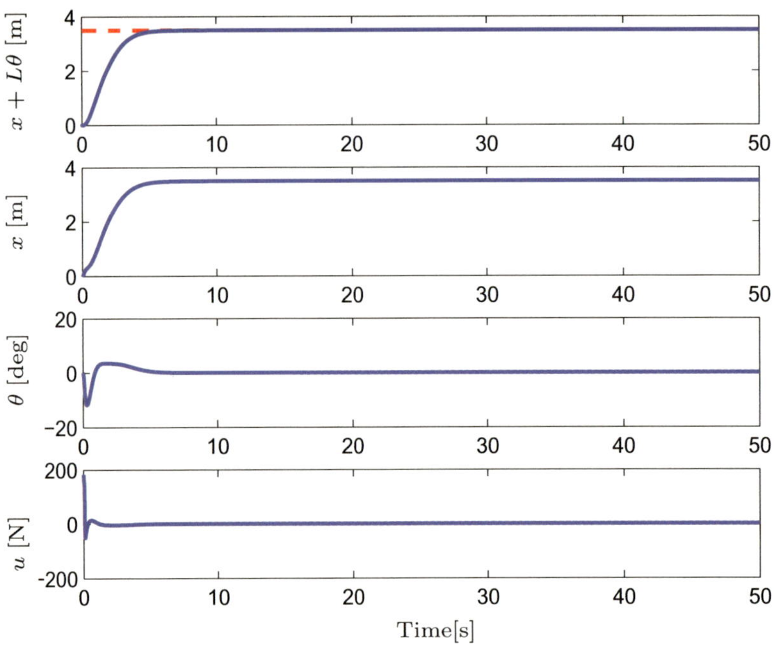

Figure 4. The simulation results of the proposed controller when $d_\theta(t) = 1.5e^{-t}$ (solid line – simulation results, dash line – desired trajectory).

and the unactuated cargo are driven to the desired destination, when there are no/negligible or vanishing perturbations $d_\theta(t) = 1.5e^{-t}$ in the unactuated dynamics.

To sum-up, the simulation results indicate that the proposed observer-based robust controller can achieve robust control in the presence of uncertainties or external perturbations, which is consistent with the theoretical analysis.

5. Concluding remarks

Considering unknown persistent perturbations in unactuated dynamics, this chapter designs an observer-based robust control method for underactuated crane systems. Specifically, a reduced-order augmented-state observer is designed to recover the lumped perturbation terms in unactuated dynamics. Further, based on the observer, a new sliding manifold is constructed to improve the robust performance of the control system. Then, the state variables are made to stay on the manifold by applying a designed robust control law in the presence of non-vanishing perturbations in unactuated dynamics. Finally, the convergence is proved in this chapter theoretically by using Lyapunov control theories. Moreover, the proposed observer-based robust controller is verified to be effective and robust by numerical simulation results.

Appendix

The system shown in Eq. (26) can be rewritten as follows:

$$\begin{aligned} \dot{\chi} &= A\chi + (e_3 - \ddot{r}_x \;\; 0 \;\; 0 \;\; \cdots \;\; 0 \;\; 0)^{\mathrm{T}}, \\ y' &= C\chi \end{aligned} \tag{56}$$

where the system variable vector $\chi(t)$ is defined as $\chi = (e_2, e_5, e_6, \cdots, e_{n+1})^{\mathrm{T}}$. The system parameter matrix $A \in \boldsymbol{R}^{(n-2)\times(n-2)}$ and $C \in \boldsymbol{R}^{(n-2)}$ are

$$\begin{aligned} A &= \begin{pmatrix} 0 & 1 & 0 & \cdots & 0 & 0 \\ 0 & 0 & 1 & \cdots & 0 & 0 \\ 0 & 0 & 0 & \cdots & 0 & 0 \\ \vdots & \vdots & \vdots & & \vdots & \vdots \\ 0 & 0 & 0 & \cdots & 1 & 0 \\ 0 & 0 & 0 & \cdots & 0 & 1 \\ 0 & 0 & 0 & \cdots & 0 & 0 \end{pmatrix}, \\ C &= (1 \;\; 0 \;\; 0 \;\; \cdots \;\; 0 \;\; 0). \end{aligned} \tag{57}$$

Then, considering the observability criteria, we can first derive the observable matrix Ψ as follows:

$$\Psi = \begin{pmatrix} C \\ CA \\ CA^2 \\ \vdots \\ CA^{(n-3)} \end{pmatrix} = \begin{pmatrix} 1 & 0 & 0 & \cdots & 0 & 0 \\ 0 & 1 & 0 & \cdots & 0 & 0 \\ 0 & 0 & 0 & \cdots & 0 & 0 \\ \vdots & \vdots & \vdots & & \vdots & \vdots \\ 0 & 0 & 0 & \cdots & 1 & 0 \\ 0 & 0 & 0 & \cdots & 0 & 1 \\ 0 & 0 & 0 & \cdots & 0 & 0 \end{pmatrix}. \tag{58}$$

It is clear that $\Psi \in \boldsymbol{R}^{(n-2)\times(n-2)}$, whose rank is rank$(\Psi) = \mathrm{A} = n-2$. Thus, the system shown in Eq. (26) is observable.

Author details

Yiming Wu[1,2*], He Chen[1,2] and Tong Yang[1,2]

*Address all correspondence to: wuyiming_1@126.com

1 Institute of Robotics and Automatic Information Systems, College of Computer and Control Engineering, Nankai University, Tianjin, China

2 Tianjin Key Laboratory of Intelligent Robotics, Nankai University, Tianjin, China

References

[1] Sun N, Wu Y, Fang Y, Chen H. Nonlinear stabilization control of multiple-RTAC systems subject to amplitude-restricted actuating torques using only angular position feedback. IEEE Transactions on Industrial Electronics. 2017;**64**(7):3084-3094

[2] Sun N, Wu Y, Fang Y, Chen H, Lu B. Nonlinear continuous global stabilization control for underactuated RTAC systems: Design, analysis, and experimentation. IEEE/ASME Transactions on Mechatronics. 2017;**22**(2):1104-1115

[3] Xin X, Liu Y. Reduced-order stable controllers for two-link underactuated planar robots. Automatica. 2013;**49**(7):2176-2183

[4] Xia D, Wang L, Chai T. Neural-network friction compensation based energy swing-up control of Pendubot. IEEE Transactions on Industrial Electron. 2014;**61**(3):1411-1423

[5] Cui R, Guo J, Mao Z. Adaptive backstepping control of wheeled inverted pendulums models. Nonlinear Dynamics. 2015;**79**(1):501-511

[6] Yang C, Li Z, Li J. Trajectory planning and optimized adaptive control for a class of wheeled inverted pendulum vehicle models. IEEE Transactions on Cybernetics. 2013;**43**(1):24-36

[7] Kato K, Wada M. Kinematic analysis and simulation of active-caster robotic drive with ball transmission (ACROBAT-S). Advanced Robotics. 2017;**31**(7):355-367

[8] Oka H, Maruki Y, Suemitsu H, Matsuo T. Nonlinear control for rotational movement of cart-pendulum system using homoclinic orbit. International Journal of Control, Automation and Systems. 2016;**14**(5):1270-1279

[9] Inoue Y, Hirama T, Wada M. Design of omnidirectional mobile robots with ACROBAT wheeled mechanisms. In: IEEE/RSJ International Conference on Intelligent Robots and Systems (IROS), 3-7 Nov, Tokyo, Japan. 2013. p. 4852-4859

[10] Sun N, Fang Y, Chen H, Fu Y, Lu B. Nonlinear stabilizing control for ship-mounted cranes with disturbances induced by ship roll and heave movements: Design, analysis, and experiments. IEEE Transactions on Systems, Man, and Cybernetics: Systems. DOI: 10.1109/TSMC.2017.2700393

[11] Sun N, Fang Y, Wu Y, Lu B. Nonlinear antiswing control of offshore cranes with unknown parameters and persistent ship-induced perturbations: Theoretical design and hardware experiments. IEEE Transactions on Industrial Electronics.

[12] Singhose W, Kim D, Kenison M. Input shaping control of double-pendulum bridge crane oscillations. Journal of Dynamic Systems, Measurement, and Control. 2008;**130**(3):1-7

[13] Grrido S, Abderrahim M, Gimnez A, Diez R, Balaguer C. Anti-swinging input shaping control of an automatic construction crane. IEEE Transactions on Automation Science and Engineering. 2008;**5**(3):549-557

[14] Maghsoudi MJ, Mohamed Z, Tokhi MO, Husain AR, Abitdin MSZ. Control of a gantry crane using input-shaping schemes with distributed delay. Transactions of the Institute of Measurement and Control. 2017;**39**(3):361-370

[15] Maghsoudi MJ, Mohamed Z, Sudin S, Buyamin S, Jaafar HI, Ahmad SM. An improved input shaping design for an efficient sway control of a nonlinear 3D overhead crane with friction. Mechanical Systems and Signal Processing. 2017;**92**:364-378

[16] Sun N, Fang Y, Zhang X. Energy coupling output feedback control of 4-DOF underactuated cranes with saturated inputs. Automatica. 2013;**49**(5):1318-1325

[17] Sun N, Fang Y. New energy analytical results for the regulation of underactuated overhead cranes: An end-effector motion-based approach. IEEE Transactions on Industrial Electronics. 2012;**59**(12):4723-4734

[18] Tuan LA, Lee SG, Dang VH, Moon SC, Kim BS. Partial feedback linearization control of a three-dimensional overhead crane. International Journal of Control, Automation and Systems. 2013;**11**(4):718-727

[19] Tuan LA, Kim GH, Kim MY, Lee SG. Partial feedback linearization control of overhead cranes with varying cable lengths. International Journal of Precision Engineering and Manufacturing. 2012;**13**(4):507-507

[20] Tuan LA, Moon SC, Lee WG, Lee SG. Adaptive sliding mode control of the overhead crane with varying cable length. Journal of Mechanical Science and Technology. 2013;**27**(3):885-893

[21] Lee LH, Huang CH, Ku SC, Yang ZH, Chang CY. Efficient visual feedback method to control a three-dimensional overhead crane. IEEE Transactions on Industrial Electronics. 2014;**61**(8):4073-4083

[22] Almutairi NB, Zribi M. Sliding mode control of a three-dimensional overhead crane. Journal of Vibration and Control. 2009;**15**(11):1679-1730

[23] Xi Z, Hesketh T. Discrete time integral sliding mode control for overhead crane with uncertainties. IET Control Theory and Applications. 2010;**4**(10):2071-2081

[24] Bartolin G, Pisano A, Usai E. Second-order sliding-mode control of container cranes. Automatica. 2002;**38**:1783-1790

[25] Ngo QH, Hong KS. Adaptive sliding mode control of container cranes. IET Control Theory and Applications. 2012;**6**(5):662-668

[26] Sun N, Fang Y, Chen H, He B. Adaptive nonlinear crane control with load hoisting/lowering and unknown parameters: Design and experiments. IEEE/ASME Transactions on Mechatronics. 2015;**20**(5):2107-2119

[27] Park MS, Chwa D, Eom M. Adaptive sliding-mode antisway control of uncertain overhead cranes with high-speed hoisting motion. IEEE Transactions on Fuzzy Systems. 2014;**22**(5):1262-1271

[28] Sun N, Fang, Lu B, Fu Y. Slew/translation positioning and swing suppression for 4-DOF tower cranes with parametric uncertainties: Design and hardware experimentation. IEEE Transactions on Industrial Electronics. 2016;**63**(10):6407-6418

[29] Zhao Y, Gao H. Fuzzy-model-based control of an overhead crane with input delay and actuator saturation. IEEE Transactions on Fuzzy Systems. 2012;**20**(1):181-186

[30] Liu D, Yi J, Zhao D, Wang W. Adaptive sliding mode fuzzy control for a two-dimensional overhead crane. Mechatronics. 2005;**15**(5):505-522

[31] Nakazono K, Ohnishi K, Kinjo H, Yamamoto T. Load swing suppression for rotary crane system using direct descent controller optimized by genetic algorithm. Transactions of the Institute of Systems, Control and Information Engineers. 2011;**22**(8):303-310

[32] Lee LH, Huang PH, Shih YC, Chiang TC, Chang CY. Parallel neural network combined with sliding mode control in overhead crane control system. Journal of Vibration and Control. 2012;**20**(5):749-760

[33] Uchiyama N, Ouyang H, Sano S. Simple rotary crane dynamics modelling and open-loop control for residual load sway suppression by only horizontal boom motion. Mechatronics. 2013;**23**(8):1223-1236

[34] Sun N, Fang Y, Zhang X, Yuan Y. Transportation task-oriented trajectory planning for underactuated overhead cranes using geometric analysis. IET Control Theory and Applications. 2012;**6**(10):1410-1423

[35] Wu X, Xia X. Optimal motion planning for overhead cranes. IET Control Theory and Applications. 2014;**8**(17):1833-1842

[36] Sun N, Fang Y, Zhang Y, Ma B. A novel kinematic coupling-based trajectory planning method for overhead cranes. IEEE/ASME Transactions on Mechatronics. 2012;**17**(1):166-173

[37] Sun N, Wu Y, Fang Y, Chen H. Nonlinear antiswing control for crane systems with double-pendulum swing effects and uncertain parameters: Design and experiments. IEEE Transactions on Automation Science and Engineering. In press. DOI: 10.1109/TASE.2017.2723539

[38] Sun N, Fang Y, Chen H, Lu B. Amplitude-saturated nonlinear output feedback antiswing control for underactuated cranes with double-pendulum cargo dynamics. IEEE Transactions on Industrial Electronics. 2017;**64**(3):2135-2146

[39] Khalil HK. Nonlinear Systems. 3rd ed. Prentice Hall: Englewood Cliffs, NJ; 2002

Chapter 16

Nonlinear Control of Flexible Two-Dimensional Overhead Cranes

Tung Lam Nguyen and Minh Duc Duong

Additional information is available at the end of the chapter

http://dx.doi.org/10.5772/intechopen.71657

Abstract

Considering gantry cable as an elastic string having a distributed mass, we constitute a dynamic model for coupled flexural overhead cranes by using the extended Hamilton principle. Two kinds of nonlinear controllers are proposed based on the Lyapunov stability and its improved version entitled barrier Lyapunov candidate to maintain payload motion in a certain defined range. With such a continuously distributed model, the finite difference method is utilized to numerically simulate the control system. The results show that the controllers work well and the crane system is stabilized.

Keywords: overhead cranes, finite difference method, Lyapunov stability, distributed modeling

1. Introduction

Nowadays, cargo transportation plays an important role in many industrial fields. For carrying the cargo in short distance or small area, such as in automotive factories and shipyards, the overhead cranes are naturally applied. To increase productivity, the overhead cranes today are required in high-speed operation. However, the fast motion of overhead cranes usually leads to the large swings of cargo and non-precise movements of trolley and bridge. The faster the cargo transport is, the larger the cargo swings. This makes dangerous and unsafe situation during the operating process. The crane itself and the concerning equipment in the factory can be damaged without proper control strategies.

In recent decades, the control problems of overhead cranes in both theory and practice have attracted many researchers. Various kinds of crane control techniques have been applied from classical methods such as linear control [1], nonlinear control [2, 5, 6], optimal approach [7], adaptive algorithms [8, 9] to modern techniques such as fuzzy logic [3, 4, 10], neural network [11], command shaping [12], and so on.

The abovementioned researches deal with crane motion modeled as pendulum or multi-section pendulum systems. As a result, their dynamics are described as an ordinary differential equation or a system of ordinary differential equations. In practice, the crane rope exhibits a certain degree of flexibility; hence, the equation of motions of the gantry crane with flexible rope is represented by a set of partial differential and ordinary differential equations. In [13–15], the authors successfully design a controller that can stabilize the system with the rope flexibility. Flexible rope also is considered in [16, 17] where coupled longitudinal-transverse motion and 3D model are investigated.

This chapter accesses the modeling and control of overhead cranes according to the other research direction. We construct a distributed model of overhead cranes in which the mass and the flexibility of payload suspending cable are fully taken into account. We utilize the analytical mechanics including Hamilton principle for constructing such the mathematical model. With the received model, we analyze and design two nonlinear control algorithms based on two versions of Lyapunov stability: one is the so-called traditional Lyapunov function and the other is the so-called barrier Lyapunov. Dissimilar to the preceding study [18, 19] whereas the problem of actuated payload positioning system is considered, the proposed controllers track the trolley to destination precisely while keeping the payload swing small during the transport process and absolutely suppressed at the payload destination with control forces exerted at the trolley end of the system. The quality of control system is investigated by numerical simulation. Since the system dynamics is characterized by a distributed mass model, the finite difference method is applied to simulate the system responses in MATLAB® environment.

The chapter content is structured as follows. Section 2 constructs a distributed mass model of overhead cranes. Section 3 analyzes and designs two nonlinear controllers based on Lyapunov direct theory. The analysis of system stability is included. Section 4 numerically simulates the system responses and analyzes the received results. Finally, the remarks and conclusions are shown in Section 5.

2. Distributed mass modeling of overhead cranes

Let us constitute a mathematical model for overhead cranes fully considering the flexibility and mass of cable. In other words, payload handling cable with length L is considered as a distributed mass string with density ρ (kg/m). An overhead crane with its physical features is depicted in **Figure 1**. The trolley with mass M (kg) handling the payload m (kg) moves along Ox which can induce the payload swing. The force Fx (N) of motor is created to push the trolley but guaranteeing the payload oscillation as small as possible. The other parameters can be seen in **Figures 1** and **2**.

Before carrying system modeling, we assume that:

1. Moving masses at the trolley end are symmetrical in X and Y directions.
2. The gantry moving in XY plane and the rope length are unchanged.

http://dx.doi.org/10.5772/intechopen.71657

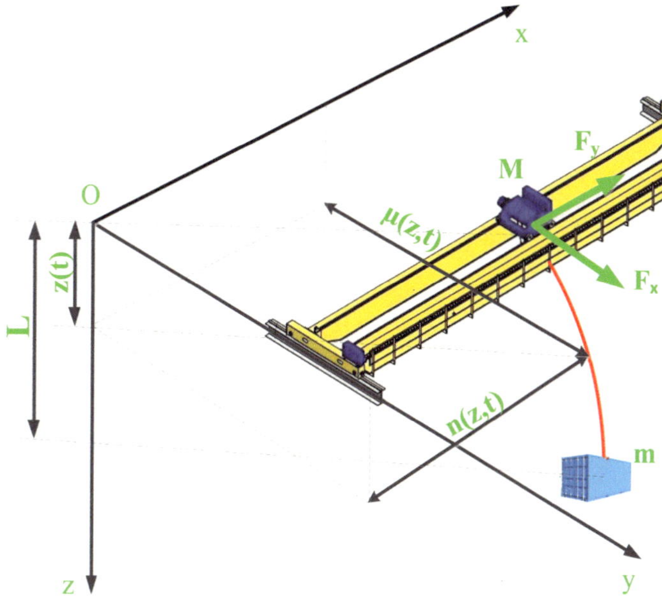

Figure 1. A practical overhead crane.

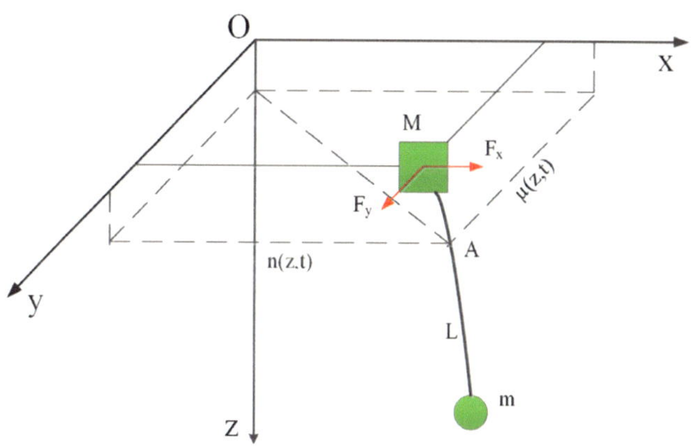

Figure 2. Physical modeling of overhead crane in OXYZ.

3. Friction and external distributed forces are neglected.

4. Longitudinal deformation of the crane rope is negligible.

From this point onward, the argument (z, t) is dropped whenever it is not confusing and $(\bullet)_t$, $(\bullet)_{tt}$, $(\bullet)_t$, and $(\bullet)_{zz}$ are used to denoted the first and second time and spatial derivatives of $(\bullet)$, respectively. We consider the physical model of an overhead crane as shown in **Figure 2**. The tension of the handing cable is of the form

$$P = g[\rho(L - z) + m] \tag{1}$$

With the differential derivation along the cable length L, the potential energy due to the elasticity of cable and gravity is determined by

$$U = \frac{1}{2}\int_0^L P(n_z^2 + \mu_z^2)dz + \frac{1}{2}EA\int_0^L \left[\frac{1}{2}(n_z^2 + \mu_z^2)\right]^2 + P_0 \tag{2}$$

where $\frac{1}{2}EA\int_0^L \left[\frac{1}{2}(n_z^2 + \mu_z^2)\right]^2$ is a potential component due to the axial deformation of the cable. The kinetic energy of system includes those of the trolley, payload, and cable motion described by

$$T = \frac{1}{2}\int_0^L \rho(n_t^2 + \mu_t^2)dz + \frac{1}{2}M(n_t^2(0,t) + \mu_t^2(0,t)) + \frac{1}{2}m(n_t^2(L,t) + \mu_t^2(L,t)) \tag{3}$$

With two force components to move trolley and bridge F_x and F_y, the total visual works of system are in the form of

$$W = F_x n(0) + F_y \mu(0) \tag{4}$$

Using the generalized form of Hamilton principle, one has the following equation:

$$H = \int_{t_1}^{t_2} (\delta T - \delta U + \delta W)dt = 0 \tag{5}$$

in which the small variations of kinematic and potential energies, respectively, are described by

$$\delta T = \delta\left[\frac{1}{2}\int_0^L \rho(n_t^2 + \mu_t^2)dz + \frac{1}{2}M(n_t^2(0,t) + \mu_t^2(0,t)) + \frac{1}{2}m(n_t^2(L,t) + \mu_t^2(L,t))\right] \tag{6}$$

$$\delta U = \delta\left[\frac{1}{2}\int_0^L P(n_z^2 + \mu_z^2)dz\right] + \delta\left\{\frac{1}{2}EA\int_0^L \left[\frac{1}{2}(n_z^2 + \mu_z^2)\right]^2 dz\right\} \tag{7}$$

and the small derivation of virtual work is written as

$$\delta W = F_x \delta n(0,t) + F_y \delta\mu(0,t) \tag{8}$$

First, one obtains

$$\delta\left[\frac{1}{2}\int_0^L \rho(n_t^2 + \mu_t^2)dz\right] - \delta\left[\frac{1}{2}\int_0^L P(n_z^2 + \mu_z^2)dz\right] - \delta\left\{\frac{1}{2}EA\int_0^L \left[\frac{1}{2}(n_z^2 + \mu_z^2)\right]^2 dz\right\}$$

We define L_c as a multivariable function

http://dx.doi.org/10.5772/intechopen.71657

$$L_c = \frac{1}{2}\rho\left(n_t^2 + \mu_t^2\right) - \frac{1}{2}P\left(n_z^2 + \mu_z^2\right) - \frac{1}{2}EA.\frac{1}{4}\left(n_z^2 + \mu_z^2\right)^2 = L_c\left(t : n_t, \mu_t, n_z, \mu_z\right) \tag{9}$$

and apply the following property:

$$\delta\int_0^L L_c dz = \int_0^L \delta L_c dz$$

with

$$\int_0^L \delta L_c dz = \int_0^L \left(\frac{\partial L_c}{\partial n_t}\delta(n_t) + \frac{\partial L_c}{\partial \mu_t}\delta(\mu_t) + \frac{\partial L_c}{\partial n_z}\delta(n_z) + \frac{\partial L_c}{\partial \mu_z}\delta(\mu_z)\right)dz \tag{10}$$

We calculate the components of (10) using the expressions of partial integration as follows:

$$\int_0^L \frac{\partial L_c}{\partial n_z}\delta(n_z)dz = \frac{\partial L_c}{\partial n_z}\delta(n)\Big|_0^L - \int_0^L \left(\frac{\partial L_c}{\partial n_z}\right)_z \delta(n)dz \tag{11}$$

$$\int_0^L \frac{\partial L_c}{\partial \mu_z}\delta(\mu_z)dz = \frac{\partial L_c}{\partial \mu_z}\delta(\mu)\Big|_0^L - \int_0^L \left(\frac{\partial L_c}{\partial \mu_z}\right)_z \delta(\mu)dz \tag{12}$$

Inserting (11) and (12) into (10) leads to

$$\begin{aligned}\int_0^L \delta L_c dz = &\int_0^L \left[\frac{\partial L_c}{\partial n_t}\delta(n_t) + \frac{\partial L_c}{\partial \mu_t}\delta(\mu_t) - \left(\frac{\partial L_c}{\partial n_z}\right)_z \delta(n) - \left(\frac{\partial L_c}{\partial \mu_z}\right)_z \delta(\mu)\right]dz \\ &+ \frac{\partial L_c}{\partial n_z}\delta(n)\Big|_0^L + \frac{\partial L_c}{\partial \mu_z}\delta(\mu)\Big|_0^L\end{aligned}$$

Integrating the abovementioned equation in term of time side by side, one has $\int_{t_1}^{t_2}\left(\int_0^L \delta L_c dz\right)dt =$ $\int_{t_1}^{t_2}\left\{\int_0^L \left[\frac{\partial L_c}{\partial n_t}\delta(n_t) + \frac{\partial L_c}{\partial \mu_t}\delta(\mu_t) - \left(\frac{\partial L_c}{\partial n_z}\right)_z \delta(n) - \left(\frac{\partial L_c}{\partial \mu_z}\right)_z \delta(\mu)\right]dz + \frac{\partial L_c}{\partial n_z}\delta(n)\Big|_0^L + \frac{\partial L_c}{\partial \mu_z}\delta(\mu)\Big|_0^L\right\}dt$ due to $\frac{\partial L_c}{\partial n_z}\delta(n_t)\Big|_{t_1}^{t_2} = 0$. Similarly, one has the following results after a series of calculation

$$\int_{t_1}^{t_2}\left[\int_0^L \frac{\partial L_c}{\partial \mu_t}\delta(\mu_t)dz\right]dt = -\int_{t_1}^{t_2}\left[\int_0^L \left(\frac{\partial L_c}{\partial \mu_t}\right)_t \delta(\mu)dz\right]dt$$

which yields

$$\begin{aligned}\int_{t_1}^{t_2}\int_0^L \delta L_c dzdt = &\int_{t_1}^{t_2}\left\{\int_0^L \left[-\left(\frac{\partial L_c}{\partial n_t}\right)_t \delta n - \left(\frac{\partial L_c}{\partial \mu_t}\right)_t \delta\mu - \left(\frac{\partial L_c}{\partial n_z}\right)_z \delta n - \left(\frac{\partial L_c}{\partial \mu_z}\right)_z \delta\mu\right]dz\right. \\ &\left.+ \frac{\partial L_c}{\partial n_z}\delta n|_0^L + \frac{\partial L_c}{\partial \mu_z}\delta\mu|_0^L\right\}dt\end{aligned} \tag{13}$$

Next, let us calculate

$$\delta\left[\frac{1}{2}M\left(n_t^2(0,t)+\mu_t^2(0,t)\right)\right]+\delta\left[\frac{1}{2}m\left(n_t^2(L,t)+\mu_t^2(L,t)\right)\right]$$

with the below notations

$$\delta\left[\frac{1}{2}M\left(n_t^2(0,t)+\mu_t^2(0,t)\right)\right]=Mn_t(0,t)\delta(n_t(0,t))+M\mu_t(0,t)\delta\left(\mu_t(0,t)\right) \tag{14}$$

and

$$\delta\left[\frac{1}{2}m\left(n_t^2(L,t)+\mu_t^2(L,t)\right)\right]=mn_t(L,t)\delta(n_t(L,t))+m\mu_t(L,t)\delta\left(\mu_t(L,t)\right) \tag{15}$$

Substituting (8), (13), (14), and (15) into (5), one obtains

$$\begin{aligned}&\int_{t_1}^{t_2}\left\{\int_0^L\left[-\left(\frac{\partial L_c}{\partial n_t}\right)_t\delta(n)-\left(\frac{\partial L_c}{\partial \mu_t}\right)_t\delta(\mu)-\left(\frac{\partial L_c}{\partial n_z}\right)_z\delta(n)-\left(\frac{\partial L_c}{\partial \mu_z}\right)_z\delta(\mu)\right]dz\right.\\&+\frac{\partial L_c}{\partial n_z}\delta(n)\Big|_0^L+\frac{\partial L_c}{\partial \mu_z}\delta(\mu)\Big|_0^L+Mn_t(0,t)\delta(n_t(0,t))+M\mu_t(0,t)\delta\left(\mu_t(0,t)\right)\\&+mn_t(L,t)\delta(n_t(L,t))+m\mu_t(L,t)\delta\left(\mu_t(L,t)\right)+F_x\delta n(0,t)+F_y\delta\mu(0,t)\}d_t=0\end{aligned} \tag{16}$$

which is simplified as

$$\left(\begin{aligned}&\int_{t_1}^{t_2}\left\{\int_0^L\left\{\left[-\left(\frac{\partial L_c}{\partial n_t}\right)_t-\left(\frac{\partial L_c}{\partial n_z}\right)_z\right]\delta(n)+\left[-\left(\frac{\partial L_c}{\partial \mu_t}\right)_t-\left(\frac{\partial L_c}{\partial \mu_z}\right)_z\right]\delta(\mu)\right\}dz\right.\\&+\frac{\partial L_c}{\partial n_z}\delta n(L,t)-\frac{\partial L_c}{\partial n_z}\delta n(0,t)+\frac{\partial L_c}{\partial \mu_z}\delta\mu(L,t)-\frac{\partial L_c}{\partial \mu_z}\delta\mu(0,t)\\&-M\delta n(0,t)n_{tt}(0,t)-M\delta\mu(0,t)\mu_{tt}(0,t)-\mathrm{m}\delta n(L,t)n_{tt}(L,t)\\&-\mathrm{m}\delta\mu(L,t)\mu_{tt}(L,t)+F_x\delta n(0,t)+F_y\delta\mu(0,t)\}dt\end{aligned}\right)=0 \tag{17}$$

Consider the following boundaries at x = 0 and x = L:

$$\begin{aligned}&\left(\frac{\partial L_c}{\partial n_t}\right)_t+\left(\frac{\partial L_c}{\partial n_z}\right)_z=0;\ \left(\frac{\partial L_c}{\partial \mu_t}\right)_t+\left(\frac{\partial L_c}{\partial \mu_z}\right)_z=0;\ \frac{\partial L_c}{\partial n_z}-mn_{tt}(L,t)=0;\\&\frac{\partial L_c}{\partial \mu_z}-m\mu_{tt}(L,t)=0;\ \frac{\partial L_c}{\partial n_z}-Mn_{tt}(0,t)+F_x=0;\ \frac{\partial L_c}{\partial \mu_z}-M\mu_{tt}(0,t)+F_y=0;\end{aligned} \tag{18}$$

which leads to

$$\left(\frac{\partial L_c}{\partial n_t}\right)_t=\rho n_{tt} \tag{19a}$$

and

http://dx.doi.org/10.5772/intechopen.71657

$$\frac{\partial L_c}{\partial n_z} = -Pn_z - \frac{1}{8}EA\left(4n_z^3 + 2.2n_z\mu_z^2\right) \tag{19b}$$

Submitting (18) into (19a) and (19b) in the interval [0, L] of z, one has

$$\rho n_{tt} - (Pn_z)_z - \frac{1}{2}EA\left(3n_z^2 n_{zz} + n_{zz}\mu_z^2 + 2n_z\mu_z\mu_{zz}\right) = 0 \tag{20}$$

and

$$\rho\mu_{tt} - \left[(P\mu_z)_z + \frac{1}{2}EA\left(3\mu_z^2\mu_{zz} + \mu_{zz}n_z^2 + 2n_z\mu_z n_{zz}\right)\right] = 0 \tag{21}$$

At boundary condition z = L, one obtains

$$Pn_z(L,t) + \frac{1}{2}EA\left(n_z^3(L,t) + n_z(L,t)\mu_z^2(L,t)\right) + mn_{tt}(L,t) = 0 \tag{22}$$

and

$$P\mu_z(L,t) + \frac{1}{2}EA\left(\mu_z^3(L,t) + \mu_z(L,t)n_z^2(L,t)\right) + m\mu_{tt}(L,t) = 0 \tag{23}$$

At boundary condition z = 0, one has

$$Pn_z(0,t) + \frac{1}{2}EA\left(n_z^3(0,t) + n_z\mu_z^2(0,t)\right) - Mn_{tt}(0,t) + F_x = 0 \tag{24}$$

and

$$P\mu_z(0,t) + \frac{1}{2}EA\left(\mu_z^3(0,t) + \mu_z(0,t)n_z^2(0,t)\right) - M\mu_{tt}(0,t) + F_y = 0 \tag{25}$$

In summary, the dynamic behavior of overhead crane governed a set of six nonlinear partial differential Eqs. (20), (21), (22), (23), (24), and (25), as follows:

$$\begin{cases} \rho n_{tt} - (Pn_z)_z - \frac{1}{2}EA\left(3n_z^2 n_{zz} + n_{zz}\mu_z^2 + 2n_z\mu_z\mu_{zz}\right) = 0 \\ \rho\mu_{tt} - (P\mu_z)_z - \frac{1}{2}EA\left(3\mu_z^2\mu_{zz} + \mu_{zz}n_z^2 + 2n_z\mu_z n_{zz}\right) = 0 \\ -Pn_z(L,t) - \frac{1}{2}EA\left(n_z^3(L,t) + n_z(L,t)\mu_z^2(L,t)\right) - mn_{tt}(L,t) = 0 \\ -P\mu_z(L,t) - \frac{1}{2}EA\left(\mu_z^3(L,t) - \mu_z(L,t)n_z^2(L,t)\right) - m\mu_{tt}(L,t) = 0 \\ Pn_z(0,t) + \frac{1}{2}EA\left(n_z^3(0,t) - n_z(0,t)\mu_z^2(0,t)\right) - Mn_{tt}(0,t) + F_x = 0 \\ P\mu_z(0,t) + \frac{1}{2}EA\left(\mu_z^3(0,t) + \mu_z(0,t)n_z^2(0,t)\right) - M\mu_{tt}(0,t) + F_y = 0 \end{cases}$$

The first and the second equations of the above system of equation represent dynamics of the gantry rope. Boundary conditions at load and trolley ends are given in the third, fourth, fifth, and sixth equations, respectively.

3. Lyapunov-based control design

Let us construct two nonlinear controllers using a traditional Lyapunov stability and its advanced version. In the first method, the control law is referred from the negative condition of a Lyapunov candidate $\dot{V} \leq 0$. In the second method, the Lyapunov function is determined so that it satisfies $0 < V \leq b$ with $b > 0$.

3.1. Conventional Lyapunov controller

The following theorem points out a nonlinear controller designed based on the second method of Lyapunov stability. The proposed control scheme tracks the outputs of a crane system approach to references asymptotically.

Theorem. Consider a mass distributed model of overhead crane that is described by six partial differential equations: (20) to (25). The following control law composed of two inputs:

$$
\begin{aligned}
F_x = {} & K_a\left[n_z(0,t) + \frac{EA}{2P(0)}\left(n_z^3(0,t) + n_z(0,t)\mu_z^2(0,t)\right)\right] \\
& - K_p\left(n(0,t) - \frac{q_d n(0,t)}{\sqrt{\mu^2(0,t) + n^2(0,t)}}\right) - K_d n_t(0,t)
\end{aligned}
\tag{26}
$$

and

$$
\begin{aligned}
F_y = {} & K_a\left[\mu_z(0,t) + \frac{EA}{2P(0)}\left(\mu_z^3(0,t) + \mu_z(0,t)n_z^2(0,t)\right)\right] \\
& -K_p\left(\mu(0,t) - \frac{q_d \mu(0,t)}{\sqrt{\mu^2(0,t) + n^2(0,t)}}\right) - K_d \mu_t(0,t)
\end{aligned}
\tag{27}
$$

pushes all state outputs of dynamic model (20)–(25) to reference q_d exponentially.

Proof. Define a positive Lyapunov candidate as follows:

$$
\begin{aligned}
V = {} & \frac{1}{2}\int_0^L\left\{\rho\left(n_t^2 + \mu_t^2\right) + P\left(n_z^2 + \mu_z^2\right) + \mathrm{EA}\left(\frac{1}{2}\left(n_z^2 + \mu_z^2\right)\right)^2\right\}dz + \frac{MP(0)}{2(P(0) + K_a)}\left(n_t^2(0,t) + \mu_t^2(0,t)\right) \\
& + \frac{1}{2}m\left(n_t^2(L,t) + \mu_t^2(L,t)\right) + \frac{P(0)K_p}{2(P(0) + K_a)}\left(\sqrt{n^2(0,t) + \mu^2(0,t)} - q_d\right)^2
\end{aligned}
\tag{28}
$$

where $P(0)$ is the tension force of cable at boundary $x = 0$. K_p and K_a are positive gains.

With the notations that $||w|^2 = \int_0^L \left\{ \left(n_t^2 + \mu_t^2\right) + \left(n_z^2 + \mu_z^2\right) + \left(n_z^2 + \mu_z^2\right)^2 \right\} dz + \left(n_t^2(0,t) + \mu_t^2(0,t)\right)$

$+\left(n_t^2(L,t) + \mu_t^2(L,t)\right) + \left(\sqrt{n^2(0,t) + \mu^2(0,t)} - q_d\right)^2,.$

one has

$$K_{\min}||w|^2 \le \mathrm{V}(\mathrm{t}) \le K_{\max}||w|^2$$

with

$$K_{\min} = \frac{1}{2}\min\left(\rho, P, \frac{EA}{4}, \frac{MP(0)}{P(0)+K_a}, m, \frac{P(0)K_p}{P(0)+K_a}\right)$$

and

$$K_{\max} = \frac{1}{2}\max\left(\rho, P, \frac{EA}{4}, \frac{MP(0)}{P(0)+K_a}, m, \frac{P(0)K_p}{P(0)+K_a}\right)$$

Differentiating Lyapunov function (28) with respect to time, one obtains

$$\dot{V} = \int_0^L \left[\rho\left(n_t n_{tt} + \mu_t \mu_{tt}\right) + P\left(n_z n_{zt} + \mu_z \mu_{zt}\right) + \frac{EA}{2}\left(n_z^3 n_{tz} + \mu_z^3 \mu_{zt} + n_z n_{zt} \mu_z^2 + \mu_z \mu_{zt} \mu_z^2\right)\right] dz$$

$$+\frac{MP(0)}{P(0)+K_a}\left\{\begin{array}{c} n_t(0,t)n_{tt}(0,t) + \mu_t(0,t)\mu_{tt}(0,t) \\ +\mu(0,t)\mu_t(0,t) + n(0,t)n_t(0,t) \\ +m\left[n_t(L,t)n_{tt}(L,t) + \mu_t(L,t)\mu_{tt}(L,t)\right] \\ -\left(\dfrac{q_d\left(\mu(0,t)\mu_t(0,t) + n(0,t)n_t(0,t)\right)}{\sqrt{\mu^2(0,t) + n^2(0,t)}}\right) \end{array}\right\} \quad (29)$$

Let us calculate the components of Lyapunov derivative (29). We refer from (20) and (21) that

$$\int_0^L \rho\left(n_t n_{tt} + \mu_t \mu_{tt}\right) dz = \int_0^L \left\{ n_t\left[\left(Pn_z\right)_z + \frac{1}{2}EA\left(3n_z^2 n_{zz} + n_{zz}\mu_z^2 + 2n_z \mu_z \mu_{zz}\right)\right] \right.$$
$$\left. +\mu_t\left[\left(P\mu_z\right)_z + \frac{1}{2}EA\left(3\mu_z^2 \mu_{zz} + \mu_{zz} n_z^2 + 2n_z \mu_z n_{zz}\right)\right] \right\} dz \quad (30)$$

Using partial integration

$$\int_0^L n_t\left(Pn_z\right)_z dz = n_t P n_z \Big|_0^L - \int_0^L P n_z n_{tz} dz$$

and

$$\int_0^L \mu_t (P\mu_z)_z dz = \mu_t P\mu_z \Big|_0^L - \int_0^L P\mu_z \mu_{tz} dz,$$

one obtains the following components of (30) as follows:

$$\int_0^L \frac{EA}{2} n_z^3 n_{tz} dz = \int_0^L \frac{EA}{2} n_z^3 d(n_t) = \frac{EA}{2} n_z^3 n_t \Big|_0^L - \int_0^L n_t \frac{EA}{2} 3n_z^2 n_{zz} dz$$

and

$$\int_0^L \frac{EA}{2} \mu_z^3 \mu_{tz} dz = \frac{EA}{2} \mu_z^3 \mu_t \Big|_0^L - \int_0^L \mu_t \frac{EA}{2} 3\mu_z^2 \mu_{zz} dz$$

Then,

$$\int_0^L \frac{EA}{2} (n_z n_{zt} \mu_z^2) dz = \frac{EA}{2} n_z \mu_z^2 n_t \Big|_0^L - \frac{EA}{2} \int_0^L n_t (n_{zz} \mu_z^2 + 2n_z \mu_z \mu_{zz}) dz$$

and

$$\int_0^L \frac{EA}{2} (\mu_z \mu_{zt} n_z^2) dz = \frac{EA}{2} \mu_z n_z^2 \mu_t \Big|_0^L - \frac{EA}{2} \int_0^L \mu_t (\mu_{zz} n_z^2 + 2n_z \mu_z n_{zz}) dz$$

The Lyapunov derivative (29) now becomes

$$\begin{aligned}
\dot{V} = {} & n_t P n_z \Big|_0^L + \mu_t P \mu_z \Big|_0^L + \frac{EA}{2} n_z^3 n_t \Big|_0^L + \frac{EA}{2} \mu_z^3 \mu_t \Big|_0^L + \frac{EA}{2} n_z \mu_z^2 n_t \Big|_0^L + \frac{EA}{2} \mu_z n_z^2 \mu_t \Big|_0^L \\
& + \frac{MP(0)}{P(0) + K_a} \big(n_t(0,t) n_{tt}(0,t) + \mu_t(0,t) \mu_{tt}(0,t)\big) + m\big(n_t(L,t) n_{tt}(L,t) + \mu_t(L,t) \mu_{tt}(L,t)\big) \\
& + \frac{P(0)K_p}{P(0) + K_a} \big(\mu(0,t)\mu_t(0,t) + n(0,t) n_t(0,t)\big) - \frac{P(0)K_p}{P(0) + K_a} \frac{q_d\big(\mu(0,t)\mu_t(0,t) + n(0,t) n_t(0,t)\big)}{\sqrt{\mu^2(0,t) + n^2(0,t)}}
\end{aligned} \tag{31}$$

Additionally, modification of (24) and (25) yields

$$\begin{aligned}
&\frac{MP(0)}{P(0)+K_a}\left(n_t(0,t)n_{tt}(0,t)+\mu_t(0,t)\mu_{tt}(0,t)\right)\\
&=\frac{P(0)}{P(0)+K_a}n_t(0,t)\left\{F_x+\left[P(0)n_z(0,t)+\frac{EA}{2}\left(n_z^3(0,t)+n_z(0,t)\mu_z^2(0,t)\right)\right]\right\}\\
&+\frac{P(0)}{P(0)+K_a}\mu_t(0,t)\left\{F_y+\left[P(0)\mu_z(0,t)+\frac{EA}{2}\left(\mu_z^3(0,t)+\mu_z(0,t)n_z^2(0,t)\right)\right]\right\}
\end{aligned} \tag{32}$$

Submitting (32) into (31) with a series of calculation, we obtain

$$\begin{aligned}
\dot{V}=&\frac{P(0)}{P(0)+K_a}n_t(0,t)\left\{-K_a\left[n_z(0,t)+\frac{EA}{2P(0)}\left(n_z^3(0,t)+n_z(0,t)\mu_z^2(0,t)\right)\right]\right.\\
&\left.+K_p\left(n(0,t)-\frac{q_d n(0,t)}{\sqrt{\mu^2(0,t)+n^2(0,t)}}\right)+F_x\right\}\\
&+\frac{P(0)}{P(0)+K_a}\mu_t(0,t)\left\{-K_a\left[\mu_z(0,t)+\frac{EA}{2P(0)}\left(\mu_z^3(0,t)+\mu_z(0,t)n_z^2(0,t)\right)\right]\right.\\
&\left.+K_p\left(\mu(0,t)-\frac{q_d\mu(0,t)}{\sqrt{\mu^2(0,t)+n^2(0,t)}}\right)+F_y\right\}
\end{aligned} \tag{33}$$

Substituting the control law (26) and (27) into (33) leads the Lyapunov function to

$$\dot{V}=-\frac{P(0)K_d}{P(0)+K_a}n_t^2(0,t)-\frac{P(0)K_d}{P(0)+K_a}\mu_t^2(0,t)\le 0 \tag{34}$$

With the negative definition of expression (34), we can conclude that the system is now exponential stability.

3.2. Barrier Lyapunov controller

We utilize an improved version of Lyapunov stability to design a control law for overhead cranes. The Lyapunov function is chosen so that its derivative is smaller than a positive constant. By this way, the Lyapunov candidate is selected similar to Eq. (28) but supplementing derivation of payload position $\frac{1}{2}\frac{P(0)}{P(0)+K_a}\ln\left(\frac{k_{b1}^2}{k_{b1}^2-z_1^2}\right)$. A modified version of Lyapunov candidate is the so-called barrier Lyapunov $V_1(t)$ being in the form of

$$\begin{aligned}
V_1=&\frac{1}{2}\int_0^L\left\{\rho\left(n_t^2+\mu_t^2\right)+P\left(n_z^2+\mu_z^2\right)+\mathrm{EA}\left(\frac{1}{2}\left(n_z^2+\mu_z^2\right)\right)^2\right\}dz\\
&+\frac{MP(0)}{2(P(0)+K_a)}\left(n_t^2(0,t)+\mu_t^2(0,t)\right)+\frac{1}{2}m\left(n_t^2(L,t)+\mu_t^2(L,t)\right)\\
&+\frac{P(0)K_p}{2(P(0)+K_a)}\left(\sqrt{n^2(0,t)+\mu^2(0,t)}-q_d\right)^2+\frac{1}{2}\frac{P_{(0)}}{P_{(0)}+K_a}\ln\left(\frac{k_{b1}^2}{k_{b1}^2-z_1^2}\right)
\end{aligned} \tag{35}$$

where $z_1 = \sqrt{n^2(L,t) + \mu^2(L,t)} - \sqrt{n^2(0,t) + \mu^2(0,t)}$ is relative position of payload in comparison with that of trolley. k_{b1} is a positive gain satisfying condition $k_{b1} > |z_1|$. The modification of (35) leads to

$$\begin{aligned}\dot{V}_1 = &\frac{P(0)}{P(0)+K_a} n_t(0,t)\left\{F_x - K_a\left[n_z(0,t) + \frac{EA}{2.P(0)}\left(n_z^3(0,t) + n_z(0,t)\mu_z^2(0,t)\right)\right]\right. \\ &\left. + K_p\left(n(0,t) - \frac{q_d n(0,t)}{\sqrt{\mu^2(0,t)+n^2(0,t)}}\right)\right\} \\ &+ \frac{P(0)}{P(0)+K_a}\mu_t(0,t)\left\{F_y - K_a\left[\mu_z(0,t) + \frac{EA}{2.P(0)}\left(\mu_z^3(0,t) + \mu_z(0,t)n_z^2(0,t)\right)\right]\right. \\ &\left. + K_p\left(\mu(0,t) - \frac{q_d \mu(0,t)}{\sqrt{\mu^2(0,t)+n^2(0,t)}}\right)\right\} + \frac{P(0)}{P(0)+K_a}\frac{z_1(z_1)_t}{k_{b1}^2 - z_1^2}\end{aligned} \tag{36}$$

Applying the following inequality

$$|(z_1)_t| \le K\left|\sqrt{n_t^2(0,t) + \mu_t^2(0,t)}\right|$$

or

$$z_1(z_1)_t \le |z_1(z_1)_t| = |z_1||(z_1)_t| \le k_{b1}K\sqrt{n_t^2(0,t) + \mu_t^2(0,t)}$$

with K being positive constant leads to

$$\frac{P(0)}{P(0)+K_a}\frac{z_1(z_1)_t}{k_{b1}^2 - z_1^2} \le \frac{P(0)}{P(0)+K_a}\frac{1}{k_{b1}^2 - z_1^2}k_{b1}K\sqrt{n_t^2(0,t) + \mu_t^2(0,t)} \tag{37}$$

Inserting (37) into (36) yields

$$\begin{aligned}\dot{V}_1 \le &\frac{P(0)n_t(0,t)}{P(0)+K_a}\left\{F_x - K_a\left[n_z(0,t) + \frac{EA}{2.P(0)}\left(n_z^3(0,t) + n_z(0,t)\mu_z^2(0,t)\right)\right]\right. \\ &\left. + K_p\left(n(0,t) - \frac{q_d n(0,t)}{\sqrt{\mu^2(0,t)+n^2(0,t)}}\right)\right\} + K_p\left(\mu(0,t) - \frac{q_d \mu(0,t)}{\sqrt{\mu^2(0,t)+n^2(0,t)}}\right)\Bigg\} \\ &+ \frac{P(0)\mu_t(0,t)}{P(0)+K_a}\left\{F_y - K_a\left[\mu_z(0,t) + \frac{EA}{2P(0)}\left(\mu_z^3(0,t) + \mu_z(0,t)n_z^2(0,t)\right)\right]\right. \\ &+ \frac{P(0)k_{b1}K}{(P(0)+K_a)(k_{b1}^2 - z_1^2)}\sqrt{n_t^2(0,t) + \mu_t^2(0,t)}\end{aligned} \tag{38}$$

Inserting the following inequality

$$\sqrt{n_t^2(0,t) + \mu_t^2(0,t)} \le |n_t(0,t)| + |\mu_t(0,t)|$$

or

$$\sqrt{n_t^2(0,t) + \mu_t^2(0,t)} \le n_t(0,t)\,\mathrm{sgn}\,(n_t(0,t)) + \mu_t(0,t)\,\mathrm{sgn}\,(\mu_t(0,t))$$

into (38), one obtains

$$\begin{aligned}
\dot{V}_1 \le & \frac{P(0)}{P(0)+K_a} n_t(0,t)\left\{F_x - K_a\left[n_z(0,t) + \frac{EA}{2P(0)}\left(n_z^3(0,t) + n_z(0,t)\mu_z^2(0,t)\right)\right]\right. \\
& \left. + K_p\left(n(0,t) - \frac{q_d n(0,t)}{\sqrt{\mu^2(0,t)+n^2(0,t)}}\right) + \frac{1}{k_{b1}^2 - z_1^2} k_{b1} K\,\mathrm{sgn}\,(n_t(0,t))\right\} \\
& + \frac{P(0)}{P(0)+K_a}\mu_t(0,t)\left\{F_y - K_a\left[\mu_z(0,t) + \frac{EA}{2.P(0)}\left(\mu_z^3(0,t) + \mu_z(0,t)n_z^2(0,t)\right)\right]\right. \\
& \left. + K_p\left(\mu(0,t) - \frac{q_d \mu(0,t)}{\sqrt{\mu^2(0,t)+n^2(0,t)}}\right) + \frac{1}{k_{b1}^2 - z_1^2} k_{b1} K\,\mathrm{sgn}\,(\mu_t(0,t))\right\}
\end{aligned} \tag{39}$$

To force the Lyapunov differentiation being negative, the control law with two components is structured as

$$\begin{aligned}
F_x = & K_a\left[n_z(0,t) + \frac{EA}{2.P(0)}\left(n_z^3(0,t) + n_z(0,t)\mu_z^2(0,t)\right)\right] - K_d n_t(0,t) \\
& - K_p\left(n(0,t) - \frac{q_d n(0,t)}{\sqrt{\mu^2(0,t)+n^2(0,t)}}\right) - \frac{1}{k_{b1}^2 - z_1^2} k_{b1} K\,\mathrm{sgn}\,(n_t(0,t))
\end{aligned} \tag{40}$$

and

$$\begin{aligned}
F_y = & K_a\left[\mu_z(0,t) + \frac{EA}{2.P(0)}\left(\mu_z^3(0,t) + \mu_z(0,t)n_z^2(0,t)\right)\right] - K_d \mu_t(0,t) \\
& - K_p\left(\mu(0,t) - \frac{q_d \mu(0,t)}{\sqrt{\mu^2(0,t)+n^2(0,t)}}\right) - \frac{1}{k_{b1}^2 - z_1^2} k_{b1} K\,\mathrm{sgn}\,(\mu_t(0,t))
\end{aligned} \tag{41}$$

which leads the Eq. (31) to

$$\dot{V}_1 \le -\frac{P(0)K_d}{P(0)+K_a}\left(n_t^2(0,t) + \mu_t^2(0,t)\right) \le 0 \tag{42}$$

for every positive gains $K_d > 0$ and $K_a > 0$. This implies that $V \le V(0)$. Hence, the system is now asymptotical stability.

4. Simulation and results

Consider the case that only the trolley motion is activated, we numerically simulate the distributed system dynamics (20)–(25) driven by either conventional Lyapunov-based input or barrier Lyapunov-based law. The finite difference method is applied for programing the control system in MATLAB environment. The system parameters used in simulation are composed of

$$m = 5\,\text{kg};\ M = 1\,\text{kg};\ L = 3,6,9\,\text{m};\ K_a = 200;\ K_p = 5; K_d = 42;$$

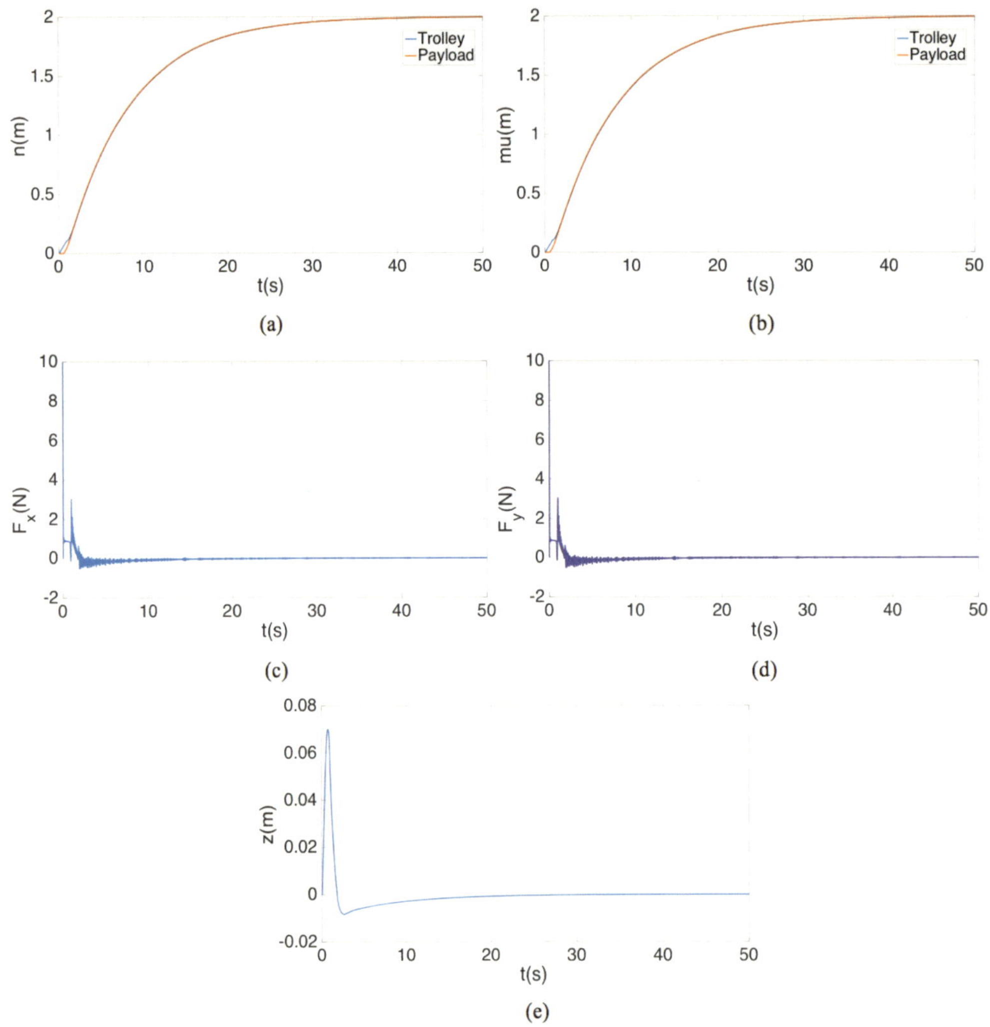

Figure 3. System responses in the case of L = 3 m and m = 3 kg.

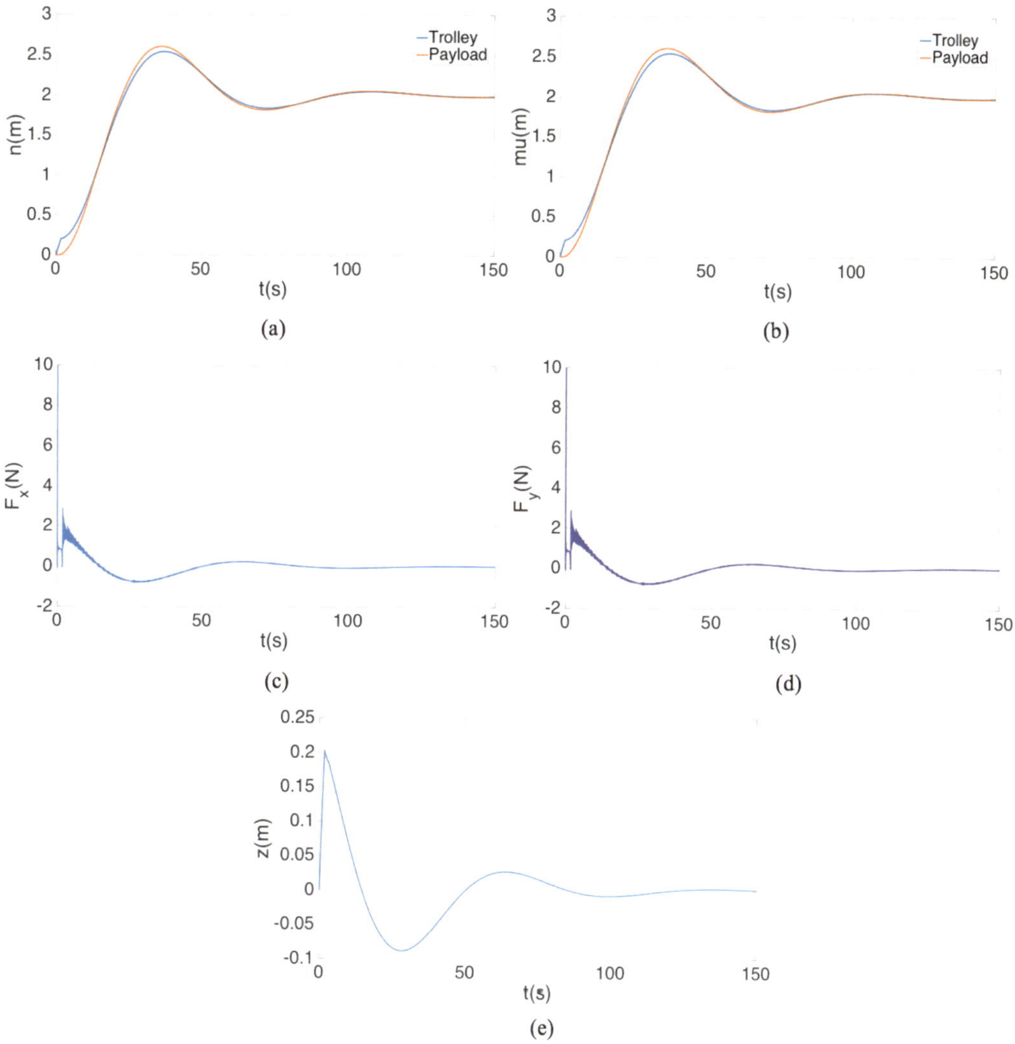

Figure 4. System responses in the case of L = 6 m and m = 6 kg.

The simulation results are depicted in **Figures 3–6**. Trolley and payload approach to destination $q_d = 2$ m precisely and speedy without maximum overshoots. The payload swing stays in a small region during the transient state and absolutely suppressed at steady state (or payload destination). However, the longer length of cable is, the lager the payload swings. The system responses show the robustness in the face of parametric uncertainty. Despite the large variation of cable length, the system responses still kept consistency as shown in **Figures 3–5**. It can be seen from **Figure 6** that with the application of the barrier Lyapunov function, payload fluctuation is controlled in an area defined by k_b. Because the motion of the trolley in X and Y directions is forced to travel the same distance to reach the desired location, system responses in X and Y directions are similar.

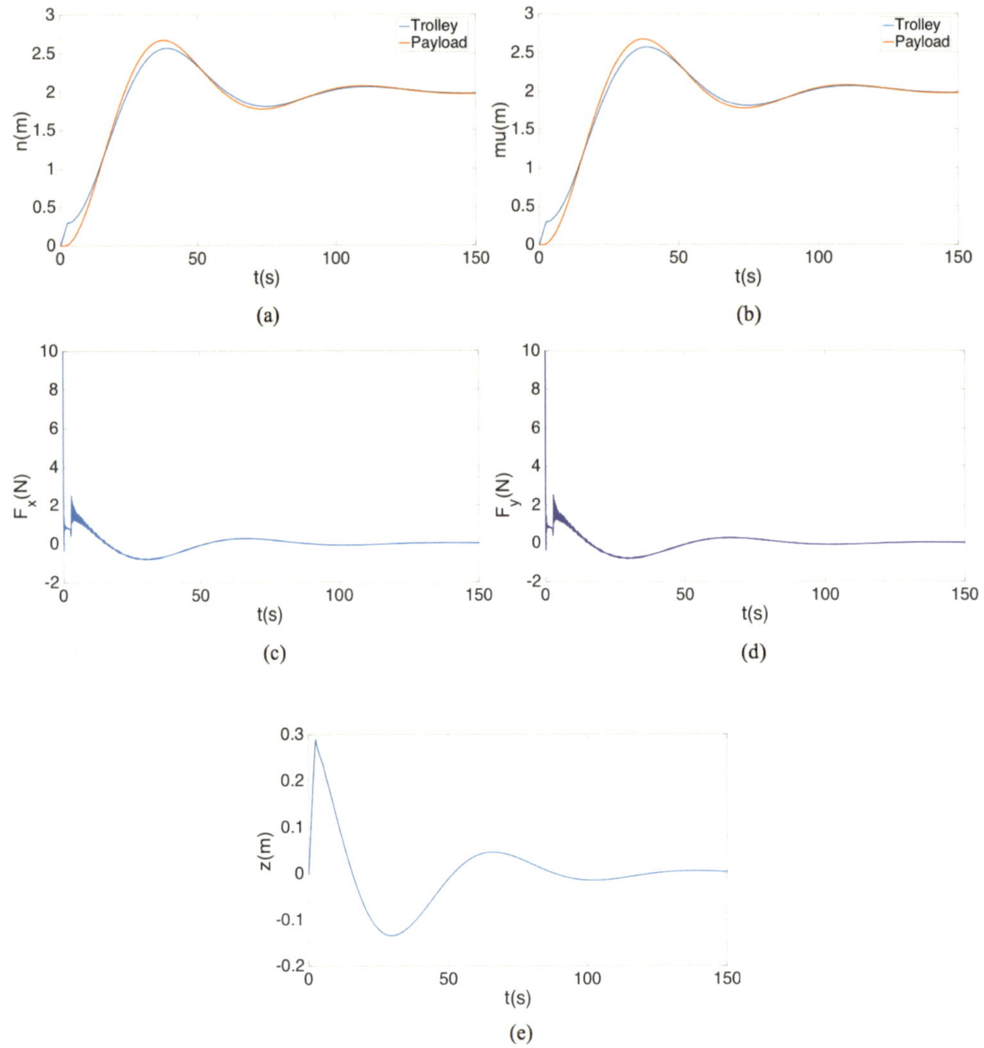

Figure 5. System responses in the case of $L = 9$ m and $m = 9$ kg with conventional Lyapunov function approach.

5. Conclusions

The dynamic model of overhead crane with distributed mass and elasticity of handling cable is formulated using the extended Hamilton's principle. Based on the model, we successfully analyzed and designed two nonlinear robust controllers using two versions of Lyapunov candidate functions. The first can steer the payload to the desired location, while the second can maintain payload fluctuation in a defined span. The proposed controllers well stabilize all

http://dx.doi.org/10.5772/intechopen.71657

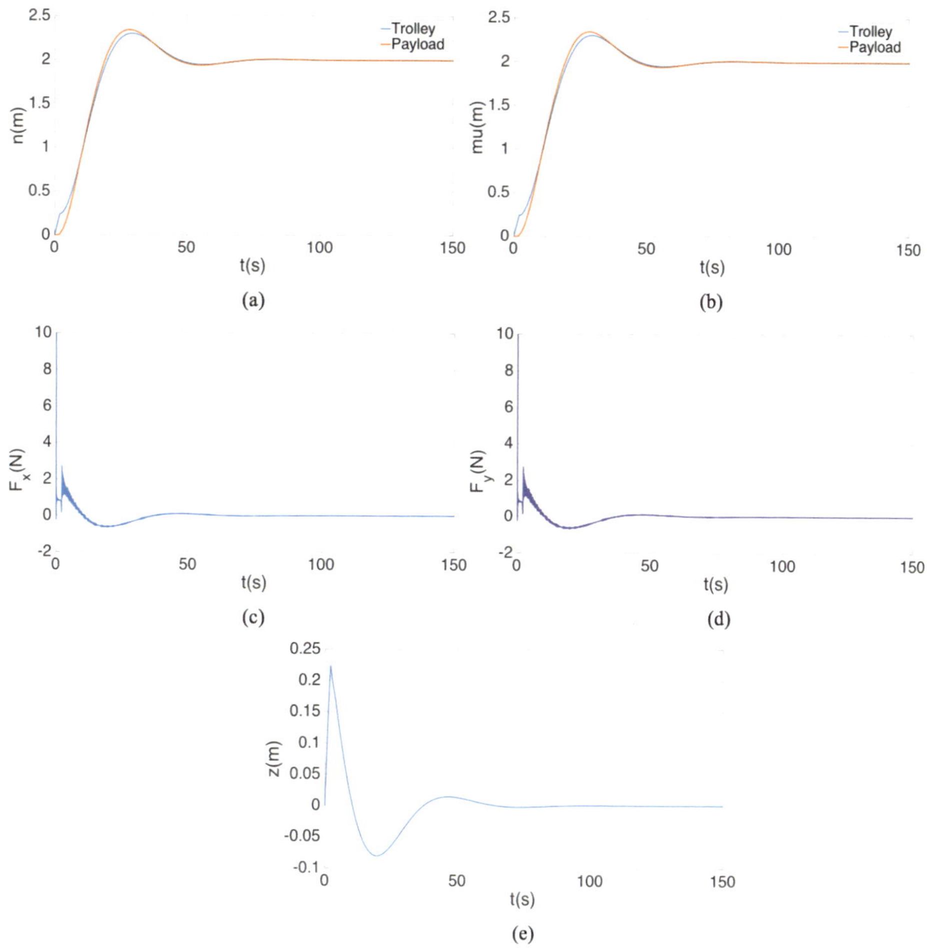

Figure 6. System responses in the case of $L = 9$ m and $m = 9$ kg with barrier Lyapunov function approach.

system responses despite the large variation of cable length and payload weight. Enhancing for 3D motion with carrying rope length will be proposed in the future studies.

Author details

Tung Lam Nguyen* and Minh Duc Duong

*Address all correspondence to: lam.nguyentung@hust.edu.vn

Department of Industrial Automation, School of Electrical Engineering, Hanoi University of Science and Technology, Hanoi, Vietnam

References

[1] Sakawa Y, Sano H. Nonlinear model and linear robust control of overhead travelling cranes. Nonlinear Analysis, Theory, Methods & Applications. 1997;**30**(4):2197-2207

[2] Su SW, Nguyen HT, Jarman R, Zhu J, Lowe DB, McLean PB, Weng K. Model predictive control of gantry crane with input nonlinearity compensation. In: International Conference on Control, Automation and Systems Engineering. 2009. pp. 312-316

[3] Park MS, Chwa D, Hong SK. Antisway tracking control of overhead cranes with system uncertainty and actuator nonlinearity using an adaptive fuzzy sliding-mode control. IEEE Transactions on Industrial Electronics. 2008;**55**(11):3972-3984

[4] Almutairi NB, Zribi M. Sliding mode control of a three-dimensional overhead crane. Journal of Vibration and Control. 2009;**15**(11):1679-1730

[5] Chwa D. Nonlinear tracking control of 3-D overhead cranes against the initial swing angle and the variation of payload weight. IEEE Transactions on Control Systems Technology. 2009;**17**(4):876-883

[6] Moustafa KAF. Reference trajectory tracking of overhead cranes. Journal of Dynamic Systems, Measurement, and Control. 2001;**123**(1):139-141

[7] Sakawa Y, Shindo Y. Optimal control of container cranes. Automatica;**18**(3):257-266

[8] Hua YJ, Shine YK. Adaptive coupling control for overhead crane systems. Mechatronics. 2007;**17**(2-3):143-152

[9] Mizumoto I, Chen T, Ohdaira S, Kumon M, Iwai Z. Adaptive output feedback control of general MIMO systems using multi-rate sampling and its application to a cart-crane system. Automatica. 2007;**43**(12):2077-2085

[10] Benhidjeb A, Gissinger GL. Fuzzy control of an overhead crane performance comparison with classic control. Control Engineering Practice. 1995;**3**(12):1687-1696

[11] Suh JH, Lee JW, Lee YJ, Lee KS. Anti-sway position control of an automated transfer crane based on neural network predictive PID controller. Journal of Mechanical Science and Technology. 2005;**19**(2):505-519

[12] Singhose W, Perter L, Kenison M, Krrikk E. Effects of hoisting on the input shaping control of gantry cranes. Control Engineering Practice. 2000;**8**(10):1159-1165

[13] Joshi S. Position control of a flexible cable gantry crane: Theory and experiment. In: Proceedings of the American Control Conference. Vol. June. 1995. pp. 2820-2824

[14] D'Andréa-Novel B, Coron JM. Exponential stabilization of an overhead crane with flexible cable via a back-stepping approach. Automatica. 2000;**36**(4):587-593

[15] Lodewijks G. Anti-sway control of container cranes as a flexible cable system. In: Proceedings of the 2004 IEEE International Conference on Control Applications. 2004. pp. 1564-1569

[16] Ge SS, Zhang S, He W. Vibration control of a coupled nonlinear string system in transverse and longitudinal directions. Proceedings of the IEEE Conference on Decision and Control. 2011;(**ii**):3742-3747

[17] Liu J-K, Qin H, He W. Modelling and vibration control for a flexible string system in three-dimensional space. IET Control Theory & Applications. 2015;**9**(16):2387-2394

[18] He W, Zhang S, Ge SS. Adaptive control of a flexible crane system with the boundary output constraint. IEEE Transactions on Industrial Electronics. 2014;**61**(8):4126-4133

[19] Tee KP, Ge SS, Tay EH. Barrier Lyapunov functions for the control of output constrained nonlinear systems. Automatica. 2009;**45**(4):918-927

Chapter 17

Robust Adaptive Control of 3D Overhead Crane System

Nga Thi-Thuy Vu, Pham Tam Thanh,
Pham Xuan Duong and Nguyen Doan Phuoc

Additional information is available at the end of the chapter

http://dx.doi.org/10.5772/intechopen.72768

Abstract

In this chapter an adaptive anti-sway controller for uncertain overhead cranes is proposed. The system model including the system uncertainties and disturbances is introduced firstly. Next, the adaptive controller which can guarantee tracking the desired position of the trolley as well as the anti-sway of the load cable is established. In this chapter, the system is proven to be input-to-state stable (ISS) which is supported by Lyapunov technique. The proposed algorithm is verified by using Matlab/Simulink simulation tool. The simulation results shown that the presented controller gives the good performances (i.e., fast transient response, position tracking, and low swing angle) when there exist system parameters variation as well as input disturbances.

Keywords: adaptive anti-swing control, input-to-state stable (ISS), overhead crane system, robust control, stability analysis, uncertainties

1. Introduction

The overhead crane system is one of the important devices in the transportation field. It includes a trolley, a driving motor, and a cable to hang the load. In the overhead crane system, there are two variables need to be controlled (the trolley position and the swing angle) but it has only one control input (acting force on the motor). This characteristic makes the control design of the overhead crane system is more difficult than full actuated system. Moreover, the operation of the system is affected by some unexpected factors such as the change of cable length and load mass, input disturbances, external disturbances. For this reason, the controller design for overhead crane system is much more challenging and attracts the consideration of many researchers.

In recent years, many controllers have been applied to the overhead crane system to move the trolley to the destination as fast as possible with acceptable swing angle. In [1–3], the PID controllers are used for the crane systems to give the good performances with simple construction. However, it is well known that PID controller is sensitive to noises and disturbances. In [4–6], the controllers based on linearized theory are introduced. Also, these controllers cannot guarantee the good performances for the system under condition of uncertain factors. In order to face with system uncertainties, many advanced controllers have been presented such as sliding mode controllers [7–13], fuzzy controllers [14–21], intelligent adaptive strategies [22] and so on.

It is well known that, robust adaptive controller is a suitable selection for the systems which are affected by working environment. In [23] an adaptive fuzzy controller is proposed for the overhead crane system to deal with nonlinear disturbances. In this scheme, the fuzzy logic controller is combined with adaptive algorithm to keep stabling for the system as well as to tune the free parameters. The given strategy is simple but robust to the variation of the system parameters (wire length and payload weight) and external disturbances. However, the stability of overall system is not presented. An adaptive sliding-mode anti-sway controller is shown in [24]. The purpose of this scheme is given the good performances for the crane system in the range of high-speed hosting motion. This algorithm includes two parts: sliding-mode controller and fuzzy observer. The first one is to keep the asymptotic stability of the sway dynamic, the other is to cope with the system uncertainties. This algorithm gives the robust anti-sway performance to overhead cranes regardless of hosting velocity and system uncertainties. The stability of the system is proven in analysis and simulation. In [25], a fuzzy sliding-mode control is incorporated with a fuzzy uncertainty observer. By this cooperation, the controller guarantees not only the anti-sway trajectory tracking of the nominal plant but also the robustness to system uncertainties as well as actuator nonlinearity. This scheme guarantees asymptotic stability and robust performances but it is quite complicated.

In this chapter a robust adaptive controller is introduced for 3D crane system. Firstly, the controller is designed based on the Euler-Lagrange model of the overhead crane system which includes the system uncertainties and external disturbances. Next, by using this controller, the error dynamic of the system is show in the form of state space model. Finally, the simulation is done to verify the effectiveness of the given algorithm. The simulation results show that the proposed controller guarantees the good tracking and no payload swing angle for the crane system even under the effect of parameters variation as well as external disturbances.

2. Robust adaptive control system design

2.1. 3D overhead crane system modeling

Figure 1 shows the structure of the 3D overhead crane. The dynamic model of the overhead crane is as follows [26]:

$$M(q)\ddot{q} + C(q,\dot{q})\dot{q} + g(q) = \tau \tag{1}$$

where

http://dx.doi.org/10.5772/intechopen.72768

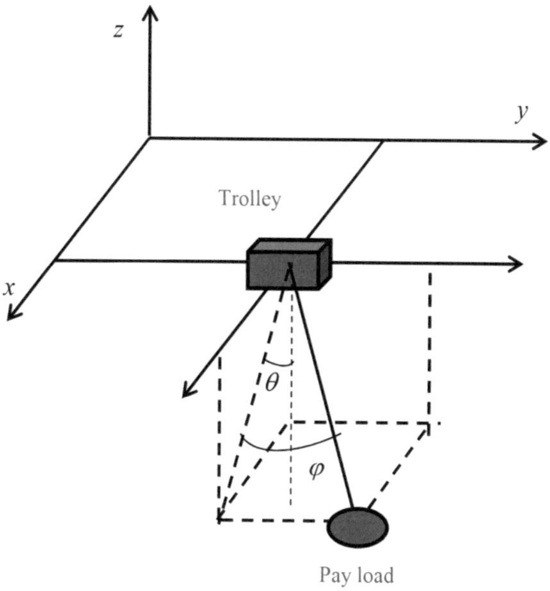

Figure 1. Structure of 3D overhead crane.

$$M(q)=\begin{pmatrix} m_c+m_h & 0 & m_h l\cos\theta\cos\varphi & -m_h l\sin\theta\sin\varphi \\ 0 & m_c+m_h+m_x & m_h l\cos\theta\sin\varphi & m_h l\sin\theta\cos\varphi \\ m_h l\cos\theta\cos\varphi & m_h l\cos\theta\sin\varphi & m_h l^2+J & 0 \\ -m_h l\sin\theta\sin\varphi & m_h l\sin\theta\cos\varphi & 0 & m_h l^2\sin^2\theta+J \end{pmatrix} \tag{2}$$

$$C(q,\dot q)=\begin{pmatrix} 0 & 0 & -m_h l\dot\theta\sin\theta\cos\varphi-m_h l\dot\varphi\cos\theta\sin\varphi & -m_h l\dot\theta\cos\theta\sin\varphi-m_h l\dot\varphi\sin\theta\cos\varphi \\ 0 & 0 & -m_h l\dot\theta\sin\theta\sin\varphi+m_h l\dot\varphi\cos\theta\cos\varphi & m_h l\dot\theta\cos\theta\cos\varphi-m_h l\dot\varphi\sin\theta\sin\varphi \\ 0 & 0 & 0 & -m_h l^2\dot\varphi\sin\theta\cos\theta \\ 0 & 0 & m_h l^2\dot\varphi\sin\theta\cos\theta & m_h l^2\dot\theta\sin\theta\cos\theta \end{pmatrix} \tag{3}$$

$$g(q)=\begin{pmatrix} 0 \\ 0 \\ m_h g l\sin\theta \\ 0 \end{pmatrix} \tag{4}$$

$$\tau=(u_1,u_2,0,0)^T,\quad q=(x,y,\theta,\varphi)^T. \tag{5}$$

In considering the system uncertainties, the model (1) is rewritten as the following:

$$M(q,d)\ddot q+C(q,\dot q,d)\dot q+g(q,d)=D(u+n(q,\dot q,\ddot q,d,t)) \tag{6}$$

where

$$D = \begin{pmatrix} I_{2\times 2} \\ 0 \end{pmatrix}, \boldsymbol{u} = (u_1, u_2)^T \tag{7}$$

The uncertain vector $d \in R^4$ includes the unknown constants in the system model and $n(q,\dot{q},\ddot{q},d,t)$ is external disturbance. In the rest of this chapter, $n(q,\dot{q},\ddot{q},d,t)$ is shorten by $n(t)$.

Model (1) is rewritten as the following:

$$\underbrace{\begin{pmatrix} M_{11}(q,d) & M_{12}(q,d) \\ M_{21}(q,d) & M_{22}(q,d) \end{pmatrix}}_{M(q,d)} \underbrace{\begin{pmatrix} \ddot{q}_1 \\ \ddot{q}_2 \end{pmatrix}}_{\ddot{q}} + \underbrace{\begin{pmatrix} C_{11}(q,\dot{q},d) & C_{12}(q,\dot{q},d) \\ C_{21}(q,\dot{q},d) & C_{22}(q,\dot{q},d) \end{pmatrix}}_{C(q,\dot{q},d)} \underbrace{\begin{pmatrix} \dot{q}_1 \\ \dot{q}_2 \end{pmatrix}}_{\dot{q}} + \underbrace{\begin{pmatrix} g_1(q,d) \\ g_2(q,d) \end{pmatrix}}_{g(q,d)} = \begin{pmatrix} u+n \\ 0 \end{pmatrix} \tag{8}$$

or

$$\begin{cases} M_{11}(q,d)\ddot{q}_1 + M_{12}(q,d)\ddot{q}_2 + C_{11}(q,\dot{q},d)\dot{q}_1 + f_1(q,\dot{q},d) = u+n \\ M_{21}(q,d)\ddot{q}_1 + M_{22}(q,d)\ddot{q}_2 + f_2(q,\dot{q},d) = 0 \end{cases} \tag{9}$$

where

$$q = \begin{pmatrix} q_1 \\ q_2 \end{pmatrix}, q_1 = (x,y)^T, \quad q_2 = (\theta,\varphi)^T \tag{10}$$

$$\begin{aligned} f_1(q,\dot{q},d) &= M_{12}(q,d)\ddot{q}_2 + C_{12}(q,\dot{q},d)\dot{q}_2 + g_1(q,d) \\ f_2(q,\dot{q},d) &= C_{21}(q,\dot{q},d)\dot{q}_1 + C_{22}(q,\dot{q},d)\dot{q}_2 + g_2(q,d) \end{aligned} \tag{11}$$

Because M(q, d) is positive definite matrix, M11(d, q) and M22(d, q) are invertible. From the second equation of (9), it can be obtained:

$$\ddot{q}_2 = -M_{22}(q,d)^{-1}\left[M_{21}(q,d)\ddot{q}_1 + f_2(q,\dot{q},d)\right] \tag{12}$$

Replacing Eq. (12) into Eq. (9) to get the following:

$$\begin{cases} M^{/}(q,d)\ddot{q}_1 + C_{11}(q,\dot{q},d)\dot{q}_1 + f^{/}(q,\dot{q},d) = u+n \\ M_{21}(q,d)\ddot{q}_1 + M_{22}(q,d)\ddot{q}_2 + f_2(q,\dot{q},d) = 0 \end{cases} \tag{13}$$

where

$$\begin{aligned} M^{/}(q,d) &= M_{11}(q,d) - M_{12}(q,d)M_{22}(q,d)^{-1}M_{21}(q,d) \\ f^{/}(q,\dot{q},d) &= f_1(q,\dot{q},d) - M_{12}(q,d)M_{22}(q,d)^{-1}f_2(q,\dot{q},d) \end{aligned} \tag{14}$$

In this paper, the following assumptions are used:

- A1: $M^{/}(\boldsymbol{q},\boldsymbol{d})$ is quadratic positive definite for all d.
- A2: $\|\boldsymbol{n}(t)\|_\infty = \sup_t |\boldsymbol{n}(t)| = \delta$ where δ is finite scalar.

http://dx.doi.org/10.5772/intechopen.72768

- A3: The relationship between the uncertainty d and the model is linear [27], i.e., the left side of Eq. (13) can be expressed as:

$$\begin{cases} M^{/}(q,d)\ddot{q}_1 + C_{11}(q,\dot{q},d)\dot{q}_1 + f^{/}(q,\dot{q},d) = F_1(q,\dot{q},\ddot{q}_1)d \\ M_{21}(q,d)\ddot{q}_1 + M_{22}(q,d)\ddot{q}_2 + f_2(q,\dot{q},d) = F_2(q,\dot{q},\ddot{q})d \end{cases} \tag{15}$$

2.2. Controller design

In this part, the following denotations are used:

$$\begin{aligned} &M^{/} = M^{/}(q,d), C_{11} = C_{11}(q,\dot{q},d), f^{/} = f^{/}(q,\dot{q},d) \\ &\widehat{M}^{/} = M^{/}\left(q,\widehat{d}\right), \widehat{C}_{11} = C_{11}\left(q,\dot{q},\widehat{d}\right), \widehat{f}^{/} = f^{/}\left(q,\dot{q},\widehat{d}\right) \\ &F_1 = F_1(q,\dot{q},\ddot{q}_1), F_2 = F_2(q,\dot{q},\ddot{q}_1) \end{aligned} \tag{16}$$

The role of the proposed controller in the system is to adapt to the constant uncertain d and robust with unknown function $n(t)$ so the error $e = q_r - q_1$, where q_r is the desired value of q_1, is bounded and converges asymptotically to 0.

The robust adaptive controller which satisfies the above requirements is obtained by the following theorem.

Theorem: Consider the system Eq. (13), the following controller:

$$u = M^{/}\left[\ddot{q}_r + K_1 e + K_2 \dot{e}\right] + C_{11}\dot{q}_1 + f^{/} + s(t) \tag{17}$$

where $K_1 = \mathrm{diag}(a), K_2 = \mathrm{diag}\left(\sqrt{(a+1)a}\right), a > 0$, and

$$\begin{cases} \dot{v} = \left(M^{/}\left(q,\widehat{d}\right)^{-1} F_1\right)^{T} (K_1, K_2)x \\ s(t) - F_1 v \end{cases} \tag{18}$$

in which $\hat{d}$, which satisfies $\max_{1 \le i \le n} \sum_{j=1}^{n} \left| m_{ij}^{/}\left(q,\widehat{d}\right)\right| \le \gamma, \forall q$ is representation of d, and $x = \mathrm{col}(e,\dot{e})$.

will converge x to the neighborhood of the are $\mathcal{O}$:

$$\mathcal{O} = \left\{ x \in \mathbb{R}^6 \middle| \, |x| < \frac{\delta\gamma}{a} \right\} \tag{19}$$

Proof: Replacing Eq. (18) into Eq. (17), the following is obtained:

$$M^{/}\ddot{q} + C_{11}\dot{q}_1 + f^{/} = u + n = \widehat{M}^{/}\left[\ddot{q}_r + K_1 e + K_2 \dot{e}\right] + \widehat{C}_{11}\dot{q}_1 + \widehat{f}^{/} + s + n \tag{20}$$

which can be rewritten as:

$$\left(M^{/}-\widehat{M}^{/}\right)\ddot{q}+\left(C_{11}-\widehat{C}_{11}\right)\dot{q}_1+\left(f^{/}-\widehat{f}^{/}\right)=\widehat{M}^{/}[\ddot{e}+K_1e+K_2\dot{e}]+s+n \tag{21}$$

By using A3, the above equation can be expressed as the following:

$$F_1\left(d-\widehat{d}\right)=\widehat{M}^{/}[\ddot{e}+K_1e+K_2\dot{e}]+s+n \tag{22}$$

or

$$\ddot{e}=-K_1e-K_2\dot{e}+\left(\widehat{M}^{/}\right)^{-1}\left[F_1\left(d-\widehat{d}\right)-s-n\right] \tag{23}$$

Equation (23) can be written in the state-space form as the following:

$$\begin{aligned}\dot{x}&=\begin{pmatrix}0 & I_{3\times3}\\ -K_1 & -K_2\end{pmatrix}x+\begin{pmatrix}0\\ \left(\widehat{M}^{/}\right)^{-1}\end{pmatrix}\left[F_1\left(d-\widehat{d}\right)-s-n\right]\\ &=Ax+B\left[F_1\left(d-\widehat{d}\right)-s-n\right]\end{aligned} \tag{24}$$

where

$$x=\begin{pmatrix}e\\ \dot{e}\end{pmatrix},\ A=\begin{pmatrix}0 & I_{3\times3}\\ -K_1 & -K_2\end{pmatrix},\ B=\begin{pmatrix}0\\ \left(\widehat{M}^{/}\right)^{-1}\end{pmatrix} \tag{25}$$

Since K_1 and K_2 are symmetric positive definite matrices, matrix A is stable, it means that all the eigenvalues of A is located in the left side of the complex plane. Consequently, the linear reference model:

$$\dot{x}_m=Ax_m \tag{26}$$

is stable. Then, $x_m(t)$ is bounded and asymptotically converges to zero as $t\rightarrow\infty$ despite the initiative value $x_m(0)$.

Next step, it will be shown that, by using the controller Eq. (17) and auxiliary controller Eq. (18), the error $(x-x_m)$ is bounded and converges to the neighborhood of the area $\mathcal{O}$ defined in Eq. (19).

From Eqs. (24) and (26), the following is obtained:

$$\begin{aligned}\dot{x}-\dot{x}_m&=A(x-x_m)+B\left[F_1\left(d-\widehat{d}\right)-s-n\right]\\ &=A(x-x_m)+B(F_1\Delta-n)\end{aligned} \tag{27}$$

where $\Delta=d-\widehat{d}-v$.

http://dx.doi.org/10.5772/intechopen.72768

Choosing the Lyapunov function as the following:

$$V = \frac{1}{2}\left[(\boldsymbol{x} - \boldsymbol{x}_m)^T \mathrm{P}(\boldsymbol{x} - \boldsymbol{x}_m) + \Delta^T \Delta\right] \tag{28}$$

The derivative of V can be expressed as:

$$\begin{aligned}\dot{V} &= \frac{1}{2}\left[(\mathrm{A}(\boldsymbol{x} - \boldsymbol{x}_m) + \mathrm{B}(\mathrm{F}_1\Delta - \boldsymbol{n}))^T \mathrm{P}(\boldsymbol{x} - \boldsymbol{x}_m) + (\boldsymbol{x} - \boldsymbol{x}_m)^T \mathrm{P}(\mathrm{A}(\boldsymbol{x} - \boldsymbol{x}_m) + \mathrm{B}(\mathrm{F}_1\Delta - \boldsymbol{n}))\right] \\ &\quad + \Delta^T \dot{\Delta} \\ &= \frac{1}{2}(\boldsymbol{x} - \boldsymbol{x}_m)^T\left(\mathrm{A}^T\mathrm{P} + \mathrm{PA}\right)(\boldsymbol{x} - \boldsymbol{x}_m) + \Delta^T\left[(\mathrm{BF}_1)^T \mathrm{P}(\boldsymbol{x} - \boldsymbol{x}_m) - \dot{\boldsymbol{v}}\right] - (\boldsymbol{x} - \boldsymbol{x}_m)^T \mathrm{PB}\boldsymbol{n}\end{aligned} \tag{29}$$

or

$$\dot{V} = -(\boldsymbol{x} - \boldsymbol{x}_m)^T \mathrm{Q}(\boldsymbol{x} - \boldsymbol{x}_m) + \Delta^T\left[(\mathrm{BF}_1)^T \mathrm{P}(\boldsymbol{x} - \boldsymbol{x}_m) - \dot{\boldsymbol{v}}\right] - (\boldsymbol{x} - \boldsymbol{x}_m)^T \mathrm{PB}\boldsymbol{n} \tag{30}$$

where

$$\begin{aligned}\mathrm{Q} &= -\frac{1}{2}\left(\mathrm{A}^T\mathrm{P} + \mathrm{PA}\right) \\ &= -\frac{1}{2}\left[\begin{pmatrix} 0 & -\mathrm{K}_1 \\ \mathrm{I}_{3\times 3} & -\mathrm{K}_2 \end{pmatrix}\begin{pmatrix} 2\mathrm{K}_1\mathrm{K}_2 & \mathrm{K}_1 \\ \mathrm{K}_1 & \mathrm{K}_2 \end{pmatrix} + \begin{pmatrix} 2\mathrm{K}_1\mathrm{K}_2 & \mathrm{K}_1 \\ \mathrm{K}_1 & \mathrm{K}_2 \end{pmatrix}\begin{pmatrix} 0 & \mathrm{I}_{3\times 3} \\ -\mathrm{K}_1 & -\mathrm{K}_2 \end{pmatrix}\right] \\ &= \begin{pmatrix} \mathrm{K}_1^2 & 0 \\ 0 & \mathrm{K}_2^2 - \mathrm{K}_1 \end{pmatrix} = \mathrm{diag}\left(a^2\right)\end{aligned} \tag{31}$$

is a symmetric positive definite matrix.

By choosing:

$$\begin{aligned}\dot{\boldsymbol{v}} = (\mathrm{BF}_1)^T \mathrm{P}(\boldsymbol{x} - \boldsymbol{x}_m) &= \left[\begin{pmatrix} 0 \\ \left(\widehat{\mathrm{M}}'\right)^{-1} \mathrm{F}_1 \end{pmatrix}\right]^T \begin{pmatrix} 2\mathrm{K}_1\mathrm{K}_2 & \mathrm{K}_1 \\ \mathrm{K}_1 & \mathrm{K}_2 \end{pmatrix}(\boldsymbol{x} - \boldsymbol{x}_m) \\ &= \left(\left(\left(\widehat{\mathrm{M}}'\right)^{-1} \mathrm{F}_1\right)^T \mathrm{K}_1, \left(\left(\widehat{\mathrm{M}}'\right)^{-1} \mathrm{F}_1\right)^T \mathrm{K}_2\right)(\boldsymbol{x} - \boldsymbol{x}_m) \\ &= \left(\left(\widehat{\mathrm{M}}'\right)^{-1} \mathrm{F}_1\right)^T (\mathrm{K}_1, \mathrm{K}_2)(\boldsymbol{x} - \boldsymbol{x}_m)\end{aligned} \tag{32}$$

then

$$\dot{V} = -(\boldsymbol{x} - \boldsymbol{x}_m)^T \mathrm{Q}(\boldsymbol{x} - \boldsymbol{x}_m) - (\boldsymbol{x} - \boldsymbol{x}_m)^T \mathrm{PB}\boldsymbol{n} \tag{33}$$

Both Eqs. (32) and (33) are always feasible with any initial values of x_m. For the simplicity, the initial value of x_m is chosen at $x_m(0) = 0$. Consequently, this leads to the following:

$$\dot{v} = \left(\left(\widehat{M}' \right)^{-1} F_1 \right)^T (K_1, K_2) \boldsymbol{x} \tag{34}$$

and

$$\begin{aligned} \dot{V} &= -\boldsymbol{x}^T Q \boldsymbol{x} - \boldsymbol{x}^T PB\boldsymbol{n} = -a^2 |\boldsymbol{x}|^2 - \boldsymbol{x}^T PB\boldsymbol{n} \\ &\leq -a^2 |\boldsymbol{x}|^2 + \|PB\| \delta |\boldsymbol{x}| \leq a[-a|\boldsymbol{x}| + \gamma\delta] |\boldsymbol{x}| \end{aligned} \tag{35}$$

This implies that as $\frac{\gamma\delta}{a} < |\boldsymbol{x}|$, i.e. when $x(t)$ is steel on the outside of the area $\mathcal{O}$, $\dot{V} < 0$ so the change of $|\boldsymbol{x}(t)|$ is monotonous decrease. This completely proves that by using the proposed controller, the trajectory x will converge to the neighborhood of the area $\mathcal{O}$.

3. Simulation verification

In order to verify the effectiveness of the proposed controller, a simulation is setup based on the MATLAB/Simulink tool. The parameters of the overhead crane system are as follow:

$m_c = 10$ kg, $m_h = 10$ kg, $m_x = 5$ kg, $l = 1.2$ m, $g = 9.8$ m/s^2.

The simulation is carried out under the three cases:

Case 1: The system parameters are nominal, no input disturbances.

Case 2: The system parameters are variation (150%), no input disturbances.

Case 3: The system parameters are nominal, existing input disturbances.

The destination positions for all cases are 1.5 m for x-axis and 2 m for y-axis, the controller gains are as follow:

$$K_1 = \begin{bmatrix} 5 & 0 \\ 0 & 5 \end{bmatrix}, \quad K_2 = \begin{bmatrix} 5.48 & 0 \\ 0 & 5.48 \end{bmatrix} \tag{36}$$

The simulation results are shown in **Figures 2–4**. In each figure, the a part is result for the x-axis and the b part is for y-axis. In addition, from top to bottom are the waveforms of trolley position, payload swing angle, and control signal, respectively.

It can be seen from **Figure 2** that, in the case of system is certainty (Case 1), the trolley reaches the destination point after 3 sec in x-axis and 4 sec in y-axis, the steady state errors are

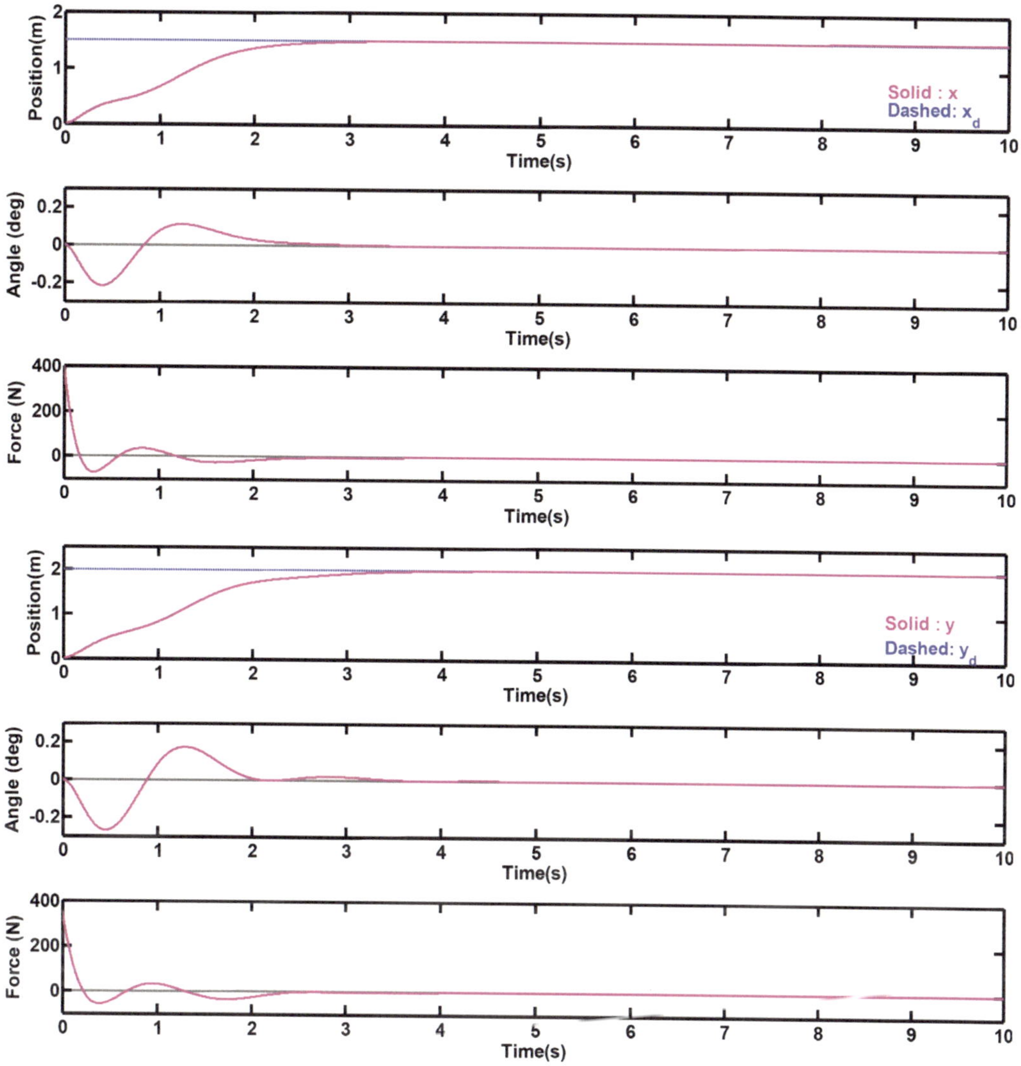

Figure 2. Simulation result of the robust adaptive controller for the case of certain system parameters.

negligible, and the payload swing quickly disappears as the trolleys finish their movements. In the **Figure 3** (Case 2), the system parameters are 150% variation but the results are nearly unchanged, i.e. the transient time is less than 5 sec, the maximum swing angle is smaller than 0.3 deg. and it is kept almost zero at the steady state.

Figure 4 is the waveform of the system under the condition of existing the external disturbances. In this case, a sinusoidal with amplitude of 2 degree is added into the inputs. The system responses are little oscillation but it is insignificant.

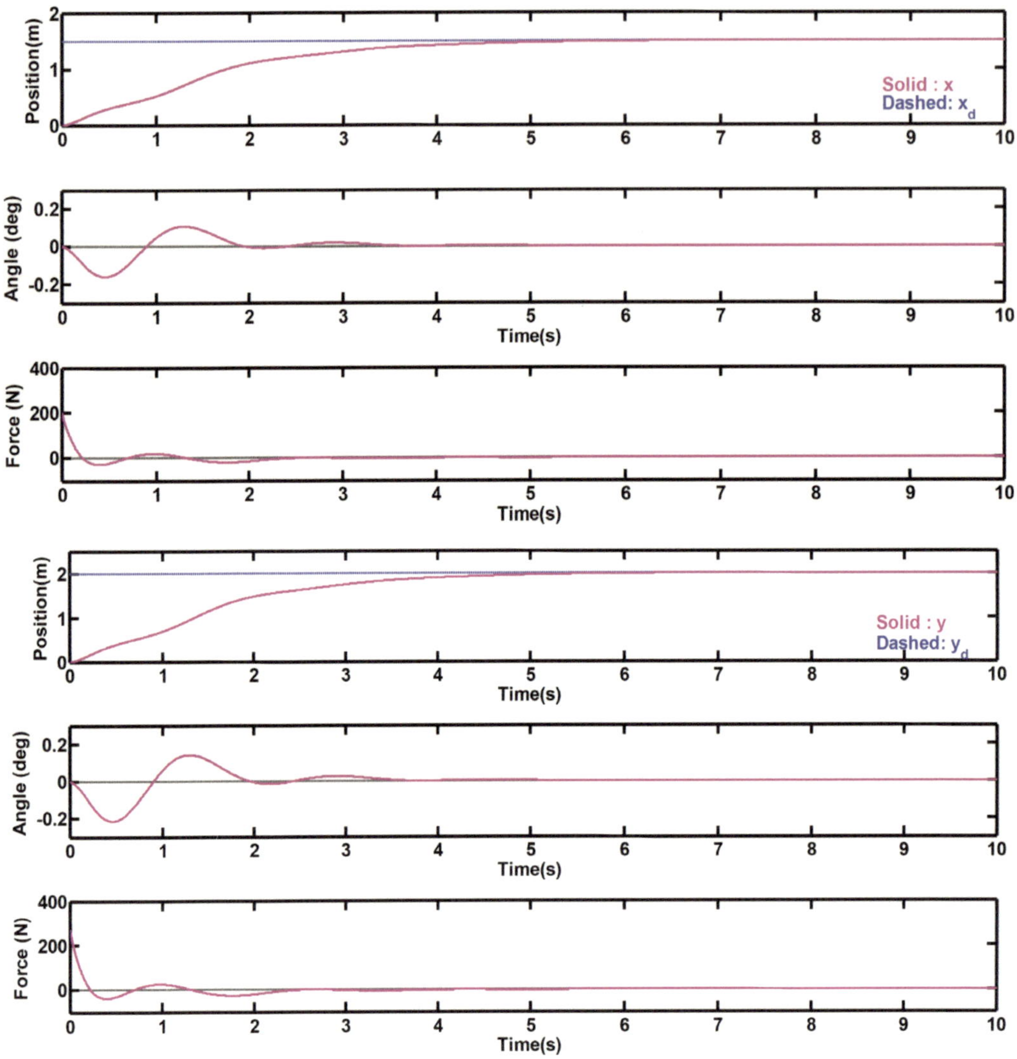

Figure 3. Simulation result of the robust adaptive controller for the case of 150% variation system parameters.

In **Table 1**, θ_{max}, φ_{max}, θ_{ss}, and φ_{ss} are maximum and steady state values of θ and φ, respectively. From the above results is can be seen that the proposed controller gives a good performance under various conditions of working. It has the ability to adapt with the uncertainties of the system such as the variation of the trolley mass, load mass, and cable length. Moreover, this controller is also robust to the external disturbance.

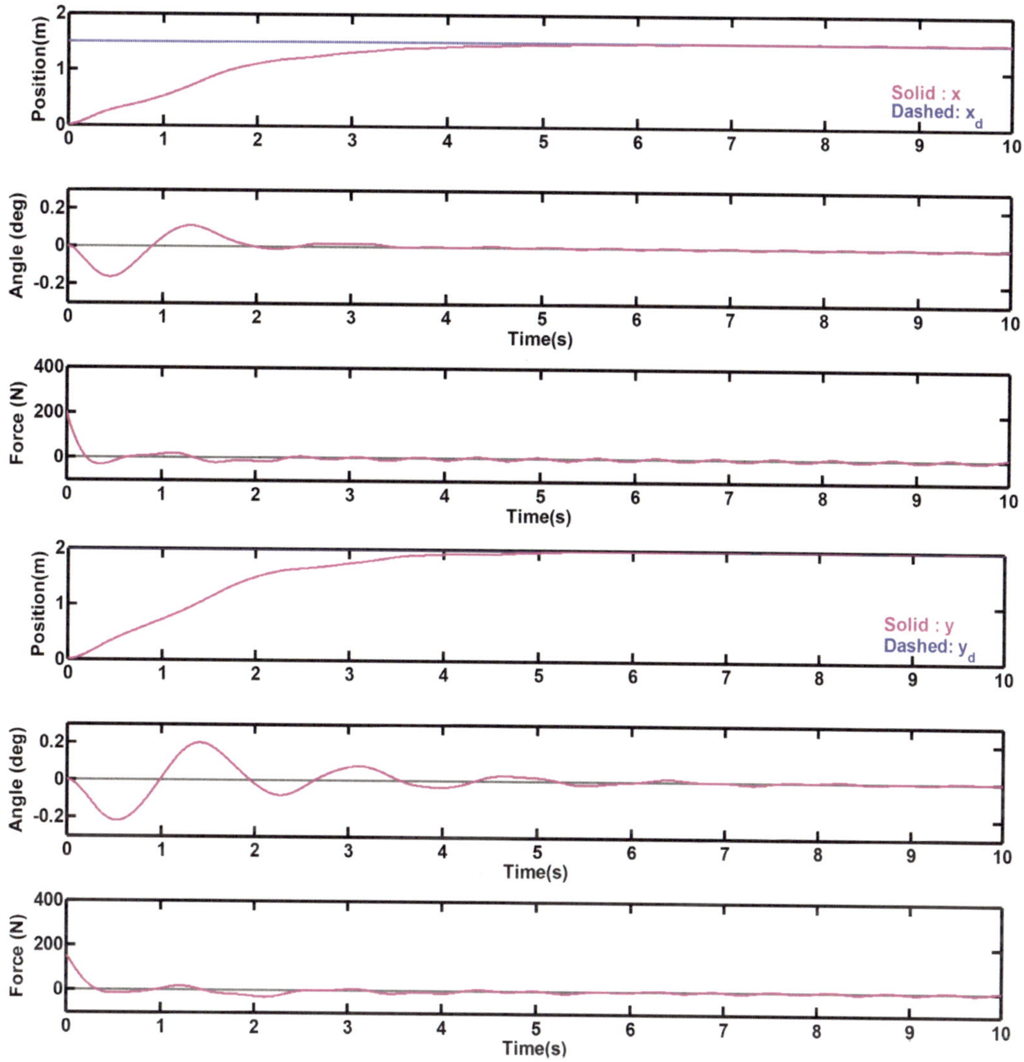

Figure 4. Simulation result of the robust adaptive controller for the case of existing external disturbances.

	θ_{max} (deg)	θ_{ss} (deg)	φ_{max} (deg)	φ_{ss} (deg)	*x*-axis settling time (sec)	*y*-axis settling time (sec)
Case 1	0.2	0	0.3	0	3	4
Case 2	0.2	0	0.25	0	5	5
Case 3	0.2	0	0.25	0	4	5

Table 1. Summarize the results for all cases.

4. Conclusion

In this chapter an adaptive robust controller which can adapt with the system uncertainties and robust to the external disturbances is establishes based on Euler–Lagrange model of the overhead crane system. Using this controller, the error dynamic of the system is show in the form of state space model. By using Lyapunov theory, it is shown that the overall system is input-to-state stable. The proposed robust adaptive controller is verified through the Matlab/Simulink toolbox under the three conditions, i.e., nominal system parameters, variation system parameters, and external disturbances. The simulation results indicate that the presented scheme gives the good performances for the overhead crane system (fast response, small swing angle in the transient time and no swing angle in the steady state, no position error) even that the system is uncertainties and existing the external disturbances.

Author details

Nga Thi-Thuy Vu[1]*, Pham Tam Thanh[2], Pham Xuan Duong[2] and Nguyen Doan Phuoc[1]

*Address all correspondence to: nga.vuthithuy@hust.edu.vn

1 Hanoi University of Science and Technology, Vietnam

2 Vietnam Maritime University, Vietnam

References

[1] Kiss B, Levine J, Mullhaupt P. A simple output feedback PD controller for nonlinear cranes. In: Proceedings of IEEE Conference Decision Control. December 2000. pp. 5097-5101

[2] Collado J, Lozano R, Fantoni I. Control of convey-crane based on passivity. In: Proceedings of American Control Conference. Chicago, IL; June 2000. pp. 1260-1264

[3] Lee H-H. A new approach for the anti-swing control of overhead cranes with high-speed load hoisting. International Journal of Control. 2003;**76**(15):1493-1499

[4] Goritov AN, Korikov AM. Optimality in robot design and control. Automation and Remote Control. 2001;**62**(7):1097-1103

[5] Hamalainen JJ, Marttinen A, Baharova L, Virkkunen J. Optimal path planning for a trolley crane: Fast and smooth transfer of load. Proceedings of Institute of Electronic Engineering Control Theory and Applications. 1995;**142**(1):51-57

[6] Piazzi A, Visioli A. Optimal dynamic-inversion-based control of an overhead crane. Proceedings of Institute of Electronic Engineering Control Theory and Applications. 2002;**149**(5): 405-411

[7] Bartolini G, Pisano A, Usai E. Second-order sliding-mode control of container cranes. Automatica. 2002;**38**(10):1783-1790

[8] Sun N, Fang Y, Chen H. A new antiswing control method for underactuated cranes with unmodeled uncertainties: Theoretical design and hardware experiments. IEEE Transactions on Industrial Electronics. 2015;**62**(1):453-465

[9] Liu D, Yi J, Zhao D, Wang W. Adaptive sliding mode fuzzy control for a two-dimensional overhead crane. Mechatronics. 2005;**15**(5):505-522

[10] Chen W, Saif M. MIMO nonlinear systems using high-order sliding-mode differentiators with application to a laboratory 3-D crane. IEEE Transactions on Industrial Electronics. 2008;**55**(11):3985-3996

[11] Ngo QH, Hong K-S. Sliding-mode antisway control of an offshore container crane. IEEE/ASME Transasction on Mechatronics. 2012;**17**(2):201-209

[12] Tuan LA, Moon S-C, Lee WG, Lee S-G. Adaptive sliding mode control of overhead cranes with varying cable length. Journal of Mechanical Science and Technology. 2013;**27**(3): 885-893

[13] Almutairi NB, Zribi M. Sliding mode control of a three-dimensional overhead crane. Journal of Vibration and Control. 2009;**15**(11):1679-1730

[14] Li C, Lee CY. Fuzzy motion control of an auto-warehousing crane system. IEEE Transactions on Industrial Electronics. 2001;**48**(5):983-994

[15] Cho SK, Lee HH. A fuzzy-logic antiswing controller for threedimensional overhead cranes. ISA Transactions. 2002;**41**(2):235-243

[16] Liang YC, Koh KK. Concise anti-swing approach for fuzzy crane control. Electronics Letters. 1997;**3**(2):167-168

[17] Chang CY. Adaptive fuzzy controller of the overhead cranes with nonlinear disturbance. IEEE Transactions on Industrial Informatics. 2007;**3**(2):164 172

[18] Zhao Y, Gao H. Fuzzy-model-based control of an overhead crane with input delay and actuator saturation. IEEE Transactions on Fuzzy Systems. 2002;**20**(1):181-186

[19] Smoczek J. Experimental verification of a GPC-LPV method with RLS and P1-TS fuzzy based estimation for limiting the transient and residual vibration of a crane system. Mechanical Systems and Signal Processing. 2015;**62-63**:324-340

[20] Park M-S, Chwa D, Hong S-K. Antisway tracking control of overhead cranes with system uncertainty and actuator nonlinearity using an adaptive fuzzy sliding-mode control. IEEE Transactions on Industrial Electronics. 2008;**55**(11):3972-3984

[21] Chang C-Y, Chiang K-H. Intelligent fuzzy accelerated method for the nonlinear 3-D crane control. Expert Systems with Applications. 2009;**36**(3):5750-5752

[22] Campeau-Lecours A, Foucault S, Laliberte T, Mayer-St-Onge B, Gosselin C. A cable-suspended intelligent crane assist device for the intuitive manipulation of large payloads. IEEE/ASME Transactions on Mechatronics. 2016;**21**(4):2073-2084

[23] Chang C-Y. Adaptive fuzzy controller of the overhead cranes with nonlinear disturbance. IEEE Transactions on Industrial Informatics. 2007;**3**(2):164-172

[24] Park M-S, Chwa D, Eom M. Adaptive sliding-mode antisway control of uncertain overhead cranes with high-speed hoisting motion. IEEE Transactions on Fuzzy Systems. 2014;**22**(5):1262-1271

[25] Park M-S, Chwa D, Hong S-K. Antisway tracking control of overhead cranes with system uncertainty and actuator nonlinearity using an adaptive fuzzy sliding-mode control. IEEE Transactions on Industrial Electronics. 2008;**55**(11):3972-3984

[26] Chen H, Fang Y, Sun N. Optimal trajectory planning and tracking control method for overhead cranes. IET Control Theory and Application. 2016;**10**(6):692-699

[27] Frank LL, Darren MD, Chaouki TA. Robot Manipulator Control Theory and Practice. 2nd ed. Revised and Expanded ed. New York: Marcel Dekker, INC.; 2004